FORD

SHOP MANUAL

More information available at Clymer.com
Phone: 805-498-6703

Haynes Publishing Group
Sparkford Nr Yeovil
Somerset BA22 7JJ England

Haynes North America, Inc
859 Lawrence Drive
Newbury Park
California 91320 USA

ISBN-10: 0-87288-517-8
ISBN-13: 978-0-87288-517-2

Printed in Malaysia

FO-47, 4S1, 14-144 **ABCDEFGHIJKLMNO**

Information and Instructions

This individual Shop Manual is one unit of a series on agricultural wheel type tractors. Contained in it are the necessary specifications and the brief but terse procedural data needed by a mechanic when repairing a tractor on which he has had no previous actual experience.

The material is arranged in a systematic order beginning with an index which is followed immediately by a Table of Condensed Service Specifications. These specifications include dimensions, fits, cleararances, capacities and tune-up information. Next in order of arrangement is the procedures section.

In the procedures section, the order of presentation starts with the front axle system and steering and proceeds toward the rear axle. The last portion of the procedures section is devoted to the power take-off and power lift systems. Interspersed where needed in this section are additional tabular specifications pertaining to wear limits, torquing, etc.

How to use the index

Suppose you wnt to know the procedure for R&R (remove and reinstall) of the engine camshaft. Your first step is to look in the index under the main heading of "Engine" until you find the entry "Camshaft." Now read to the right. Under the column covering the tractor you are repairing, you will find a number which indicates the beginning paragraph pertaining to the camshaft. To locate this paragraph in the manual, turn the pages until the running index appearing on the top outside corner of each page contains the number you are seeking. In this paragraph you will find the information concerning the removal of the camshaft.

Common spark plug conditions

NORMAL

Symptoms: Brown to grayish-tan color and slight electrode wear. Correct heat range for engine and operating conditions.
Recommendation: When new spark plugs are installed, replace with plugs of the same heat range.

WORN

Symptoms: Rounded electrodes with a small amount of deposits on the firing end. Normal color. Causes hard starting in damp or cold weather and poor fuel economy.
Recommendation: Plugs have been left in the engine too long. Replace with new plugs of the same heat range. Follow the recommended maintenance schedule.

CARBON DEPOSITS

Symptoms: Dry sooty deposits indicate a rich mixture or weak ignition. Causes misfiring, hard starting and hesitation.
Recommendation: Make sure the plug has the correct heat range. Check for a clogged air filter or problem in the fuel system or engine management system. Also check for ignition system problems.

ASH DEPOSITS

Symptoms: Light brown deposits encrusted on the side or center electrodes or both. Derived from oil and/or fuel additives. Excessive amounts may mask the spark, causing misfiring and hesitation during acceleration.
Recommendation: If excessive deposits accumulate over a short time or low mileage, install new valve guide seals to prevent seepage of oil into the combustion chambers. Also try changing gasoline brands.

OIL DEPOSITS

Symptoms: Oily coating caused by poor oil control. Oil is leaking past worn valve guides or piston rings into the combustion chamber. Causes hard starting, misfiring and hesitation.
Recommendation: Correct the mechanical condition with necessary repairs and install new plugs.

GAP BRIDGING

Symptoms: Combustion deposits lodge between the electrodes. Heavy deposits accumulate and bridge the electrode gap. The plug ceases to fire, resulting in a dead cylinder.
Recommendation: Locate the faulty plug and remove the deposits from between the electrodes.

TOO HOT

Symptoms: Blistered, white insulator, eroded electrode and absence of deposits. Results in shortened plug life.
Recommendation: Check for the correct plug heat range, over-advanced ignition timing, lean fuel mixture, intake manifold vacuum leaks, sticking valves and insufficient engine cooling.

PREIGNITION

Symptoms: Melted electrodes. Insulators are white, but may be dirty due to misfiring or flying debris in the combustion chamber. Can lead to engine damage.
Recommendation: Check for the correct plug heat range, over-advanced ignition timing, lean fuel mixture, insufficient engine cooling and lack of lubrication.

HIGH SPEED GLAZING

Symptoms: Insulator has yellowish, glazed appearance. Indicates that combustion chamber temperatures have risen suddenly during hard acceleration. Normal deposits melt to form a conductive coating. Causes misfiring at high speeds.
Recommendation: Install new plugs. Consider using a colder plug if driving habits warrant.

DETONATION

Symptoms: Insulators may be cracked or chipped. Improper gap setting techniques can also result in a fractured insulator tip. Can lead to piston damage.
Recommendation: Make sure the fuel anti-knock values meet engine requirements. Use care when setting the gaps on new plugs. Avoid lugging the engine.

MECHANICAL DAMAGE

Symptoms: May be caused by a foreign object in the combustion chamber or the piston striking an incorrect reach (too long) plug. Causes a dead cylinder and could result in piston damage.
Recommendation: Repair the mechanical damage. Remove the foreign object from the engine and/or install the correct reach plug.

SHOP MANUAL

FORD

MODELS 3230, 3430, 3930, 4630, 4830

The tractor identification plate is located under the tractor hood. Serial and model numbers of the tractor, engine, transmission, rear axle, hydraulic pump and the hydraulic lift are recroded on this identification plate. If equipped with Front-Wheel Drive, a similar plate is attached to the rear surface of the front drive axle housing.

INDEX (By Starting Paragraph)

INDEX (Cont.)

CONDENSED SERVICE DATA

	3230	3430	3930	4630	4830
GENERAL					
Engine Make			Own		
Number of Cylinders	3	3	3	3	4
Bore			111.8 mm (4.4 in.)		
Stroke	106.7 mm (4.2 in.)	106.7 mm (4.2 in.)	111.8 mm (4.4 in.)	111.8 mm (4.4 in.)	106.7 mm (4.2 in.)
Displacement	3147 cc (192 cid)	3147 cc (192 cid)	3294 cc (201 cid)	3294 cc (201 cid)	4186 cc (256 cid)
Compression Ratio			16.3:1		
Firing Order	1-2-3	1-2-3	1-2-3	1-2-3	1-3-4-2
Valve Clearance (Cold):					
Inlet			0.36-0.46 mm (0.014-0.018 in.)		
Exhaust			0.43-0.48 mm (0.017-0.019 in.)		
Valve Face Angle, Degrees					
Inlet and Exhaust			44.25-44.5		
Valve Seat Angle, Degrees					
Inlet and Exhaust			45-45.5		

CONDENSED SERVICE DATA (Cont.)

	3230	3430	3930	4630	4830
GENERAL (Cont.)					
Injection Timing			See Paragraph 88		
Engine Low Idle, rpm	600-700	600-850	600-850	600-850	600-850
Engine High Idle, rpm			See Paragraph 92		
Engine Rated Speed, rpm	2000	2000	2000	2200	2200
Battery Terminal Grounded			Negative		
SIZES					
Crankshaft Main Journal Diameter			See Paragraph 62		
Crankshaft Crankpin Diameter			See Paragraph 61		
Camshaft Journal Diameter			See Paragraph 54		
Piston Pin Diameter			38.092-38.100 mm (1.4997-1.5000 in.)		
Valve Stem Diameter- Inlet			9.426-9.444 mm (0.3711-0.3718 in.)		
Exhaust			9.400-9.418 mm (0.3701-0.3708 in.)		
CLEARANCES					
Main Bearing Diametral Clearance			See Paragraph 62		
Rod Bearing Diametral Clearance - Copper-Lead Material			0.043-0.096 mm (0.0017-0.0038 in.)		
Aluminum-Tin Alloy			0.053-0.107 mm (0.0021-0.0042 in.)		
Camshaft Bearing Diametral Clearance			0.025-0.076 mm (0.001-0.003 in.)		
Crankshaft End Play			0.10-0.20 mm (0.004-0.008 in.)		
Piston Skirt to Cylinder Clearance			0.069-0.094 mm (0.0027-0.0037 in.)		
CAPACITIES					
Cooling System- With Cab & Heater	11.4 L (12 qts.)	11.4 L (12 qts.)	11.4 L (12 qts.)	11.6 L (13 qts.)	13.5 L (14 qts.)
Without Cab	10.4 L (11.2 qts.)	10.4 L (11.2 qts.)	10.4 L (11.2 qts.)	10.4 L (11.2 qts.)	12.5 L (13.2 qts.)
Crankcase With Filter	6.6 L (7 qts.)	6.6 L (7 qts.)	6.6 L (7 qts.)	6.6 L (7 qts.)	7.7 L (8 qts.)
Transmission, Constant Mesh (8 × 2)			12 L* (3.2 gal.*)		
Transmission, Synchronized (8 × 8)	....		9.4 L (2.5 gal.)		
Transmission, Synchronized with Dual Power (16 × 8)	....		8.4 L (2.2 gal.)		

CONDENSED SERVICE DATA (Cont.)

CAPACITIES (Cont.)	3230	3430	3930	4630	4830
Final Drive & Hydraulic, with Constant Mesh Transmission			45.7 L** (12 gals.**)		
Final Drive & Hydraulic, with Synchronized Transmission			32.5 L** (8.5 gals.**)		
Steering Gear, Manual			0.6 L (1.3 pints)		
Hydrostatic Steering			2.2 L (2.3 qts.)		
Front Drive Axle Hubs (Each Side)			0.9 L (1 qt.)		
Front Drive Axle Housing			5.5 L (5.8 qts.)		

* Add 3 liters (3.3 qts.) if equipped with reduction (creeper) gearbox.
** Add 1.4 liters (1.5 qts.) if equipped with four-wheel drive.

DUAL DIMENSIONS

This shop manual provides specifications in both the Metric (SI) and U.S. Customary systems of measurement. The first specification is given in the measuring system perceived by us to be the preferred system when servicing a particular component; the second specification (given in parentheses) is the converted measurement. For instance, a specification of "0.28 mm (0.011 inch)" would indicate that we feel the preferred measurement, in this instance, is the metric system of measurement and the U.S. system equivalent of 0.28 mm is 0.011 inch.

FRONT AXLE SYSTEM (TWO-WHEEL DRIVE)

FRONT AXLE ASSEMBLY AND STEERING LINKAGE

Two-Wheel-Drive Models

1. WHEELS AND BEARINGS. To remove front wheel hub and bearings, raise and support the front axle extension, then unbolt and remove the tire and wheel assembly. Remove cap (13—Fig. 1), cotter pin (12), slotted nut (11), washer (10) and outer bearing cone (8). Slide hub assembly from spindle axle shaft. Remove grease retainer (2) and inner bearing cone (3). Hub is slotted to facilitate removal of bearing cups (5 and 7). Pack wheel bearings liberally with Ford M1C137-A, M1C75-B or equivalent grease. Reassemble by reversing disassembly procedure. Grease retainer (2) and inner bearing cone (3) should be positioned on spindle. Tighten slotted nut (8) to a torque of 27-40 N•m (20-30 ft.-lbs.), then rotate hub several turns. Retighten nut (8) to 61-74 N•m (45-50 ft.-lbs.), then back nut off less than two flats and install cotter pin (12). Be sure to install cap (13) securely.

2. TIE ROD, DRAG LINK AND TOE-IN. Models without hydrostatic steering are equipped with one drag link from the steering gear arm to the steering arm of the left spindle. A tie (connecting) rod connects the steering arms of the left and right spindles. Models with hydrostatic steering are equipped with one tie rod extending between left and right steering arms. The hydrostatic steering cylinder is attached between the front axle and the center of the tie rod. On all models, ends of tie rod, drag link and steering cylinder are automotive-type and should be renewed if wear is excessive. The procedure for removing and installing the ends is self-evident.

Recommended toe-in is 0-13 mm (0-½ inch) for all models. Spindle arms and axle extensions are marked as shown at (M—Fig. 2) to assist setting toe-in. To adjust length of tie rod when changing axle width, remove track adjusting screws (13—Fig. 3), slide rod inside tube until the appropriate hole is aligned, then install screw (13). To make small toe-in adjustments, loosen clamp bolt (B), remove screw (13), then turn the threaded section (T) until toe-in is correct. Reinstall screw (13) and tighten clamp bolt. On models with drag link, remove bolt from clamp (9—Fig. 4) and extend or retract drag link as required to correspond to axle width, then reinstall clamp bolt.

On all models, clamp bolts (B—Fig. 3) in tie rod ends (used to adjust toe-in) should be tightened to 45 N•m (33 ft.-lbs.) torque. Track adjusting screws (13) should be tightened to 105 N•m (78 ft.-lbs.) torque. Drag link adjusting clamp bolt should be tightened to 13 N•m (9 ft.-lbs.) torque if equipped with manual steering; 35 N•m (25 ft.-lbs.) torque if equipped with power steering. Tighten the track adjusting clamp screws (28—Fig. 4 or Fig. 5) that attach axle extensions inside the axle center section to 210 N•m (155 ft.-lbs.) torque.

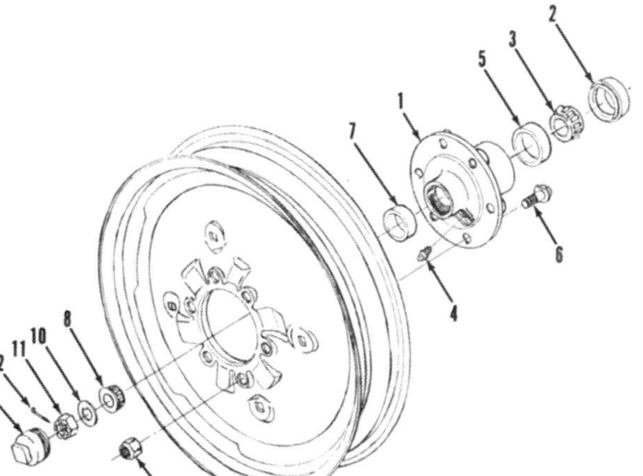

Fig. 1—Exploded view of front wheel hub used on two-wheel-drive models.

1. Hub
2. Grease retainer
3. Inner bearing cone
4. Grease fitting
5. Inner bearing cup
6. Wheel stud
7. Outer bearing cup
8. Outer bearing cone
9. Wheel lug
10. Tang washer
11. Slotted nut
12. Cotter pin
13. Cap

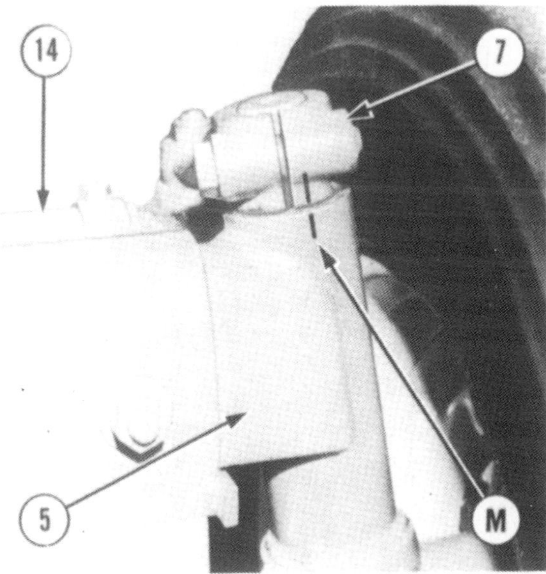

Fig. 2—Marks (M) are provided on the axle extensions and the steering arms to assist in setting toe-in.

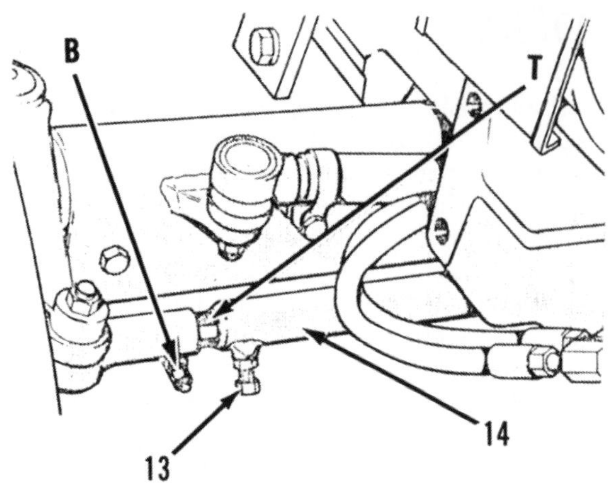

Fig. 3—View of track and toe-in adjusting points for two-wheel-drive models. Hydrostatic steering model is shown, but adjustment is similar for models with manual or power steering.

B. Clamp bolt
T. Threaded rod
13. Track bolt
14. Tie rod tube

3. SPINDLES, AXLE EXTENSIONS AND BUSHINGS. To remove spindle (1—Fig. 4 or Fig. 5), first remove the wheel and hub. Disconnect tie rod end from steering arm (7 or 18). On models with manual or power steering, detach drag link if left side is being removed. On all models, remove clamp screw from steering arm, then remove steering arm. Remove key (2) and washers (6) from top of spindle, then lower spindle out of axle extension (5). Remove thrust bearing (3) from spindle. Clean and inspect parts for wear or other damage and renew as necessary.

When reassembling, install thrust bearing (3) on spindle so that numbered side of bearing is facing upward and insert spindle through axle extension. Install washer (6) and key (2), then locate steering arm on top of spindle. Tighten steering arm retaining clamping screw to a torque of 170 N•m (125 ft.-lbs.) for models with cab, 65 N•m (50 ft.-lbs.) torque for models without. Refer to paragraph 2 for track and toe-in adjustment and recommended torques. Balance of reassembly is the reverse of disassembly.

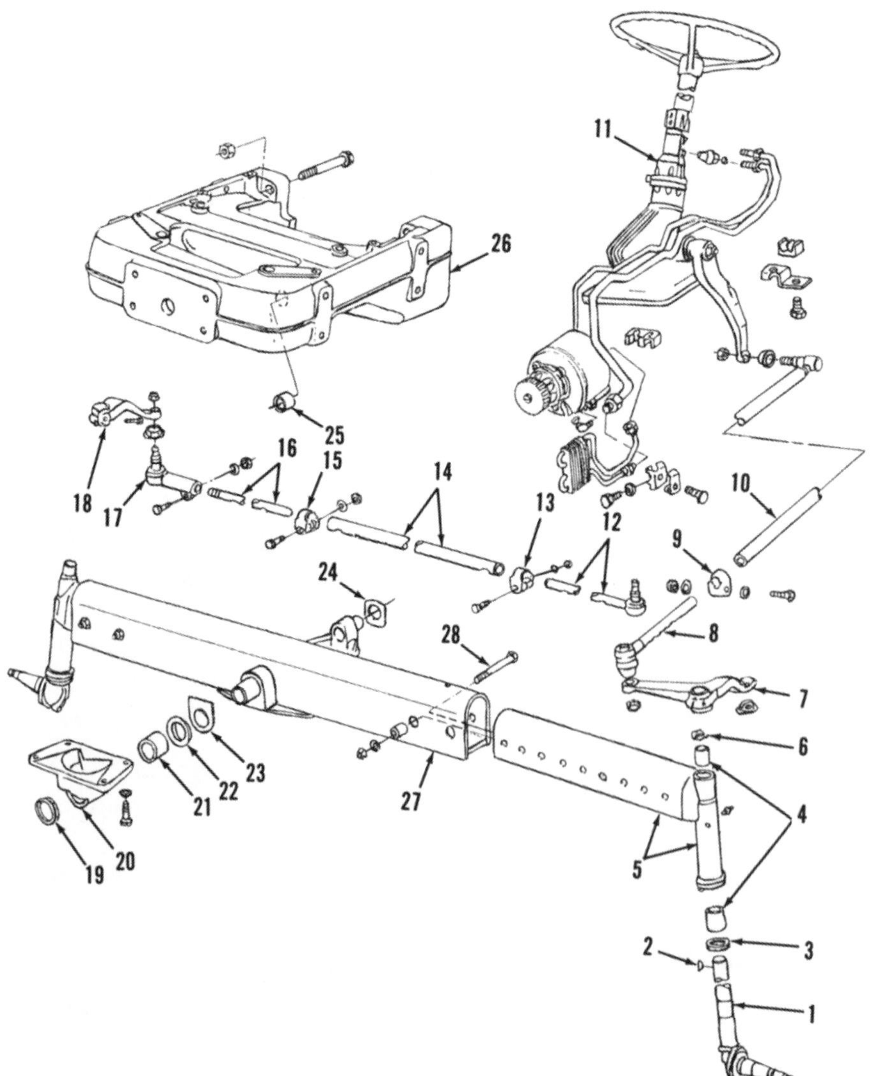

Fig. 4—Exploded view of row crop, two-wheel-drive front axle, typical of models with power steering (shown) and manual steering.

1. Spindle
2. Woodruff key
3. Thrust bearing
4. Bushings
5. Axle extension
6. Seal
7. Left steering arm
8. Drag link end
9. Clamp
10. Drag link tube
11. Steering gear
12. Tie rod end
13. Clamp
14. Tie rod tube
15. Clamp
16. Tie rod adjusting rod
17. Tie rod end
18. Right steering arm
19. Washer
20. Axle pivot bracket
21. Bushing
22. Shim
23. Washer
24. Washer
25. Pivot bushing
26. Front support
27. Axle center member
28. Track bolt

4. AXLE CENTER MEMBER, PIVOT PIN AND BUSHINGS. To remove front axle assembly, raise front of tractor in such a way that it will not interfere with the removal of the axle. A hoist may be attached to front support or special stands can be attached to sides. Removal of the axle center member may be easier if the axle extension and spindle assembly is first removed from each side; however, the complete assembly can be removed as a unit. Remove front wheels and weights, then support the axle with a suitable jack or special safety stand to prevent tipping while permitting the axle to be lowered and moved safely. Disconnect drag link from steering arm of models without hydrostatic steering. On models with hydrostatic steering, disconnect hoses from steering cylinder and cover openings to prevent entry of dirt. On all models, unbolt and remove pivot bracket (20—Fig. 4 or Fig. 5). Move axle forward until axle rear pivot is free from pivot in front support (26). Lower axle assembly and carefully roll axle away.

Check axle pivot bushings and renew if necessary. Reverse removal procedure when assembling. Tighten screws retaining the axle pivot bracket (20) to 90 N.m (65 ft.-lbs.) torque. End clearance of axle center member should be 0.254 mm (0.010 inch) and is adjusted by changing thickness of shim (22).

5. FRONT SUPPORT. To remove the front support, the axle must be removed, the radiator must be removed from the support and the front support must be unbolted from the front of engine. The front axle, the front support and the remainder of the tractor must each be supported separately while removing, while separated and while assembling. Be sure that sufficient equipment is available before beginning.

Remove hood, drain cooling system and disconnect radiator hoses. Disconnect air inlet hose from the front-mounted air cleaner. Remove the grille and disconnect lines from oil coolers, if so equipped. Disconnect battery ground cable and wires to the headlights. Unbolt radiator shell upper support bracket from the shell. Disconnect steering linkage and steering hoses that would interfere with removal of the front axle or the front support. Support front of tractor in such a way that it will not interfere with removal of either the front support or the front axle.

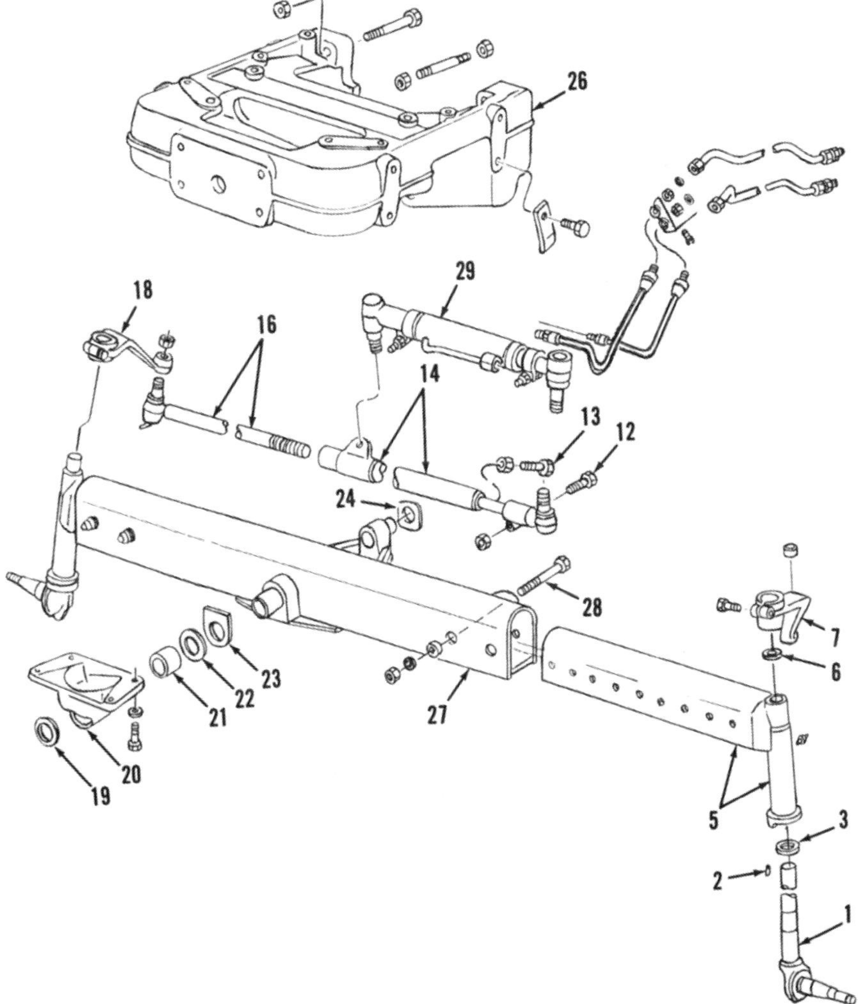

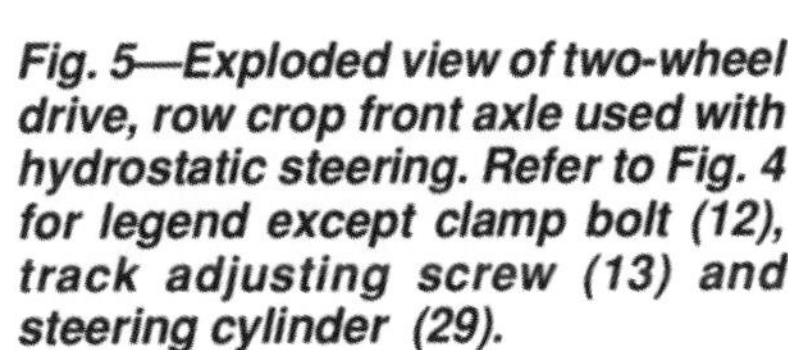

Fig. 5—Exploded view of two-wheel drive, row crop front axle used with hydrostatic steering. Refer to Fig. 4 for legend except clamp bolt (12), track adjusting screw (13) and steering cylinder (29).

Remove axle as outlined in paragraph 4. Attach a hoist or other supporting device to the front support, then unbolt and separate front support from front of engine.

Reattach front support to engine. Tighten retaining screws to 339-420 N•m (250-310 ft.-lbs.) for models with stamped oil pan; to 240-298 N•m (180-220 ft.-lbs.) torque for models with cast oil pan. Complete assembly by reversing removal procedure.

6. STEERING GEAR. Refer to paragraph 18 for manual steering gear or 23 for the power steering gear. Refer to paragraph 30 and following for testing and service to the hydrostatic steering system used on some models in place of a gear unit.

FRONT-WHEEL DRIVE

7. The mechanical front-wheel drive available on these models uses a front drive axle unit manufactured by Carraro. There are some differences between the front-wheel drive systems used on these models, which will be referred to in the following servicing instructions.

The transfer gearbox is engaged on some models by moving mechanical linkage shown in Fig. 18. On 4630 and 4830 models, engagement is controlled by an electric solenoid/hydraulic valve that directs oil pressure to move the dog clutch (35—Fig. 20) and engage the front-wheel drive. The transfer gearbox is attached to the bottom of the rear axle center housing of all models. A drive shaft with two "U" joints connects the transfer gearbox to the front axle.

TIE RODS AND TOE-IN

All Models So Equipped

8. Tie rod ends may be one of several different types, but none are adjustable for wear; faulty units must be renewed.

To check toe-in, first turn steering wheel so front wheels are in straight-ahead position. Measure distance at front and rear of front wheels from rim flange to rim flange at hub height. Toe-in should be 0-6 mm (0-¼ inch).

To adjust toe-in of narrow tread models, loosen the lock nuts at each end of the tie rod, then turn the tie rod tube to set the toe-in. Tighten lock nuts at each end when adjustment is correct.

To adjust toe-in of all other models (standard tread four-wheel drive), loosen clamp bolt (2—Fig. 6), then turn tie rod (1) in or out of tie rod end (3) as required. Adjust both sides evenly. When adjustment is correct, tighten clamp bolt (2) to 25 N•m (18 ft.-lbs.) torque.

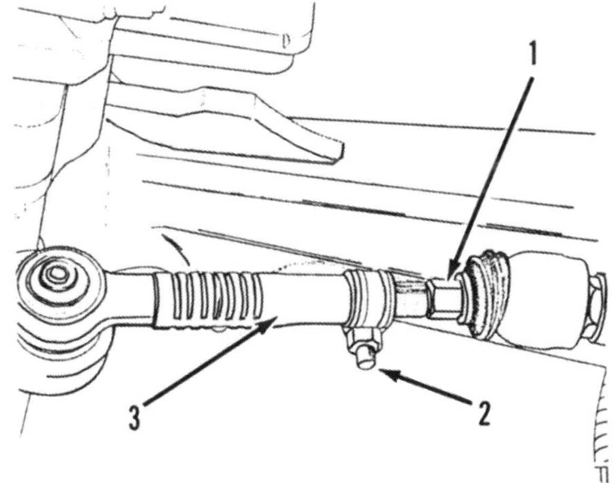

Fig. 6—View of toe-in adjustment points for four-wheel-drive models. Be sure to adjust both sides equally.

1. Adjuster
2. Clamp bolt
3. Tie rod end

DRIVE SHAFT

All Models So Equipped

9. REMOVE AND REINSTALL. Standard and waterproof drive shafts have been installed. Refer to Fig. 7 or Fig. 8 and the appropriate following paragraphs.

To remove the standard drive shaft, first unbolt and remove the shield assembly (1—Fig. 7). Remove clamp bolts (3) from coupling, unbolt center bearing (9) from bracket, then slide couplings onto drive shaft and remove shaft. Grease couplings through fittings

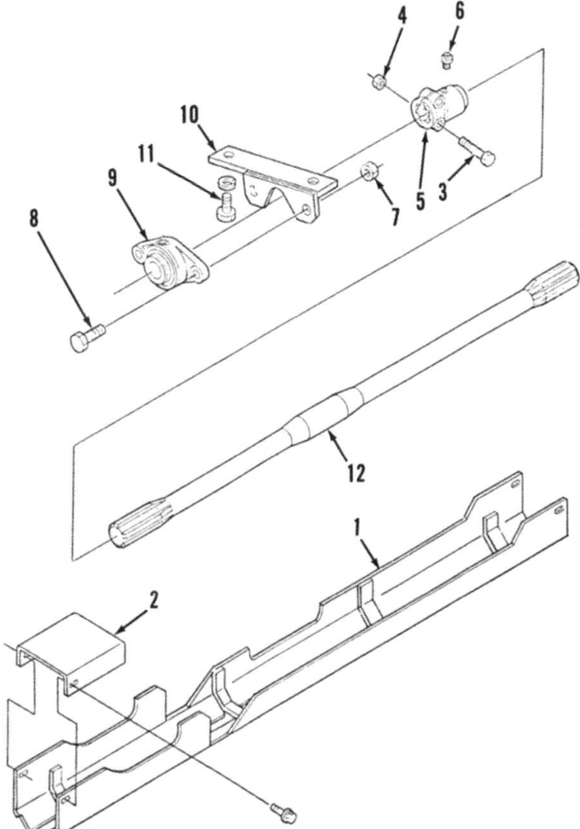

Fig. 7—View of drive shaft and shield used on most four-wheel-drive models. Waterproof drive shaft is shown in Fig. 8.

1. Shield
2. Cover
3. Clamp screw
4. Nut
5. Coupling
6. Grease fitting
7. Nut
8. Screw
9. Center bearing
10. Center bracket
11. Screw
12. Drive shaft

(6) after assembling. Tighten coupling clamp bolts (3 and 4) to 60 N•m (44 ft.-lbs.) torque.

To remove waterproof drive shaft, first clean the four parts of drive shaft housing (1, 2 and 13—Fig. 8) thoroughly. Unbolt front and rear sections (2) and slide front and rear sections onto center parts (1 and 13). Drive pins (3) from front and rear couplings (5), then slide couplings onto drive shaft. Unbolt center bearing housing (16) and drive shaft housings (1 and 13) from bracket (10), and remove shaft and housing. Grease caps and shields (14) and "O" rings (6 and 15) before assembling. Tighten screws (S) to 352 N•m (260 ft.-lbs.) torque and center bearing attaching bolts (7 and 8) to 57 N•m (42 ft.-lbs.) torque.

FRONT DRIVE AXLE

All Models So Equipped

10. R&R ASSEMBLY. First remove drive shaft and shield as outlined in paragraph 9. Raise front of tractor in such a way that it will not interfere with axle removal. A hoist may be attached to front support or special stands can be attached to sides. Remove front wheels and weights, then support the axle with a suitable jack or special safety stand to prevent tipping while permitting the axle to be lowered and moved safely. Disconnect hoses from the steering cylinder and cover openings to prevent the entry of dirt. Unbolt and remove front pivot bracket (1—Fig. 13) and drop housing (17). Lower axle until free, then carefully roll axle away.

Reinstall front drive axle by reversing the removal procedure. Tighten screws attaching front pivot (1) and drop housing (17) to 325 N•m (240 ft.-lbs.) torque.

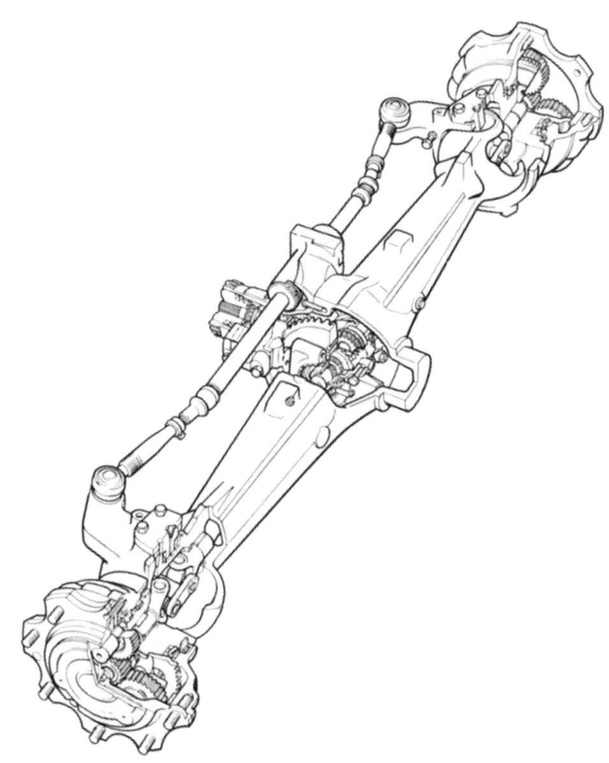

Fig. 9—Cross section of standard tread front drive axle.

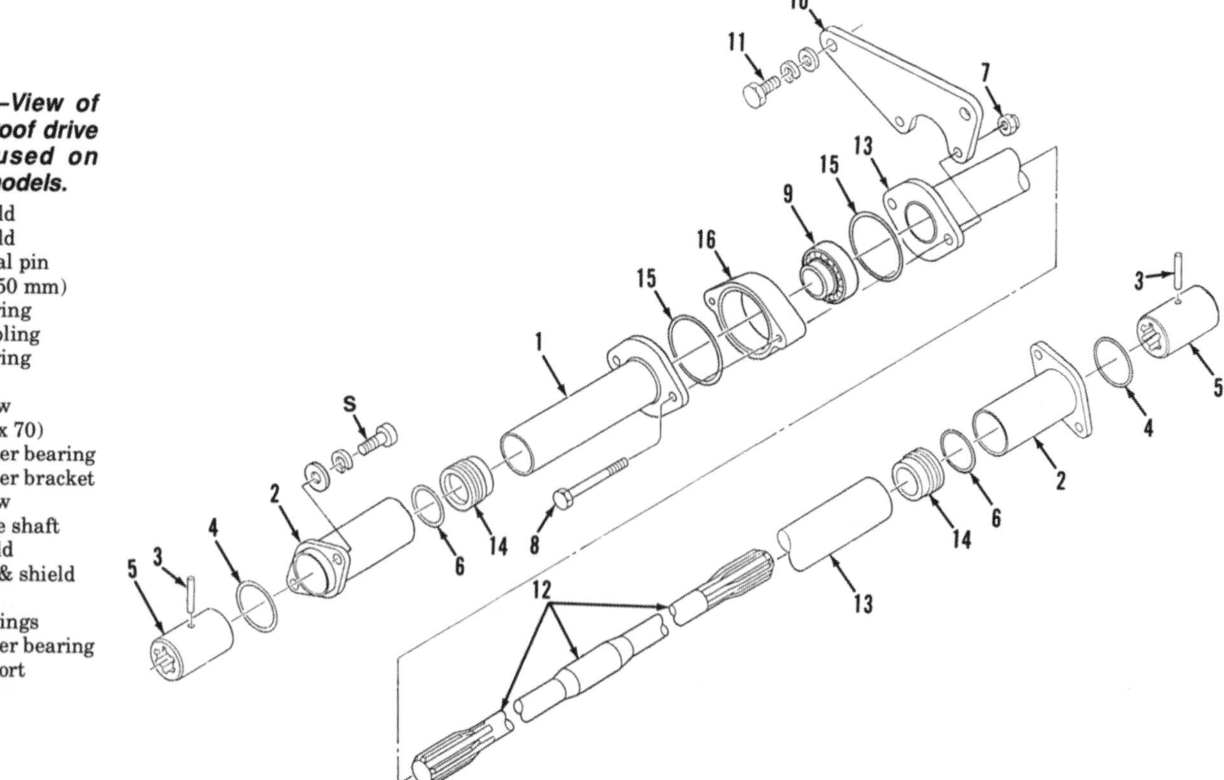

Fig. 8—View of waterproof drive shaft used on some models.

1. Shield
2. Shield
3. Spiral pin (6 x 50 mm)
4. "O" ring
5. Coupling
6. "O" ring
7. Nut
8. Screw (M8 x 70)
9. Center bearing
10. Center bracket
11. Screw
12. Drive shaft
13. Shield
14. Cap & shield assy.
15. "O" rings
16. Center bearing support

WHEEL HUB AND PLANETARY

All Models So Equipped

11. R&R AND OVERHAUL. Refer to Fig. 9. Either front wheel hub and planetary can be serviced without removing the steering knuckle housing (26—Fig. 10). Support front axle housing and remove front wheel. Remove drain plug (1) and drain oil from hub assembly. Remove the two Allen screws (2) and lift off planetary carrier (3). Remove snap rings (11) and retainer plate (9). Mark shafts (10) and gears (7), then keep bearings (8) and thrust washers (6) separate so that they can be reinstalled as a set.

Before removing the ring gear (14), mark relative position of ring gear (14) and steering knuckle housing (26) to facilitate installation in same location. Unscrew the six self-locking cap screws (15). Install three screws in holes around driven gear carrier (17). Use these screws as jack screws to remove driven gear carrier (17) and ring gear (14). It may be necessary to heat area around bushings (18) to remove gear. Bushings (18) may be installed with Loctite. Use a slide hammer puller and remove bushings (18).

Bump hub (22) and bearings (20 and 23) from steering knuckle (26). Remove cups for bearings (20 and 23) and seal (24) from hub. If necessary, remove wear ring (25) and inner bearing cone from knuckle housing.

Clean and inspect all parts for excessive wear or other damage and renew as necessary.

When reassembling, drive cups for bearings (20 and 23) and oil seal (24) into hub (22). If removed, install new wear ring (25) and cone for inner bearing (23) on knuckle housing (26). Install hub (22) and cone for outer bearing (20) onto knuckle housing. It may be necessary to heat bearing cones before install-

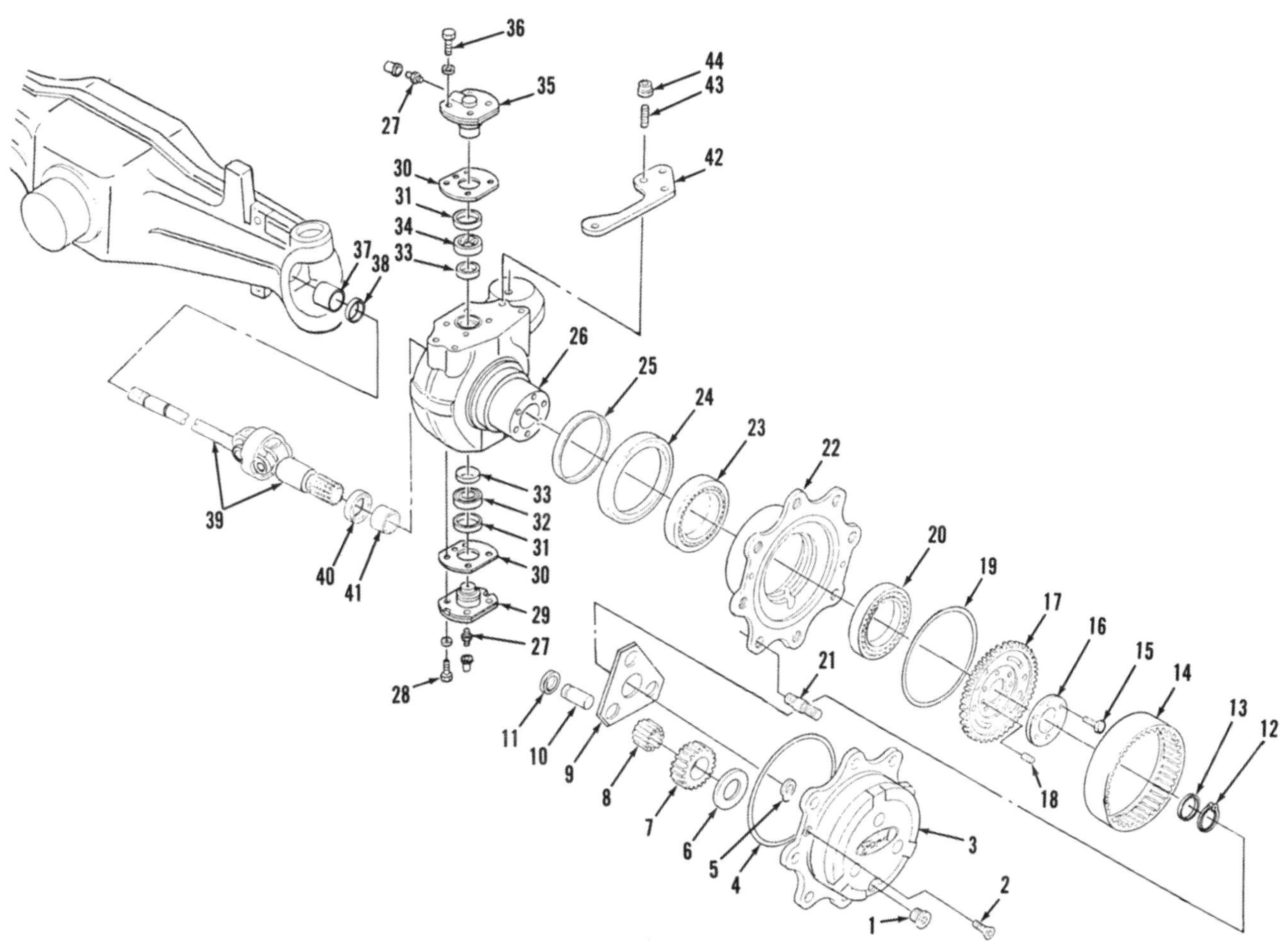

Fig. 10—Exploded view of front-wheel-drive steering knuckle and planetary assembly.

1. Magnetic drain plug
2. Allen screws
3. Planet carrier
4. "O" ring
5. Thrust pin
6. Thrust washer
7. Planet gear
8. Rollers (30/gear)
9. Planet carrier plate
10. Planet gear shaft
11. Snap ring
12. Snap ring
13. Washer
14. Ring gear
15. Screws (M10 x 30)
16. Retaining plate
17. Driven gear carrier
18. Bushing
19. Retaining ring
20. Bearing cup & cone
21. Wheel stud
22. Front hub
23. Bearing cup & cone
24. Seal
25. Wear ring
26. Steering knuckle
27. Grease fitting
28. Screw
29. Lower retainer
30. Shims (0.10, 0.19, 0.35 mm)
31. Seals
32. Lower bearing
33. Caps
34. Upper bearing
35. Upper retainer
36. Screw
37. Bushing
38. Seal
39. Axle shaft assy.
40. Seal
41. Bushing

ing. Coat bushings (18) with Loctite 638 and drive into knuckle housing. If bushings (18) are spring pin type, split should be toward direction of rotation or away from direction of rotation. **Split of pin should never be toward center of axle or away from center of axle.** Use cap screws to force ring gear (14) and driven gear carrier (17) on bushings. Secure ring gear with cap screws (15), tightened to 250 N•m (184 ft.-lbs.) torque. Make sure thrust pin (5) is in place and install new "O" ring (4) in groove of hub (3). Install bearings (8), planetary gears (7) and retainer plate (9), then secure with snap rings (11). Install planetary carrier assembly. Install the two Allen screws (2). Fill hub and planetary to the level of opening for plug (1) with "OIL LEVEL" mark (L—Fig. 11) horizontal. Fill with Ford M2C134-D/C or equivalent. Install front wheel and tighten disc to hub nuts to 270 N•m (200 ft.-lbs.) torque. Disc to wheel rim nuts should be tightened to 240 N•m (177 ft.-lbs.) torque.

STEERING KNUCKLE HOUSING

All Models So Equipped

12. R&R AND OVERHAUL. To remove either steering knuckle housing, first remove wheel hub and planetary as outlined in paragraph 11. Disconnect tie rod from steering knuckle arm. Unbolt and remove upper and lower retainers (29 and 35—Fig. 10) and shims (30). Measure and note thickness and number of shims (30) under each retainer for aid in reassembly. Carefully remove steering knuckle housing (26). Remove snap ring (12) and washer (13). Axle shaft and double "U" joint assembly (39) may be withdrawn with knuckle housing. Lower bearing cone (32) may fall from lower retainer (29). Remove upper bearing cone (34), cap (33) and seal (31). Bearing cups can be removed if necessary.

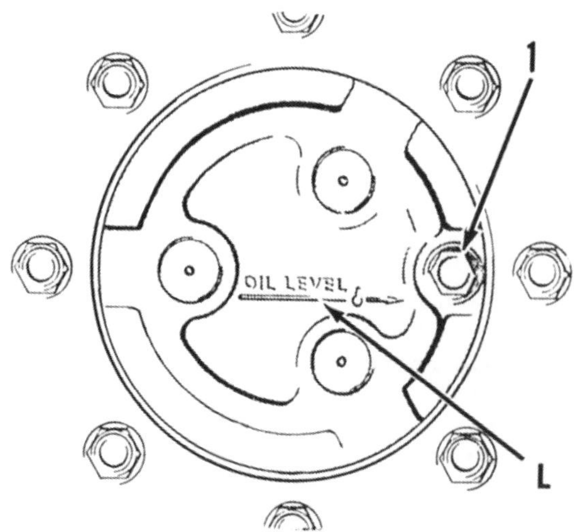

Fig. 11—View of "OIL LEVEL" mark (L) and drain plug (1) of front-wheel-drive front hub.

If desired, axle shaft and "U" joint assembly can be withdrawn for inspection or repair. If renewal is required, oil seal (40) and bushing (41) can be removed from knuckle housing (26). Oil seal (38) and bushing (37) can be removed from axle housing. Bushings (37 and 41) should be pressed into position with external groove toward top and internal arrow-shaped grooves pointing toward inside of oil-filled housing (away from seal). Be careful not to damage seals (38 and 40) when installing axle and knuckle housing.

Reassemble by reversing the disassembly procedure. Sufficient shims (30) should be installed to provide king pin with 0.2-0.4 mm (0.008-0.016 inch) preload. To set correct preload, add enough shims (30) to provide some end play and tighten cap screws (28 and 36) to 96 N•m (71 ft.-lbs.) torque. Position a dial indicator as shown in Fig. 12 and check vertical play of king pin bearings. Be sure that king pin has some end play. If measured end play is 0.2 mm (0.008 inch), it will be necessary to remove 0.4-0.6 mm (0.016-0.024 inch) to have preload within desired limits. Divide shims (30—Fig. 10) equally at top and bottom, to center the assembly. Shims are available in varying thicknesses.

Refer to paragraph 11 when reassembling hub and planetary.

DROP HOUSING

All Models So Equipped

13. R&R AND OVERHAUL. To remove the drop housing, refer to paragraph 9 and remove the drive shaft, then refer to paragraph 10 and lower the front axle assembly. The drop housing (17—Fig. 13) can be withdrawn from the axle. Gears in the drop housing are lubricated by the oil contained in the front axle assembly. Driven gear (5) will remain on differential bevel pinion shaft.

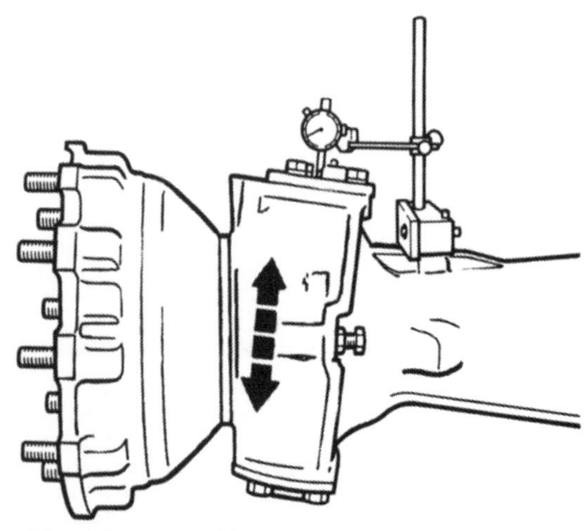

Fig. 12—Adjust the king pin bearings to a preload of 0.2-0.4 mm (0.008-0.016 inch).

Remove snap ring (6) and withdraw driven gear (5) from pinion shaft. Remove bolts (26) and drive shaft shield bracket (27), then withdraw cover (25) from drop housing. Remove input shaft (21), gear (20) and bearing (18). Remove screw (12) and remove oscillation bushing (10).

Clean all parts, complete disassembly, if required, and inspect all parts for wear or damage. Use new "O" rings and seals when assembling. Gears (5 and 20) are available with different numbers of teeth to match the tire size and rear axle reduction ratio.

DIFFERENTIAL

All Models So Equipped

14. R&R AND OVERHAUL. To remove the differential assembly, refer to paragraph 9 and remove the drive shaft, then paragraph 10 and remove the front axle assembly. The drop housing (17—Fig. 13) can be withdrawn from the axle. Refer to paragraph 12 and remove steering knuckle and axle shaft assembly (39—Fig. 10) from both sides. Some mechanics prefer to remove the steering knuckles and axle shafts before removing the axle from the tractor, to reduce the weight of the unit. Unbolt and remove the differential housing from the axle center housing. The differential housing of standard width axles contains the hydrostatic steering cylinder.

Before disassembling, mark both bearing caps and housing as shown at (M—Fig. 14) to facilitate alignment when reassembling. Straighten tabs of both lock plates (13—Fig. 15), then remove both cap screws (12), washers and lock plates (13). Loosen, but do not remove, the four screws attaching caps (C—Fig. 14), then unscrew adjusting rings (14). Remove

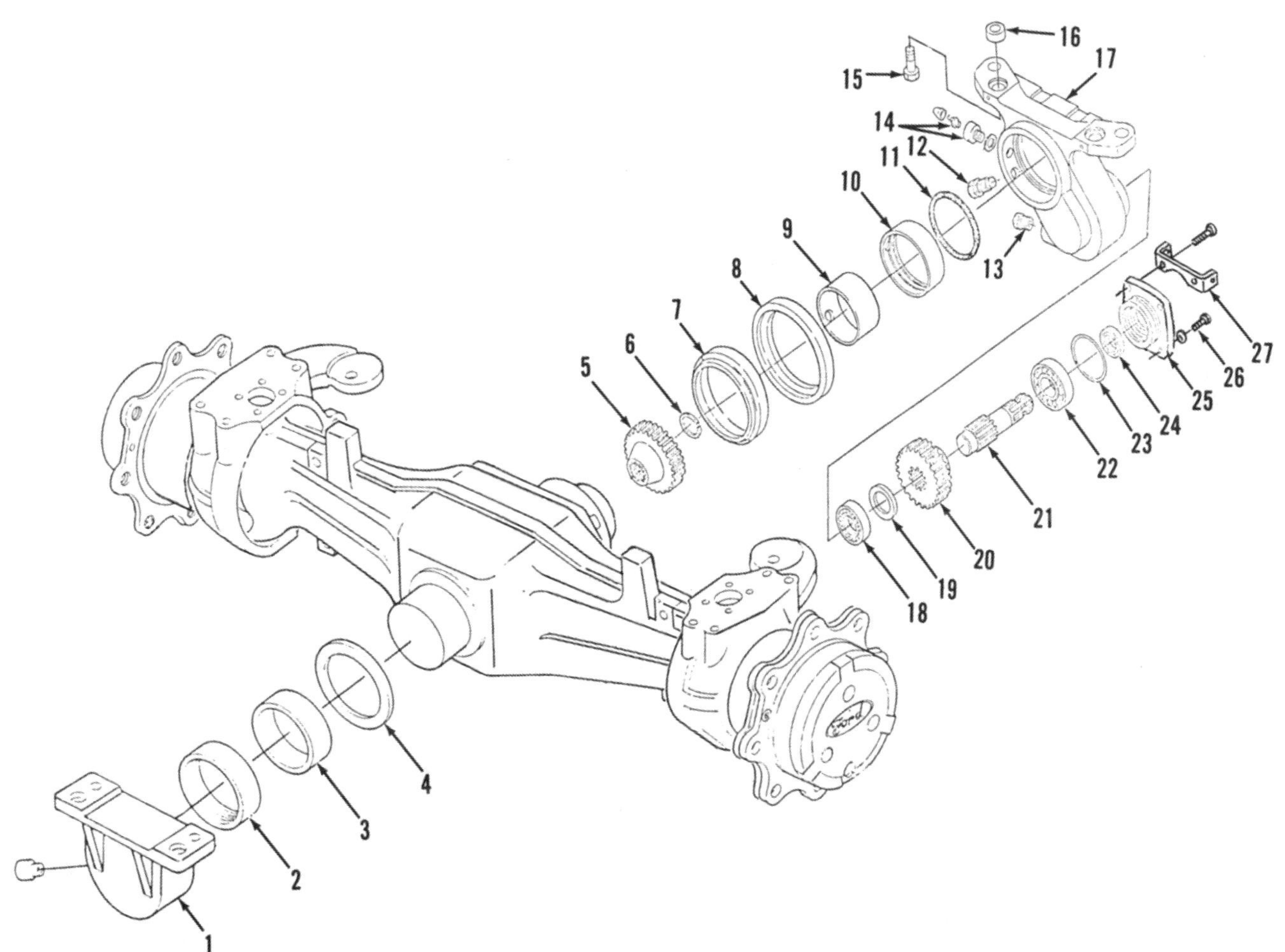

Fig. 13—Partially exploded view of front-wheel-drive drop housing. Some parts (7, 8 & 12) may be used only with waterproof axles.

1. Front pivot bracket
2. Bushing
3. Sleeve
4. Thrust washer
5. Driven gear
6. Snap ring
7. Seal
8. Seal support
9. Cover
10. Bushing
11. "O" ring
12. Retainer screw
13. Drain plug
14. Grease fitting & adapter
15. Screw
16. Locating spacer
17. Drop housing
18. Bearing
19. Washer
20. Input gear
21. Input shaft
22. Bearing
23. "O" ring
24. Seal
25. Cap
26. Screw
27. Shield support

both bearing caps and lift differential, bearing cups and adjusting rings from housing.

To remove the bevel pinion (10—Fig. 15), remove nut (2) and washer (3), then push pinion out toward inside. Cone of bearing (4) will slide from shaft as pinion is removed.

The differential case (16) may be either one or two pieces. If two-piece case is used, mark halves to facilitate alignment when reassembling. Remove screws (17) and remove ring gear (11). If two-piece case, the case halves can be separated after removing screws (17). If one-piece case, drive pin (21) from shaft (20), toward ring gear face, then remove shaft. Rotate pinions (19) and thrust washers (18) so that they can be removed. Side gears (22) and clutch (23, 24 and 25) can be removed from one-piece differential housing after pinions (19) are removed. Regardless of case type, assembly of original parts is easier if side gears (22) and clutch parts (23, 24 and 25) are kept together for each side and not mixed.

Clean and inspect all parts for wear or other damage. Lubricate all parts while assembling. Alternate the five external splined plates (24) and four internal splined discs (25). The internally splined thrust disc (23) is thicker than other discs and should be assembled next to side gear (22). Install thrust washers (18), pinions (19) and shaft (20). Drive pin (21) in from ring gear face of differential case. On two-piece differential case, align previously affixed marks. On all models, install ring gear (11) and screws (17). Tighten screws (17) to 78 N•m (58 ft.-lbs.) torque. Press bearing cones (15) onto case (16) until seated.

Thickness of shims (9) should be selected to change mesh position of bevel pinion (10) if gear set (10 and 11), pinion bearings (4 and 8) and/or differential housing (1 or 1A) is renewed or if previous mesh position is questioned. Install cups for bearings (4 and 8) in housing (1 or 1A), then position bearing cones in cups as shown in Fig. 16. Use a retainer as shown at (H) to hold cones tight in cups while measuring. Install both carrier bearing caps and tighten the retaining nuts securely. Measure the diameter of the bearing bores (A). Measure the distance from the bearing bore to the inner flange of bearing (8) as shown at (B). Both bearing bores should be the same diameter. Determine distance (C) from the center of the bearing bore to the bearing flange as follows. Divide bearing bore diameter (A) by 2 then add the result to the measured distance (B). To determine the thickness of shims to be installed at (9—Fig. 15), subtract the amount stamped on pinion face from the calculated distance (C—Fig. 16). Select a shim (9—Fig. 15) of the correct thickness and install on pinion shaft (10) with chamfer toward pinion gear. Remove carrier bearing caps and bearing cones (4 and 8) after selecting shims (9).

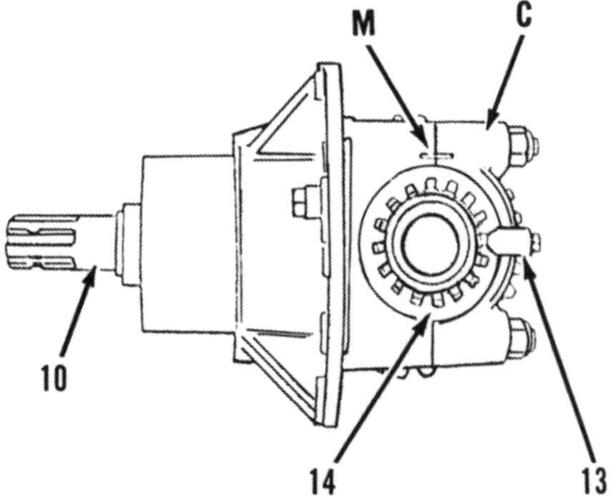

Fig. 14—View of removed differential assembly showing marks (M) on one of the caps (C) and housing. Mark the other cap so that caps can be quickly identified for assembly to the correct side and in the correct position.

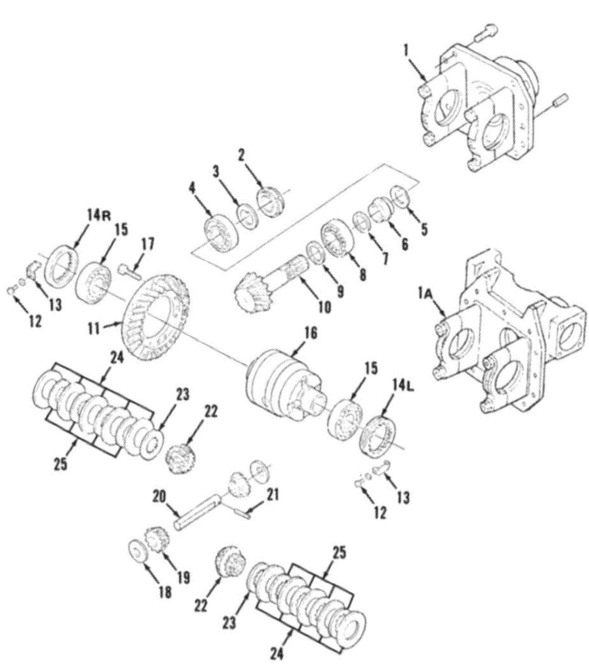

Fig. 15—Exploded view of differential assembly. Housing (1) is used for narrow axle models. Housing (1A) is for standard width axle and contains hydrostatic steering cylinder. Gears (10 & 11) are available only as matched sets. Different ratio ring gear and pinion is used depending upon other tractor drive ratios.

1. Housing
1A. Housing
2. Nut
3. Lock washer
4. Bearing cup & cone (Same as 8)
5. Washer (Same as 7)
6. Spacer
7. Washer (Same as 5)
8. Bearing cup & cone (Same as 4)
9. Shim (2.5-3.4 mm)
10. Bevel pinion
11. Ring gear
12. Screws
13. Locking clips
14L. Adjusting ring
14R. Adjusting ring
15. Bearing cup & cone
16. Differential case (1 or 2 pieces)
17. Screws
18. Thrust washers
19. Pinion gears
20. Pinion shaft
21. Roll pin
22. Side gears
23. Spacer (1 each side)
24. External spline plates (5 each side)
25. Internal spline discs (4 each side)

Install inner bearing cone (8) on pinion (10) against selected shims.

Install washer (7), new (not yet collapsed) spacer (6) and washer (5) on pinion shaft (10) and insert into housing through the installed bearing cups. Install cone for bearing (4), lock washer (3) and nut (2). Tighten nut (2) until all play in bearings is just taken up, then measure rotating drag with a spring scale and string wrapped around the pinion shaft as shown in Fig. 17. Do not measure starting torque (force necessary to start pinion shaft turning), only measure force necessary to keep pinion turning. Correct amount of rotational force measured by the spring scale should be within the range of 9.3-13.9 kg. (21-30 lbs.). Tighten nut (2) to tighten bearings and increase rotational force. Bearing adjustment is accomplished by crushing spacer (6) and nut (2) should not be loosened, unless a new spacer is installed.

TRANSFER GEARBOX

3230, 3430, 3930 Models So Equipped

15. R&R AND OVERHAUL. To remove the transfer gearbox, remove drain plug (24—Fig. 18) and allow oil to drain. Remove drive shaft shield (1—Fig. 7), remove clamp bolt (3 and 4—Fig. 7) or pin (3—Fig. 8) from rear coupler (5—Fig. 7 or Fig. 8), then slide

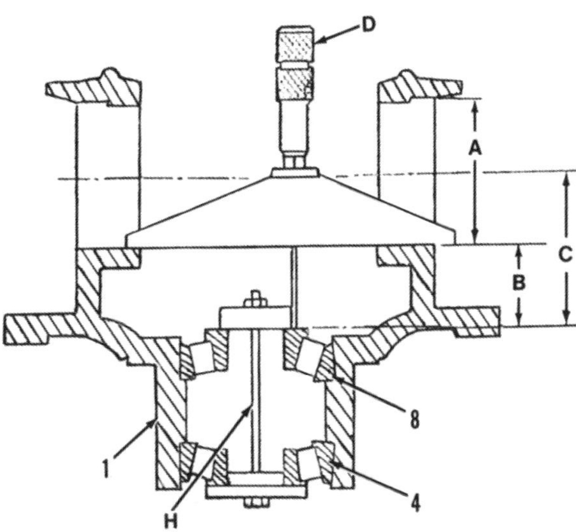

Fig. 16—Cross section of differential housing showing measurements required for accurately setting pinion position. Refer to text for measuring distance (C) from bearing inner race to center of bearing bore.

D. Depth gage
H. Bearing retainer
1. Differential housing
4 & 8. Pinion bearings

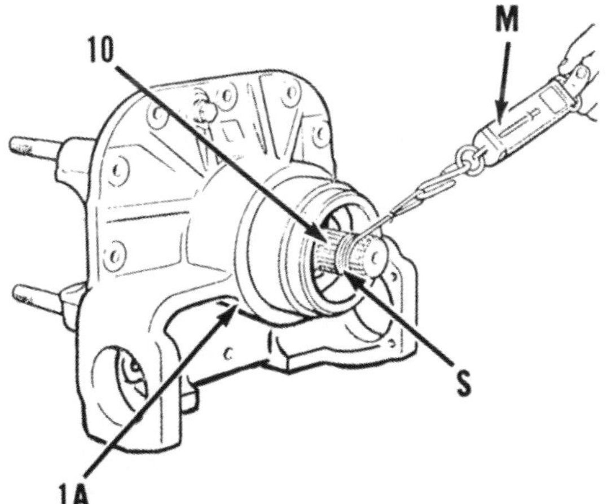

Fig. 17—Measure rolling drag using a spring scale (M) and string (S) as shown. Differential housing is shown at (1A) and pinion shaft at (10).

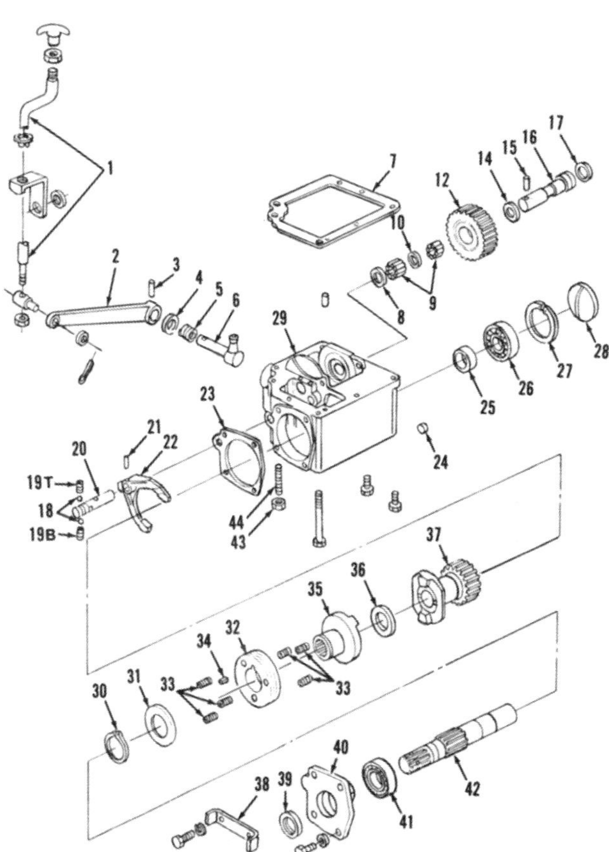

Fig. 18—Exploded view of transfer case used on 3230, 3430 and 3930 models. Refer to Fig. 20 for other models.

1. Shift rod assy.
2. Lever
3. Pin
4. Oil seal
5. Spring
6. Shifter shaft
7. Gasket
8. Washer
9. Bearing needles
10. Spacer
12. Idler gear
14. Washer
15. Roll pin
16. Idler support shaft
17. "O" ring
18. Ball
19B. Detent springs
19T. Detent springs
20. Shift rail
21. Pin
22. Shift fork
23. Gasket
24. Drain plug
25. Spacer
26. Bearing
27. Snap ring
28. Plug
29. Housing
30. Snap ring
31. Washer
32. Collar
33. Springs (6 used)
34. Key
35. Dog clutch
36. Washer
37. Gear & clutch coupler
38. Bracket
39. Oil seal
40. End plate
41. Bearing
42. Output shaft
43. Locknut
44. Detent adjuster screw

the coupling forward off transfer case output shaft (42—Fig. 18). Disconnect shift rod (1) from lever (2). Position floor jack under transfer gearbox, remove retaining screws and lower transfer gearbox from the tractor.

To disassemble unit, drive pin (3) out and remove lever (2). Unbolt and remove bracket (38) and end plate (40). Use a suitable puller and remove output shaft (42) and bearing (41). Loosen lock nut (43) and unscrew detent adjuster screw (44). Drive pin (21) out and remove shift rail (20), detent springs (19) and balls (18). Remove shift fork (22) and clutch components (30 through 37) from housing. Remove plug (28), snap ring (27), bearing (26) and spacer (25). Remove shifter shaft (6), spring (5) and oil seal (4). Drive out roll pin (15), withdraw support shaft (16) and remove idler gear and shaft assembly (8 through 14).

Clean and inspect all parts and renew any showing excessive wear or other damage. When reassembling, renew all "O" rings, seals and gaskets. Reassemble by reversing disassembly procedure. Adjust detent screw (44) until a pull of 12 kg. (26 lbs. or 115 N) on a spring scale will move lever (2) to the engaged position. Tighten lock nut (43) to 9 N•m (6 ft.-lbs.) torque.

Reinstall transfer gearbox by reversing the removal procedure. Tighten retaining screws to 48 N•m (35 ft.-lbs.) torque. Fill rear axle center housing through plug (P—Fig. 19) to level of dipstick (D) with Ford M2C134-D/C or equivalent.

Models 4630 and 4830 So Equipped

16. R&R AND OVERHAUL. To remove the transfer gearbox, remove drain plug (24—Fig. 20) and allow oil to drain. Remove drive shaft shield (1—Fig. 7), remove clamp bolt (3 and 4—Fig. 7) or pin (3—Fig. 8) from rear coupler (5—Fig. 7 or Fig. 8), then slide the coupling forward off transfer case output shaft (42—Fig. 20). Remove cover (4), then disconnect wire (1) from the solenoid (2). Disconnect housing for wire (1) and hydraulic line (5) from housing (29). Position floor jack under the transfer gearbox, remove retaining screws and lower transfer gearbox from the tractor.

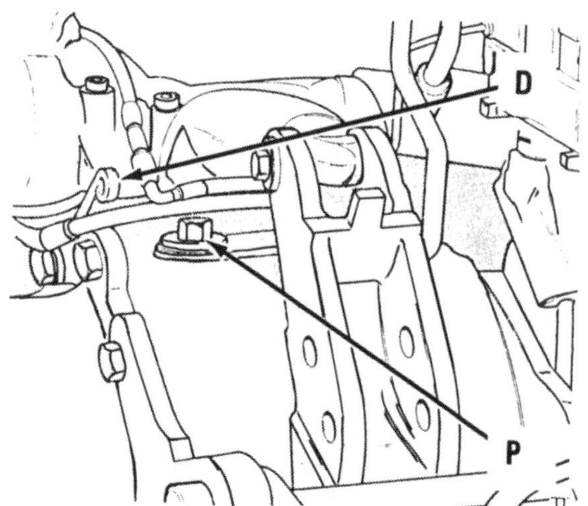

Fig. 19—Rear axle center housing should be filled to level marked on dipstick (D). Fill through hole for plug (P).

To disassemble the unit, remove the external oil line (8), remove the nut from the solenoid and remove solenoid coil. After the solenoid coil is removed, the core and valve assembly can be removed from the housing. Drive pin (15) out and remove idler support shaft (16), retainer (13) and needle bearing rollers (9). Remove idler gear and shaft assembly (12, 17 and 18). Unbolt and remove bracket (38) and both end plates (28 and 40). Use a suitable puller and remove output shaft (42) and bearing (41). Washer (31) must be used to compress springs (33) when removing snap ring (30).

Clean and inspect all parts and renew any showing excessive wear or other damage. Pilot valve assembly (27) can be removed by blowing into the cross passage with compressed air. When reassembling, renew all "O" rings, seals and gaskets. Reassemble by reversing disassembly procedure. Be sure that pilot valve assembly (27) is installed in end of output shaft (42) so that oil will pass through valve from transfer tube in end cover (28) into the shaft toward dog clutch (35), but not the other direction. Refer to Fig. 21. Tighten end plate cap screws to 40 N•m (29.5 ft.-lbs.) torque. Refer to Fig. 22 for brake installed on some models.

Reinstall transfer gearbox by reversing the removal procedure. Tighten retaining screws to 80 N•m (59 ft.-lbs.) torque. Fill rear axle center housing through plug (P—Fig. 19) to level of dipstick (D) with Ford M2C134-D/C or equivalent.

FRONT SUPPORT

All Models So Equipped

17. REMOVE AND REINSTALL. To remove the front support, the axle must be removed, the radiator must be removed from the support and the front support must be unbolted from the front of engine. Each of these heavy components and the tractor must be supported separately while removing, while separated and while assembling. Be sure that sufficient equipment is available before beginning.

Remove hood, drain cooling system and disconnect radiator hoses. Disconnect the air inlet hose from the front mounted air cleaner. Remove the grille and disconnect lines from oil coolers, if so equipped. Disconnect battery ground cable and wires to the headlights. Unbolt radiator shell upper support bracket from the shell. Remove the front drive shaft and shield as outlined in paragraph 9. Disconnect steering hoses that would interfere with the removal of the front axle or the front support. Support front of tractor in such a way that it will not interfere with removal of either the front support or the front axle. Remove axle as outlined in paragraph 10. Attach a hoist or other supporting device to the front support, then unbolt and separate the front support from the front of the engine.

Reattach front support to engine, tightening the retaining screws to 339-420 N•m (250-310 ft.-lbs.) for models with stamped oil pan; to 240-298 N•m (180-220 ft.-lbs.) torque for models with cast oil pan. Complete assembly by reversing the removal procedure.

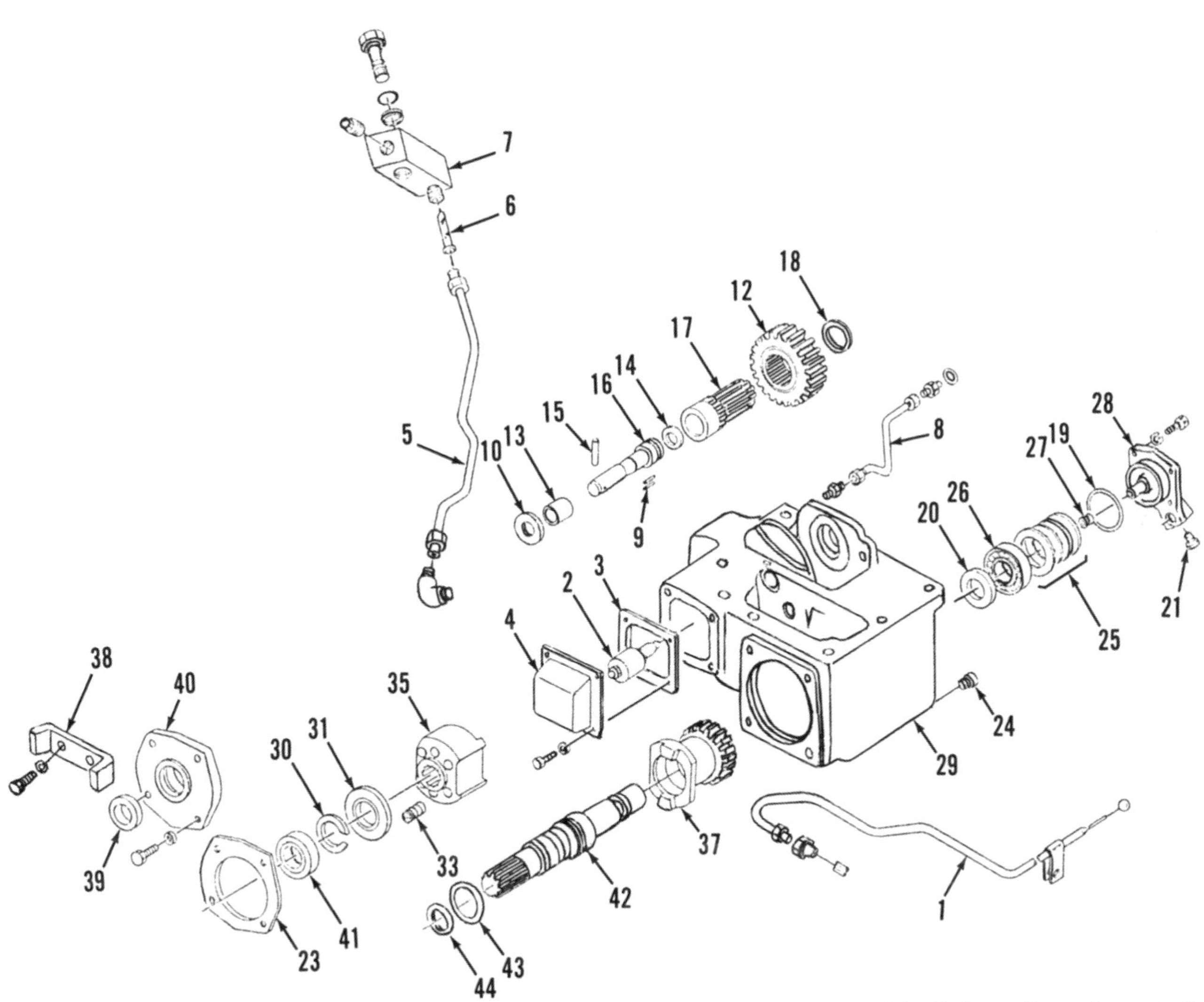

Fig. 20—Exploded view of four-wheel-drive transfer gearbox used on 4630 and 4830 models. Refer to Fig. 22 for parts that are different with four-wheel braking.

1. Electrical wire & protective tube
2. Control solenoid
3. Gasket
4. Cover
5. Oil pressure line
6. Filter screen
7. Fitting
8. External oil line
9. Bearing needles (48 used)
10. Thrust bearing
12. Idler gear
13. Retainer
14. Seal
15. Pin
16. Idler support shaft
17. Idler gear shaft
18. Snap ring
19. "O" ring
20. Washer
21. Plug
23. Gasket
24. Drain plug
25. Shims
26. Tapered roller bearing
27. Pilot valve assy.
28. Output shaft end plate
29. Housing
30. Snap ring
31. Washer
33. Springs (6 used)
35. Dog clutch coupler
37. Gear & clutch
38. Bracket
39. Oil seal
40. End plate
41. Tapered roller bearing
42. Output shaft
43. "O" ring
44. "O" ring

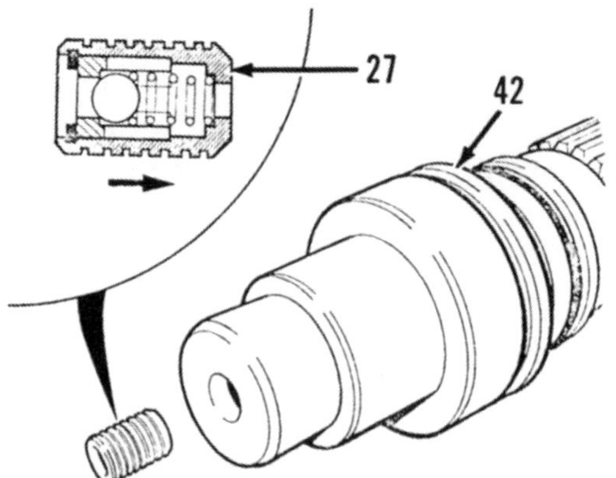

Fig. 21—The pilot valve assembly (27) must be installed as shown so that pressurized oil can pass through the check valve in the direction of the arrow to engage the front drive. Valve should slide easily in bore of shaft (42) during normal operation.

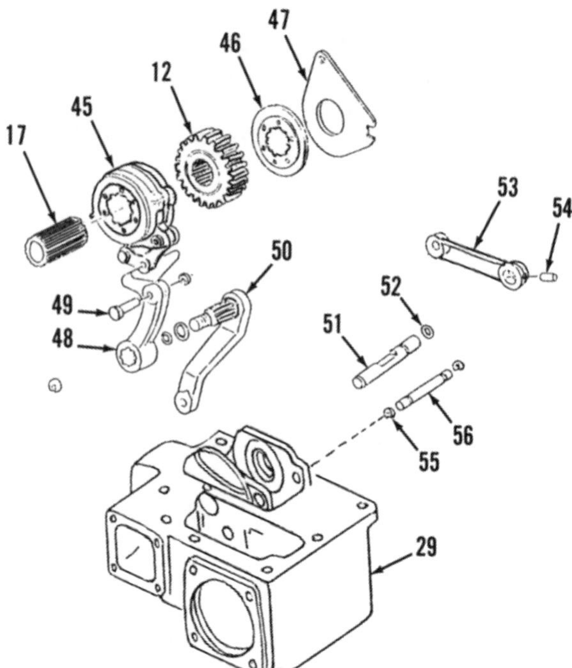

Fig. 22—Exploded view of parts used with transfer gearbox when equipped with four-wheel braking. Refer also to Fig. 20.

12. Idler gear
17. Idler gear shaft
29. Housing
45. Brake assembly
46. Disc
47. Plate
48. Lever
49. Pin
50. Lever
51. Operating shaft
52. "O" ring
53. Lever
54. Pin
55. "E" rings
56. Plate retaining shaft

MANUAL STEERING

STEERING GEAR

All Models So Equipped

18. ADJUSTMENT. Adjustment of the steering gear can be considered correct if:

1. There is no perceptible end play of either the steering (worm) shaft (6—Fig. 23) or rocker shaft (22).

2. A pull at the outer edge of the steering wheel of 0.45-1.13 kg. (4.5-12.2 N or 1-2½ lbs.) is required to pull the steering wheel past the mid-position (with the drag link disconnected from the steering arm). Adjustments can be made to correct excessive end play or turning effort, but adjustments are usually performed during reassembly of the gear unit after overhauling the unit.

19. WORMSHAFT END PLAY. Remove unit as outlined in paragraph 21. Remove side cover (25—Fig. 23), clean inside unit and inspect for damage or excessive wear. If no signs of wear or damage are noticed, wormshaft end play can be adjusted by adding or removing shims (14) and gaskets (15).

Unbolt steering column (4) from gear housing (26), then add or remove shims (14) and gaskets (15) so wormshaft (6) has zero end play, but turns freely. The approximate thickness of shims and gaskets (14 and 15) can be determined as follows: Install the wormshaft and associated parts (6 through 13). Install upper bearing (5) and steering column (4), but tighten retaining nuts only finger tight. Measure gap between lower flange of steering column and face of gear housing (26). Install paper gaskets and steel shims (14 and 15) equal to the measured gap. A paper gasket should be located on each side of shims to assure proper sealing. Tighten steering column retaining screws to 34 N•m (25 ft.-lbs.) torque if not equipped with cab; 41 N•m (30 ft.-lbs.) torque if equipped with cab.

20. ROCKER SHAFT END PLAY. First remove the unit as outlined in paragraph 21, inspect the unit for wear or damage and adjust the wormshaft end play as outlined in paragraph 19.

Be sure the rocker shaft (22—Fig. 23) and the ball nut (7 through 11) are in the center position, then install side cover (25) without gaskets or shims (23 and 24). Install retaining nuts and screws evenly, finger tight, then measure gap between housing (26) and cover (25). Install paper gaskets and steel shims (23 and 24) equal to the measured gap. A paper gasket should be located on each side of shims to ensure proper sealing. Tighten screws retaining cover (25) to 34 N•m (25 ft.-lbs.) torque if not equipped with cab; 18 N•m (14 ft.-lbs.) torque if equipped with cab.

21. REMOVE AND REINSTALL. To remove the steering gear assembly, disconnect battery, remove engine hood and disconnect drag link from steering arm. Remove cab from models so equipped. Unbolt steering gear assembly from transmission of all models, then lift assembly from the tractor.

When reinstalling, tighten retaining bolts to 88 N•m (65 ft.-lbs.) torque. Tighten steering arm to drag link retaining nut to 55 N•m (40 ft.-lbs.) torque on models without cab; 88 N•m (65 ft.-lbs.) torque on models with cab.

22. OVERHAUL. Remove steering gear unit as outlined in paragraph 21, refer to exploded view in Fig. 23 and proceed as follows:

Remove the nut retaining the steering drop arm (16) and use a suitable puller to remove arm from

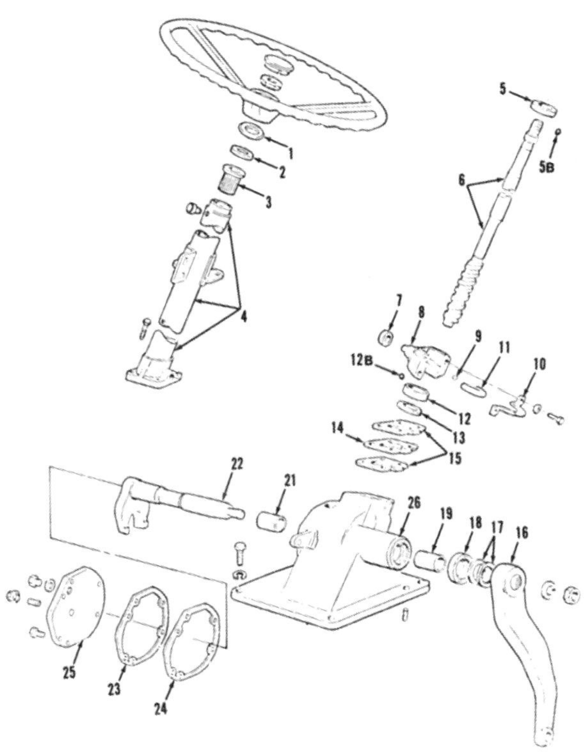

Fig. 23—Exploded view of the manual steering gear unit used on some models.

1. Grommet
2. Dust seal
3. Bushing
4. Steering column
5. Upper bearing race
5B. Steel balls-3/8 in. (10 used)
6. Wormshaft
7. Roller
8. Ball nut
9. Steel balls-3/8 in. (14 used)
10. Retainer
11. Tube
12. Lower bearing race
12B. Steel balls-3/8 in. (10 used)
13. Spacer
14. Shims
15. Gaskets
16. Steering arm
17. Dust seal
18. Oil seal
19. Bushing
21. Bushing
22. Rocker shaft
23. Shims
24. Gasket
25. Side cover
26. Gear housing

rocker shaft (22). Unbolt and remove the side cover (25), shims (24) and gaskets (23). Refer to Fig. 24. Remove roller (7—Fig. 23) from ball nut (8) and slide rocker shaft (22) from housing. Unbolt and remove steering column (4), shims (14) and gaskets (15). Remove bushing (3) from upper end of steering column if worn. Pull wormshaft (6) upward, then remove the 10 bearing balls (5B). Remove the wormshaft and ball nut assembly from gear housing as shown in Fig. 25, then remove the 10 bearing balls (12B—Fig. 23). Unscrew ball nut assembly from the wormshaft and remove the 14 recirculating balls (9) from the nut. Tube (11) can be removed if necessary. Remove lower bearing race (12), spacer (13), bushings (19 and 21) and oil seal (18) from housing if renewal is required.

To reassemble, first install new bushings (19 and 21) using piloted driver, then install new seal (18) with lip toward inside. Install spacer (13) and lower bearing race in gear housing, then stick the 10 loose bearing balls (12B) in race with grease (Fig. 26). Assemble tube (11—Fig. 23) to ball nut (8) if removed, then stick the 14 recirculating balls (9) in the tube and groove of nut with grease. Thread the ball nut assembly onto wormshaft, then install shaft and nut assembly in gear housing. Insert wormshaft into lower bearing; avoid dislodging loose bearing balls (12B). Hold wormshaft in bearing, place upper bearing race (5) over shaft. Invert the assembly, allowing the gear housing to rest against the end of shaft. Stick the 10 bearing balls (5B) in upper race with grease, then push bearing assembly up into housing. Hold upper bearing race in position while turning the assembly upright. Install column (4) over wormshaft and install retaining screws finger tight. Refer to paragraph 19 and install sufficient shims and gaskets (14 and 15) to remove end play without causing wormshaft to bind in bearings. Refer to Fig. 27.

Insert rocker shaft (22—Fig. 23), place roller (7) on end of ball nut (8) and install side cover (25). Tighten retaining nuts finger tight, then refer to paragraph 20. Install sufficient thickness of shims and gaskets (23 and 24) to remove rocker shaft end play without causing binding. Install steering arm (16) and tighten retaining nut to 162 N•m (120 ft.-lbs.) torque. Steering gearbox contains 0.6 liter (1.3 pints) of Ford M2C134-D/C or equivalent gear lubricant.

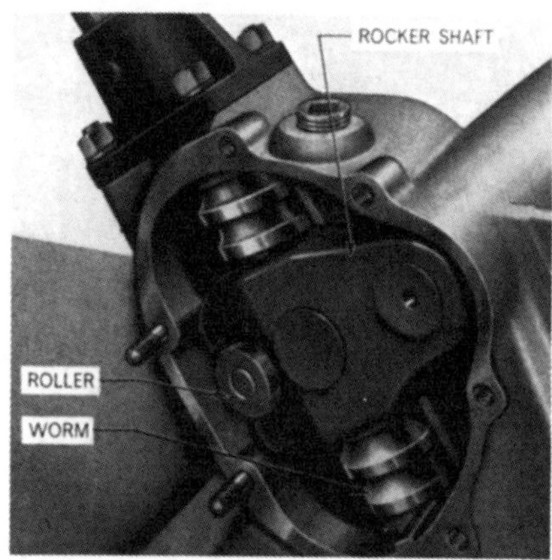

Fig. 24—View of manual steering gear assembly with side cover removed. Roller moves in slot located in side cover.

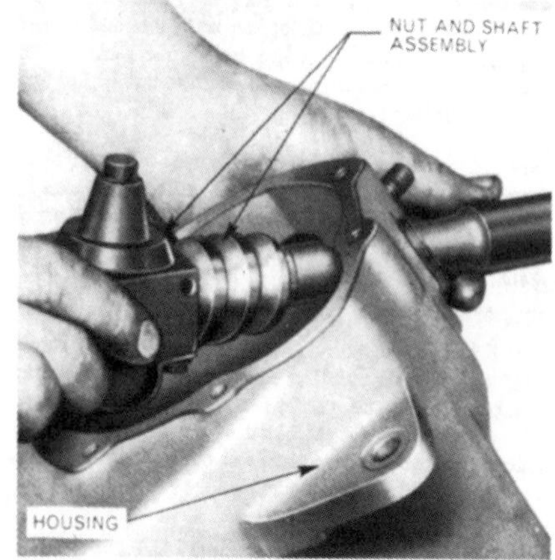

Fig. 25—Removing ball nut and steering shaft assembly from gear housing.

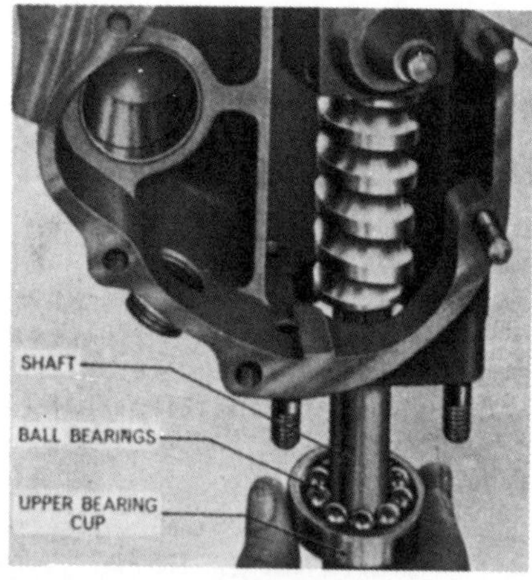

Fig. 26—Use grease to hold loose ball bearings in upper bearing race while installing.

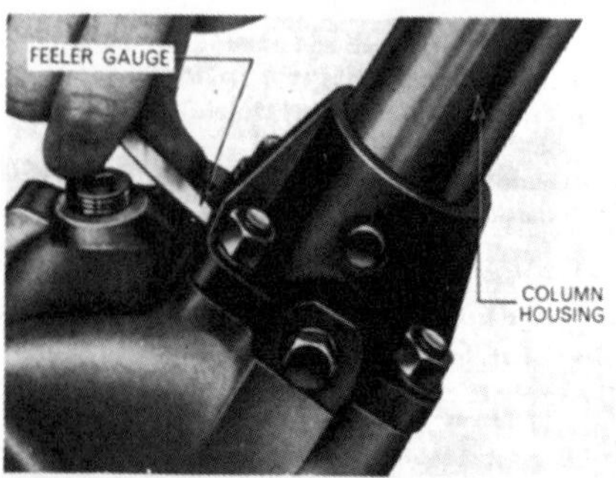

Fig. 27—The correct thickness of shims and gaskets to be installed can be determined by measuring gap between column and gearcase with a feeler gage as shown.

POWER STEERING

All Models So Equipped

23. OPERATION. Some models are equipped with a power assisted steering system. The control valve and assist piston are integral with the gear unit. A piston is built into the shaft ball nut and a cylinder is machined into the gear case housing. Control is by a rotary valve attached to the wormshaft. The lower end of the piston is pressurized through the external tube (24—Fig. 31) and the upper end is pressurized through an internal passage. The piston and ball nut assembly moves upward for right turns. Manual operation is possible because of check valve (7) located in the valve housing.

24. FLUID AND BLEEDING. Recommended power steering fluid is Ford M2C134-D/C or equivalent. Maintain fluid level at bottom of reservoir filler hole with tractor level. Fluid and filter should be changed if steering unit is overhauled, if fluid is suspected of contamination or at least after 1200 hours of operation.

The filter (3—Fig. 29) can be renewed as follows. Disconnect cooler lines at unions (U—Fig. 28) Remove screw (S) and move the oil cooler (C) and lines out of the way. Catch oil in a suitable container, remove screw (1) and pull reservoir from pump unit. Remove the filter assembly (3—Fig. 29) and sealing ring (4). Clean the inside of the reservoir and outside of the pump. Install new sealing ring (4) and filter, then install reservoir, aligning breather (B—Fig. 28) with lug (L) before tightening screw (1) to 16 N•m (12 ft.-lbs.) torque.

The power steering system is self bleeding. When the unit has been disassembled, refill with new oil, then start engine and cycle the system several times by turning the steering wheel from lock to lock. Recheck fluid level and add fluid as required. System is fully bled when no more air bubbles appear in reservoir as system is cycled.

SYSTEM PRESSURE

All Models So Equipped

25. CHECK AND ADJUST. The power steering pump incorporates a pressure relief valve (32—Fig. 30). System relief pressure should be 4137-4826 kPa (600-700 psi) for models without cab; 5861-6550 kPa (850-950 psi) for models equipped with cab. Normal pump flow with engine operating at 1000 rpm is 9.4 L/min. (2.5 gal./min.) for models without cab; 13.6 L/min. (3.6 gal./min.) for models equipped with cab.

To check the system relief pressure, install a "T" fitting in pump pressure line at pump (14). Connect a 0-7000 kPa (0-1000 psi) test gage to the fitting and operate engine at 1000 rpm. Turn the front wheels to one extreme against lock, and observe gage reading.

CAUTION: When checking system relief pressure, hold the steering wheel against lock only long enough to observe pressure indicated by gage. Pump may be damaged if steering wheel is held in this position too long or if flow is otherwise stopped.

Pump reservoir must be removed as described in paragraph 24 to adjust opening pressure. Remove relief valve assembly (24—Fig. 29). Thickness of

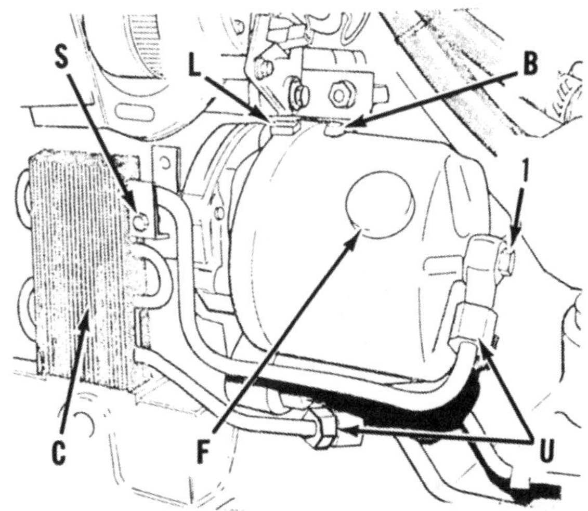

Fig. 28—The power steering pump and oil cooler typical of some models.

1. Reservoir retaining screw
B. Breather
C. Cooler
F. Filler cap
L. Lug
S. Screw
U. Fitting unions

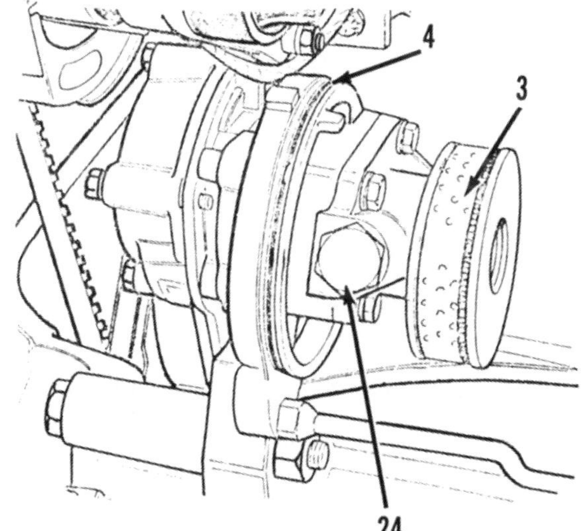

Fig. 29—View of power steering pump with reservoir removed.

shims (25—Fig. 30) controls pressure of relief valve spring (26). Valve seat (30) is threaded into valve body (24). Changing thickness of shim pack (25) 0.127 mm (0.005 inch) will change relief pressure approximately 345 kPa (50 psi). Be sure valve (29), seat (30), spring (26) and remaining parts (27, 28 and 31) are in new or serviceable condition before assembling. Tighten relief valve assembly (24—Fig. 29) to 22 N•m (17 ft.-lbs.) torque.

PUMP

All Models So Equipped

26. R&R AND OVERHAUL. Clean pump and area around pump thoroughly, then disconnect lines from pump and allow fluid to drain. Cover all openings to prevent dirt from entering pump or lines, then unbolt and remove pump from the engine front plate.

Remove screw (1—Fig. 30), then remove reservoir (2) and filter (3). The relief valve assembly (32) can be removed without further disassembly. Refer to paragraph 25 for checking and adjusting pressure.

Bend locking tab away, then remove nut (23), gear (22) and key (12). Mark relative position of flange housing (18), body (13) and cover (6), then remove bolts (5). Pump gears (10 and 11) are available only as matched set. Check condition of bearing blocks (9), gears (10 and 11) and body (13) for wear or other damage.

Always use new filter and seals when assembling. Tighten pump body through-bolts (5) to 21 N•m (15 ft.-lbs.) torque, nut (23) to 78 N•m (58 ft.-lbs.) torque and relief valve body (24) to 22 N•m (17 ft.-lbs.) torque.

Tighten pump retaining screws to 36 N•m (26 ft.-lbs.) torque. Refer to paragraph 24 for filling and bleeding.

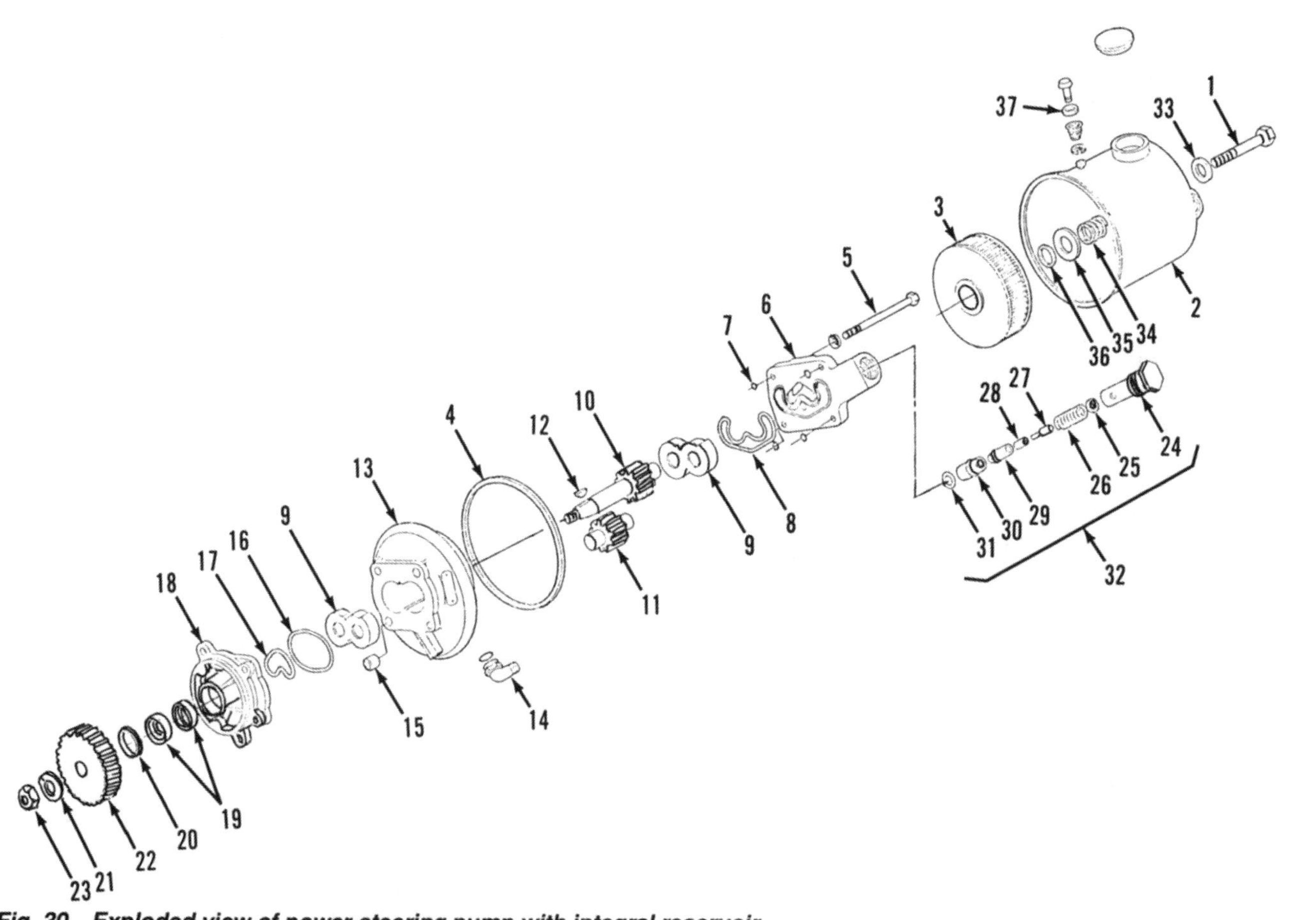

Fig. 30—Exploded view of power steering pump with integral reservoir.

1. Screw
2. Reservoir
3. Filter
4. Seal ring
5. Through-bolt
6. Cover
7. "O" ring
8. Seal ring
9. Bearing blocks
10. Drive gear & shaft
11. Driven gear
12. Woodruff key
13. Body
14. Outlet elbow
15. Dowel
16. Seal ring
17. Seal ring
18. Flange housing
19. Oil seal
20. Snap ring
21. Washer
22. Drive gear
23. Nut
24. Valve body
25. Shim pack
26. Spring
27. Spring guide
28. Seal
29. Valve head
30. Valve seat
31. "O" ring
32. Relief valve assembly
33. Seal
34. Spring
35. Washer
36. Retainer
37. Breather seal

STEERING GEAR, CONTROL VALVE AND CYLINDER

All Models So Equipped

27. REMOVE AND INSTALL UNIT. To remove the steering gear assembly, disconnect battery, remove the engine hood and disconnect drag link from the steering arm. Remove cab from models so equipped. Disconnect oil lines, unbolt steering gear assembly from the transmission of all models, then lift assembly from the tractor.

When reinstalling, tighten gear assembly to transmission retaining bolts to 88 N•m (65 ft.-lbs.) torque on models without cab; 122 N•m (90 ft.-lbs.) torque on models with cab. Tighten steering arm to drag link retaining nut to 55 N•m (40 ft.-lbs.) torque on models without cab; 88 N•m (65 ft.-lbs.) torque on models with cab. Refer to paragraph 24 for filling and bleeding procedures.

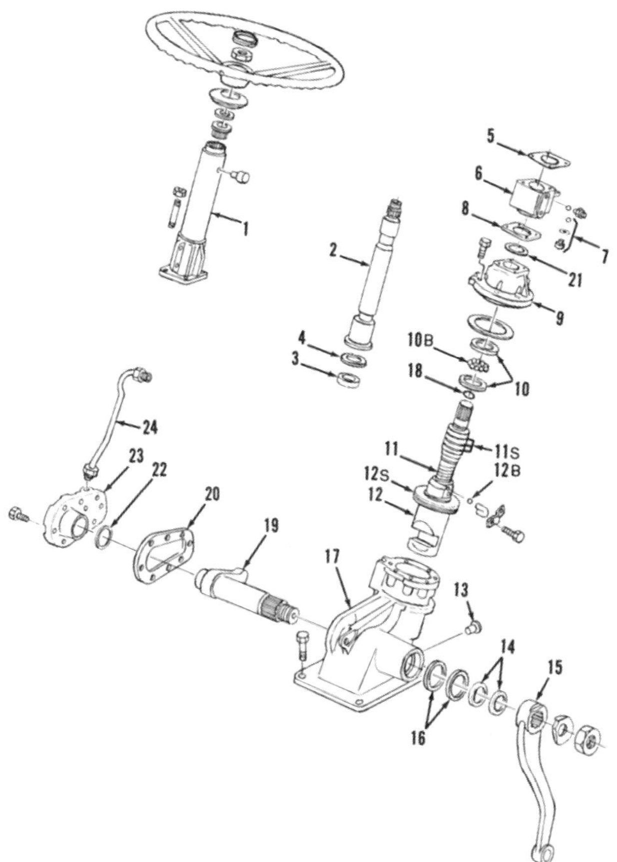

Fig. 31—Exploded view of power steering gearbox used on some models.

1. Steering column housing
2. Steering shaft
3. Housing seal
4. Washer
5. Shim pack
6. Valve housing
7. Check valve (3/8 in. ball)
8. Shim pack
9. Bearing housing
10. Bearing races
10B. Balls-5/16 in. (15 used)
11. Wormshaft
11S. Seal rings
12. Piston and ball nut
12B. Balls-5/16 in. (28 used)
12S. Seal rings
13. Guide peg
14. Dust seals
15. Steering arm
16. Oil seals
17. Gear housing
18. "O" ring
19. Rocker shaft
20. Gasket
21. Seal
22. Shims
23. Side cover
24. Pressure tube

28. OVERHAUL. Refer to paragraph 27 and remove steering gear unit. Temporarily reinstall steering wheel and disconnect external steering tube (24—Fig. 31). Turn steering wheel from lock to lock several times to pump as much fluid as possible from the housing.

Remove the steering arm (15), using a suitable puller. Remove the side cover (23), gasket (20) and shim (22). Turn steering shaft until rocker arm shaft arm is centered in the housing opening, then withdraw the rocker shaft (Fig. 32).

Remove the four stud nuts that attach steering column (1—Fig. 31), then lift off the column and shaft (2). Carefully remove and save shim pack (5). Install shim protector over steering shaft spline as shown in Fig. 33, then gently tap valve housing (6) away from bearing housing (9). Lift the valve housing off and carefully save the shims (8).

Remove screws retaining bearing housing (9—Fig. 31), then turn splined end of steering shaft (11) counterclockwise to force bearing housing (9) and bearing (10) up and out. Shaft bearing (10) contains 15 (5/16

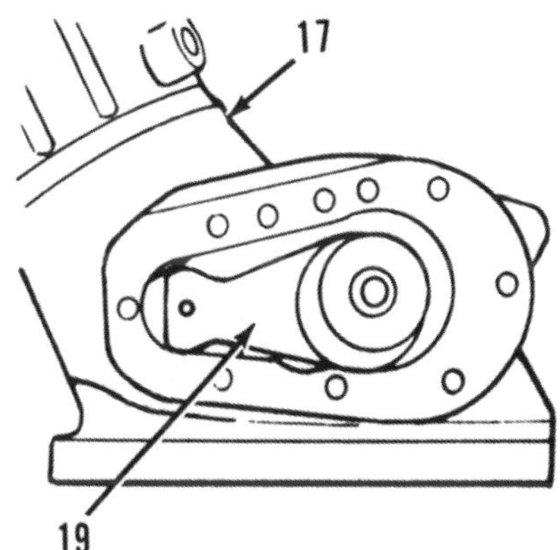

Fig. 32—Rocker shaft can be withdrawn when arm is centered in housing opening as shown.

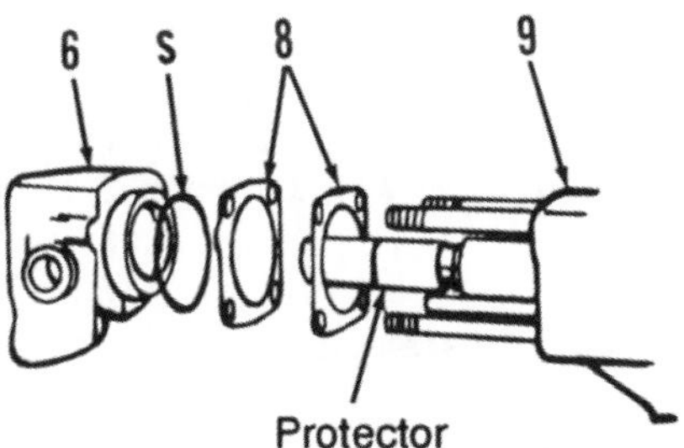

Fig. 33—Oil seal protector sleeve should be used to cover splines as shown. "O" ring seal is shown at (S).

inch diameter) loose steel balls. Balls can fall free as parts are removed. Bearing balls (10B) are the same as the 28 steel balls (12B) used in the steering nut, but should not be mixed when disassembling. Work through side opening and push ball nut (N—Fig. 34), piston and associated parts from the main housing. Groove (G) fits over guide peg (13) to prevent ball nut and piston from turning. Be careful not to damage piston rings (12S) as piston is withdrawn. Remove clamp bracket (B), transfer tube and the 28 bearing balls from ball nut.

Examine all parts for wear or scoring and make sure all parts are clean. Renew all seals, gaskets and "O" rings when assembling. "O" rings are located on piston guide peg (13) and between bushing sleeves of control valve as shown at (11S—Fig. 35). Carefully push bushings out top end of housing as shown in Fig. 35. Seal (3) must be installed from underside (chamfered) end of bushing (N) with a special tool that grips seal between the two seal lips. Coat sealing "O" rings sparingly with lubricant and install carefully using Fig. 35 as a guide.

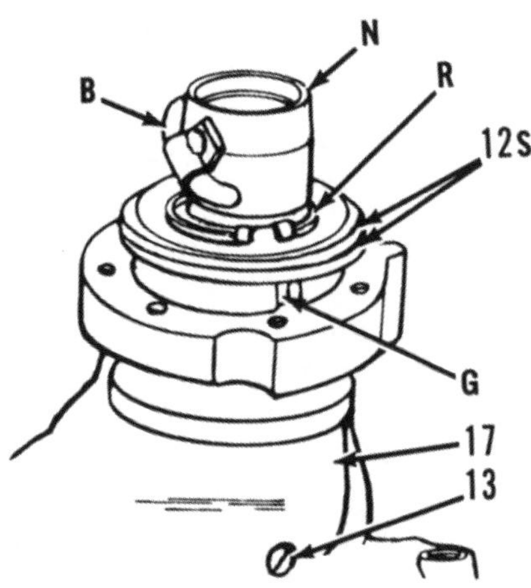

Fig. 34—Partially disassembled view of steering gear. Refer to text.

B. Clamp bracket
G. Locating groove
N. Ball nut
R. Snap ring

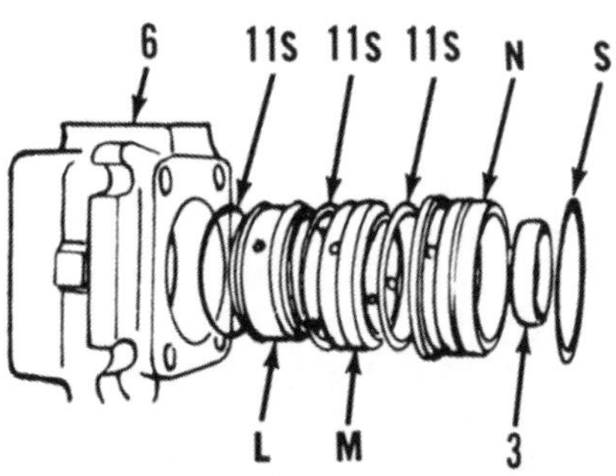

Fig. 35—Partially disassembled view of control valve.

3. Oil seal
6. Valve housing
11S. Seal rings
L. Lower sleeve
M. Middle sleeve
N. Upper sleeve
S. "O" ring

Install piston rings (12S—Fig. 34), if removed, and position end gaps 180 degrees apart. Align groove (G) with locating pin (13) and carefully install piston using a suitable ring compressor. Position ball nut assembly so that rocker shaft arm slot is aligned with main housing side of opening as shown in Fig. 32, then install rocker shaft. Temporarily install side cover (23—Fig. 31), using new gasket (20), but omitting shims (22). Tighten screws retaining side cover (23) to 54 N•m (40 ft.-lbs.) torque, then use a dial indicator to measure rocker shaft end float. Remove side cover (23) and install shims (22) equal to the measured end float minus 0.203 mm (0.008 inch). Reinstall side cover and tighten screws to 55 N•m (40 ft.-lbs.) torque.

Turn rocker shaft until ball nut and piston unit is at top of its stroke, then install wormshaft (11—Fig. 36) and lower half of transfer tube (T). Feed the 28 loose bearing balls (12B) into ball nut and transfer tube with grease, install the upper half of tube and

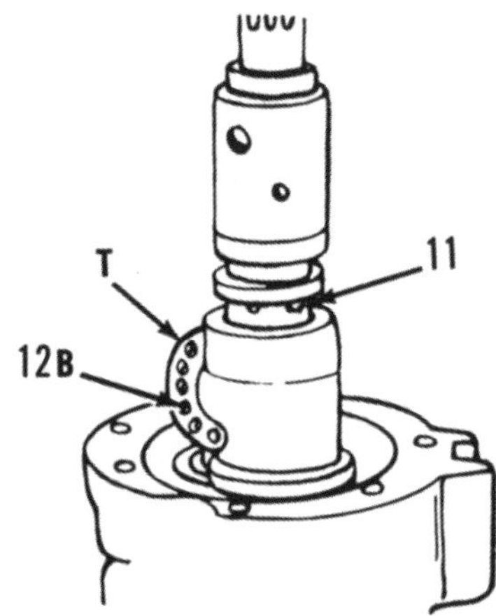

Fig. 36—Install wormshaft (11) and lower half of transfer tube (T) with piston at top of stroke. Feed the 28 loose balls (12B) into passage using clean grease. Refer to text.

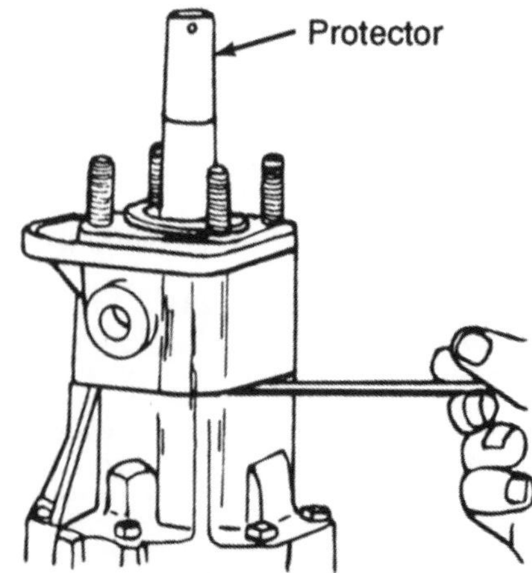

Fig. 37—Use protector sleeve to cover splines and install bottom valve housing without shims. Measure clearance as shown, then install shims equal to measured clearance minus 0.076 mm (0.003 inch).

clamp bracket (B—Fig. 34). Install bearing housing (9—Fig. 31) and tighten retaining screws to 24 N•m (18 ft.-lbs.) torque. Slide lower race of bearing (10) into bearing housing bore, grooved side up, then install 15 steel balls (10B) in bearing groove using clean grease. Install upper bearing race. Position seal protector over splines as shown in Fig. 37, then temporarily install control valve housing (6—Fig. 31) omitting shims (8). Be sure housing is bottomed, then measure gap between valve housing (6) and bearing housing (9) as shown in Fig. 37. Remove valve housing and install shim pack at (8—Fig. 31) equal to measured clearance **minus 0.076 mm (0.003 inch).** Be sure to measure clearance and shim pack thickness accurately. Shim pack (8) controls preload of worm gear bearing (10 and 10B). After installing correct thickness of shims (8), temporarily install the steering column (1), steering shaft (2), seal (3) and washer (4), omitting all shims (5). Be sure all parts are bottomed, then measure clearance between steering column and valve housing as shown in Fig. 38. Install shim pack (5—Fig. 31) equal to measured clearance **plus 0.127 mm (0.005 inch).** Be sure to measure clearance and shim pack thickness accurately. See Fig. 39. Tighten steering column retaining nuts to 34 N•m (25 ft.-lbs.) torque if not equipped with cab; 45 N•m (33 ft.-lbs.) torque if equipped with cab. Tighten nut retaining steering arm to 162 N•m (120 ft.-lbs.) torque if not equipped with cab; 305 N•m (225 ft.-lbs.) torque if equipped with cab.

Complete assembly by reversing disassembly procedure. Refer to paragraph 27 when installing. Tighten steering wheel retaining nut to 95 N•m (70 ft.-lbs.) torque.

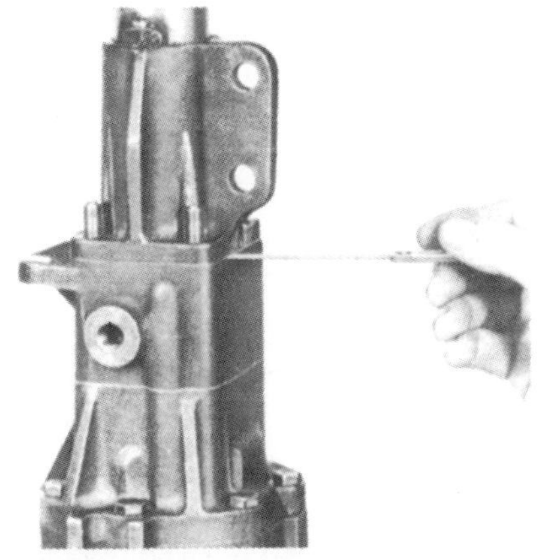

Fig. 38—Make sure steering column is bottomed in steering shaft splines, then measure clearance for upper shim pack as shown. Install shims equal to measured clearance PLUS 0.127 mm (0.005 inch).

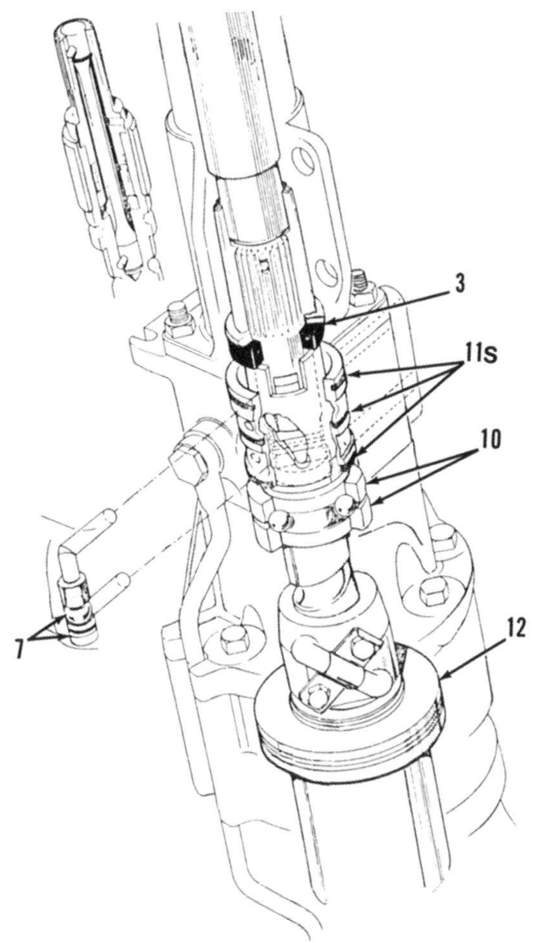

Fig. 39—Cross section of power steering control valve. Refer also to Fig. 31.

HYDROSTATIC STEERING

29. Hydrostatic steering system is used on some models and consists of a steering valve assembly, and one double acting steering cylinder. In the event of hydraulic failure or engine stoppage, manual steering can be accomplished by the gerotor pump in the steering valve. Refer to paragraph 18 and following for the manual steering used on some models or paragraph 23 and following for the power steering system used on other models.

FLUID AND BLEEDING

All Models With Hydrostatic Steering

30. Recommended steering fluid is Ford M2C134-D/C or equivalent. Maintain fluid level at bottom of reservoir filler hole with tractor level. Fluid and filter should be changed if steering unit is overhauled, if fluid is suspected of contamination or at least after 1200 hours of operation.

The filter (3—Fig. 41) can be renewed as follows. Disconnect cooler lines at unions (U—Fig. 40). Remove screw (S) and move oil cooler (C) and lines out of the way. Catch oil in a suitable container, remove screw (1) and pull reservoir from pump unit. Remove filter assembly (3—Fig. 41) and sealing ring (4). Clean inside the reservoir and outside the pump. Install new sealing ring (4) and filter, then install reservoir, aligning breather (B—Fig. 40) with lug (L) before tightening screw (1) to 16 N•m (12 ft.-lbs.) torque.

The hydrostatic steering system is self bleeding. Capacity is approximately 2.2 L (2.3 qts.). When the unit has been disassembled, refill with new oil, then start engine and cycle the system several times by turning the steering wheel from lock to lock. Recheck fluid level and add fluid as required. System is fully bled when no more air bubbles appear in reservoir as system is cycled.

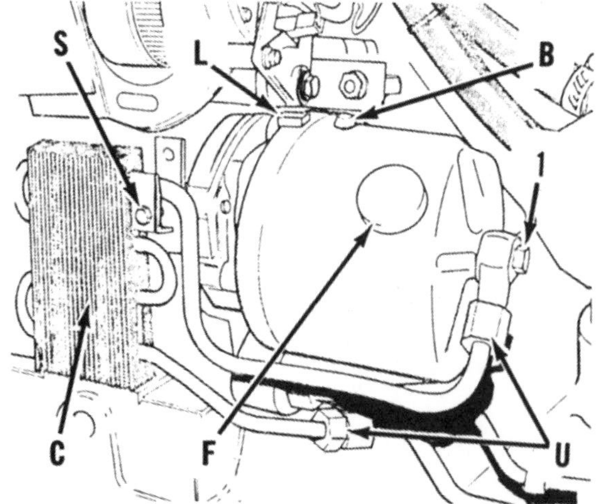

Fig. 40—The power steering pump and oil cooler typical of some models.

1. Reservoir retaining screw
B. Breather
C. Cooler
F. Filler cap
L. Lug
S. Screw
U. Fitting unions

TROUBLESHOOTING

All Models With Hydrostatic Steering

31. Some problems that may occur during operation of power steering and their possible causes are as follows:

1. Steering wheel hard to turn. Could be caused by:
   a. Defective power steering pump.
   b. Leaking or missing recirculating ball.
   c. Mechanical parts of front steering system binding.
   d. Ball bearings in steering column damaged.
   e. Leaking steering cylinder.
   f. Control valve spool and sleeve binding.

2. Steering wheel turns on its own. Could be caused by:
   a. Leaf springs in control valve weak or broken.

3. Steering wheel does not return to neutral position. Could be caused by:
   a. Control valve spool and sleeve jammed.
   b. Leakage between valve sleeve and housing.
   c. Dirt or metal chips between valve spool and sleeve.

4. Excessive steering wheel play. Could be caused by:
   a. Inner teeth of rotor or drive shaft teeth worn.
   b. Upper flange of drive shaft worn.
   c. Leaf springs in control valve weak or broken.
   d. Drive shaft teeth worn.

5. Steering wheel rotates at steering cylinder stops. Could be caused by:
   a. Excessive leakage in steering cylinder.
   b. Rotor and stator excessively worn.
   c. Excessive leakage between valve spool and sleeve.
   d. Excessive leakage between sleeve and housing.

6. Steering wheel "kicks" violently. Could be caused by:
   a. Incorrect adjustment between drive shaft and rotor.

7. Steering wheel responds too slowly. Could be caused by:
   a. Not enough oil.
   b. Steering control valve worn.

8. Tractor steers in wrong direction. Could be caused by:
   a. Hoses to steering cylinder incorrectly connected.
   b. Incorrect timing of drive shaft to rotor.

32. SYSTEM PRESSURE AND FLOW. The power steering pump incorporates a flow control valve (32—Fig. 41). Normal pump flow with engine operating at 1000 rpm is 13.6 L/min. (3.6 gal./min.). Pressure relief valve for the hydrostatic steering system is located in the steering control valve (32 through 36—Fig. 46). System relief pressure should be 9652-10,342 kPa (1400-1500 psi).

To check the system relief pressure, install a "T" fitting in pump pressure line at pump (14—Fig. 41). Connect a 0-21,000 kPa (0-3000 psi) test gage to the fitting and operate engine at 1000 rpm. Turn the front wheels to one extreme against lock, and observe gage reading.

CAUTION: When checking system relief pressure, hold the steering wheel against lock only long enough to observe pressure indicated by gage. Pump may be damaged if steering wheel is held in this position too long or if flow is otherwise stopped.

Turning the relief valve spring seat (34—Fig. 46) in, increases the pressure against spring (35) and increases system relief pressure. Spring seat plug (34) is accessible after removing plug (32).

Pump flow should be 32 L/minute (7.0 gpm) with engine operating at 1820 rpm. If flow is incorrect, valve can be disassembled and cleaned or renewed. Pump reservoir must be removed as described in paragraph 30 to remove and disassemble the flow control valve assembly (32—Fig. 41).

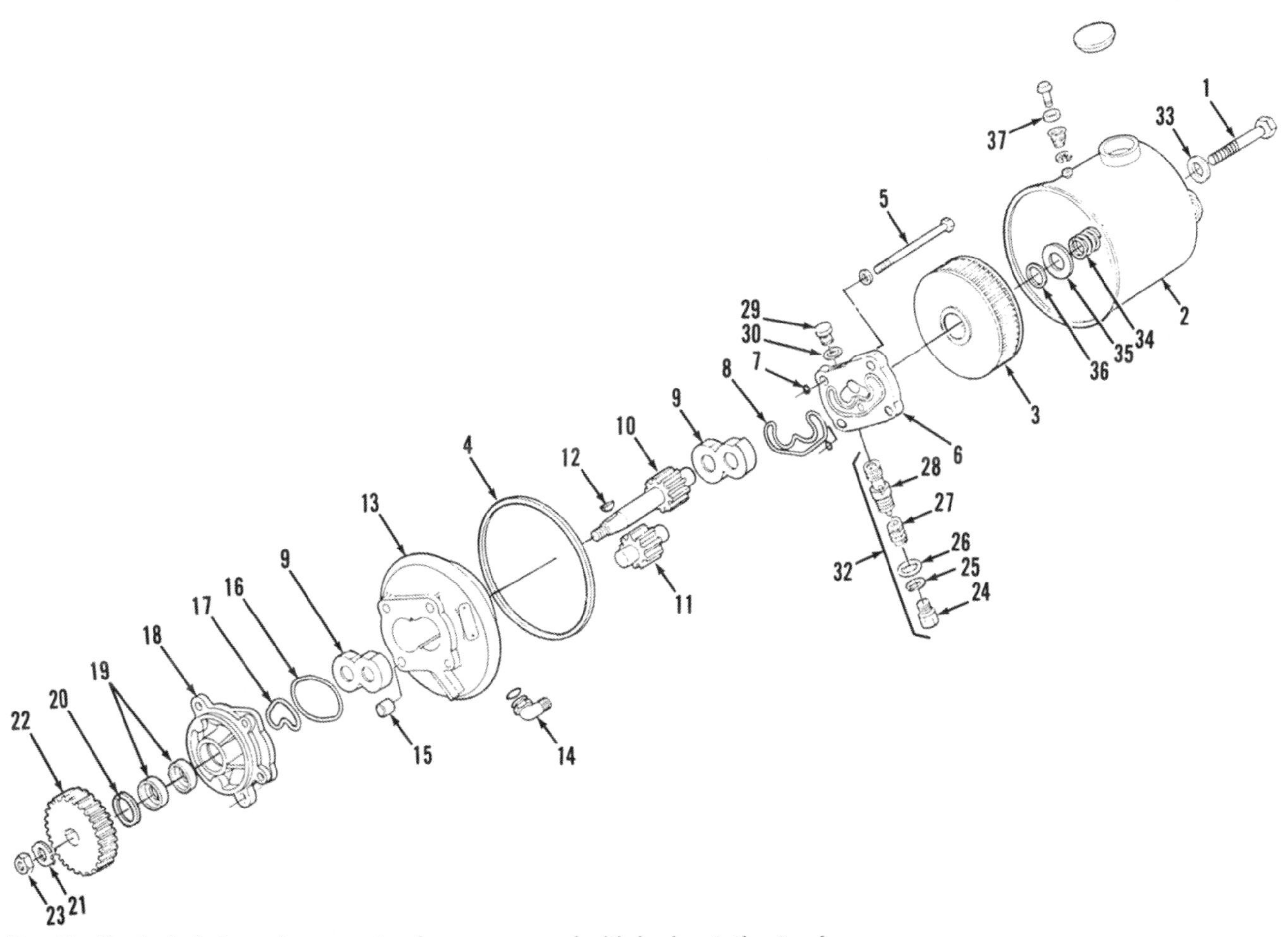

Fig. 41—Exploded view of power steering pump used with hydrostatic steering.

1. Screw
2. Reservoir
3. Filter
4. Seal ring
5. Through bolt
6. Cover
7. "O" ring
8. Seal ring
9. Bearing blocks
10. Drive gear & shaft
11. Driven gear
12. Woodruff key
13. Body
14. Outlet elbow
15. Dowel
16. Seal ring
17. Seal ring
18. Flange housing
19. Oil seal
20. Snap ring
21. Washer
22. Drive gear
23. Nut
24. Plug
25. Seal (same as 30)
26. Seal
27. Spring
28. Spool
29. Plug
30. Seal (same as 25)
32. Flow control valve
33. Seal
34. Spring
35. Washer
36. Retainer
37. Breather seal

PUMP

All Models With Hydrostatic Steering

33. REMOVE AND REINSTALL. Clean pump and area around pump thoroughly, then disconnect lines from pump and allow fluid to drain. Cover openings to prevent dirt from entering, then unbolt and remove pump from engine front plate.

Tighten pump retaining screws to 36 N•m (26 ft.-lbs.) torque. Refer to paragraph 30 for filling and bleeding.

34. OVERHAUL. Remove screw (1—Fig. 41), then remove reservoir (2) and filter (3). The flow control valve assembly (32) can be removed without further disassembly.

Bend locking tab away, then remove nut (23), gear (22) and key (12). Mark relative position of flange housing (18), body (13) and cover (6), then remove bolts (5). Pump gears (10 and 11) are available only as matched set. Check condition of bearing blocks (9), gears (10 and 11) and body (13) for wear or other damage.

Always use new filter and seals when assembling. Tighten pump body through bolts (5) to 21 N•m (15 ft.-lbs.) torque, nut (23) to 78 N•m (58 ft.-lbs.) torque.

POWER CYLINDER

Two-Wheel-Drive Models With Hydrostatic Steering

35. R&R AND OVERHAUL. The cylinder is attached between axle center member and the tie rod. To remove the cylinder, disconnect hydraulic hoses at steering cylinder and cover openings to prevent entrance of dirt, then detach both ends of steering cylinder.

Refer to Fig. 42 and Fig. 43. Clamp rod end (2) in a vise or holding fixture, remove clamp screw (1) and unscrew piston rod (3) from rod end. Clamp cylinder in a holding fixture that will not crush or deform cylinder, then unscrew extension tube and end cap (12). Pull piston and piston rod (3) from the cylinder.

When assembling, refer to Fig. 43 for installation of seals. Tighten end cap (12) to cylinder to 271 N•m (200 ft.-lbs.) torque.

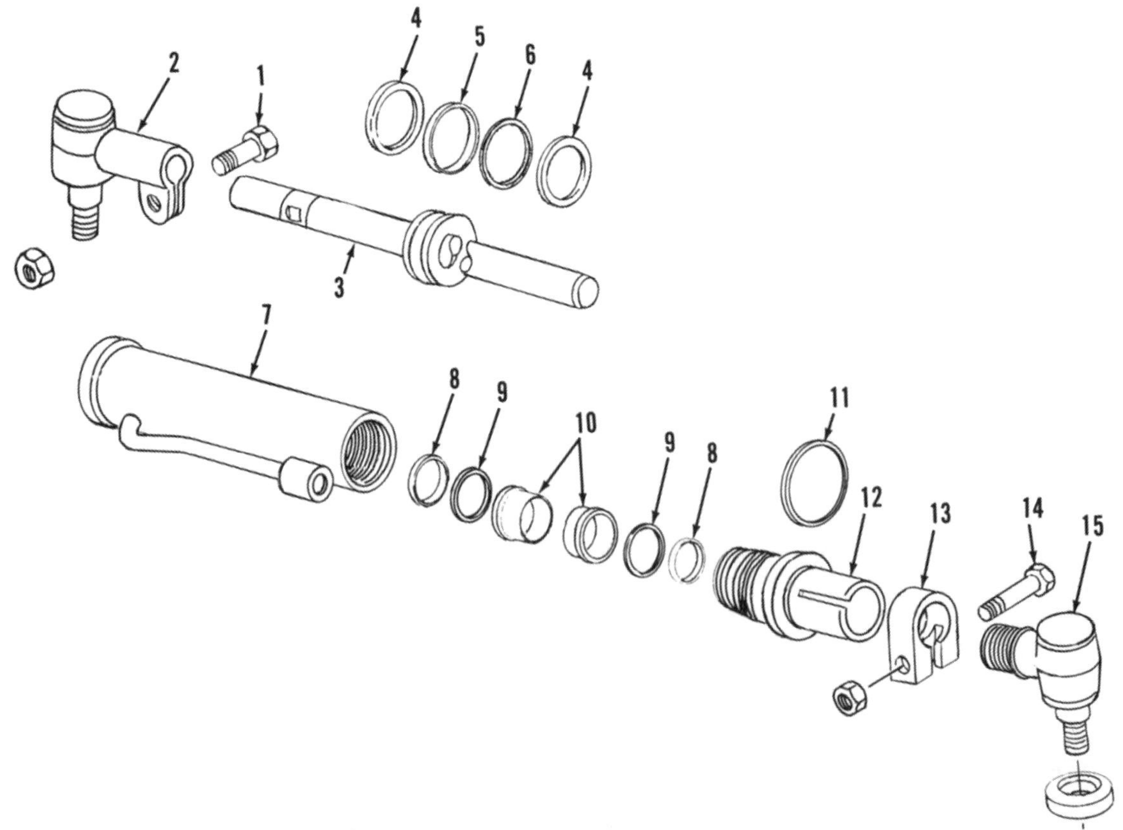

Fig. 42—Exploded view of separate steering cylinder used on two-wheel-drive models with hydrostatic steering.

1. Screw
2. Rod end
3. Piston & rod
4. Wear ring
5. Piston seal
6. "O" ring
7. Cylinder
8. Rod wiper
9. Rod seal
10. Bushing
11. "O" ring
12. Extension tube
13. Clamp
14. Clamp screw
15. Rod end

Narrow Tread Four-Wheel-Drive Models

36. R&R AND OVERHAUL. The cylinder is attached between axle and steering arm. To remove the cylinder, disconnect hydraulic hoses at steering cylinder and cover openings to prevent entrance of dirt, then detach both ends of the steering cylinder.

Refer to Fig. 44. Clamp the connector end of cylinder (4) in a holding fixture and unscrew cap (10). Withdraw piston rod, cap and piston assembly from cylinder. Clamp connector end of piston rod (12) in a holding fixture and remove nut (5). Remove piston (7) and end cap (10) from the piston rod. Seals (11) can be removed and new seals installed after piston rod is removed.

When assembling, tighten end cap (10) to cylinder to 271 N•m (200 ft.-lbs.) torque.

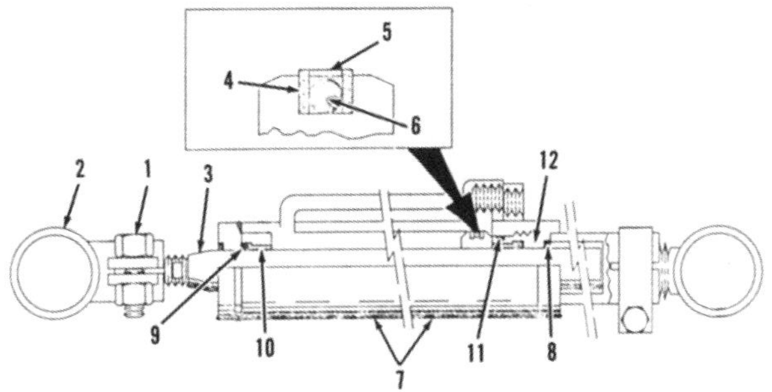

Fig. 43—Cross section of hydrostatic steering cylinder shown exploded in Fig. 42.

Standard Tread Four-Wheel-Drive Models

37. R&R AND OVERHAUL. The hydrostatic steering cylinder used on standard tread width four-wheel-drive models is integral with the front axle differential housing as shown in Fig. 45. The double-acting steering cylinder used on these models must be disassembled for removal. Housing containing the steering cylinder sleeve is an integral part of front drive axle differential housing.

To disassemble steering cylinder, first disconnect tie rod ends from steering arms. Loosen screws (12) from clamps (14) and unscrew left and right tie rods (15).

NOTE: Tie rod joints (13) may be locked with Loctite and heat must be applied to area before unscrewing from piston rod (1).

Disconnect hydraulic hoses at steering cylinder and cover openings to prevent entrance of dirt. Remove the four cap screws (4), then pull on end of piston rod (1) to remove internal parts from differential housing. Remove end plate (3) from cylinder (7), then pull piston rod and piston assembly out of cylinder (7). Remove the piston seal (10) and expander (9) from piston. Remove wiper seal (6) and piston rod seal (5).

Clean and inspect all parts for excessive wear, scoring or other damage and renew as necessary. Reassemble cylinder by reversing the disassembly procedure, observing the following. Always use new

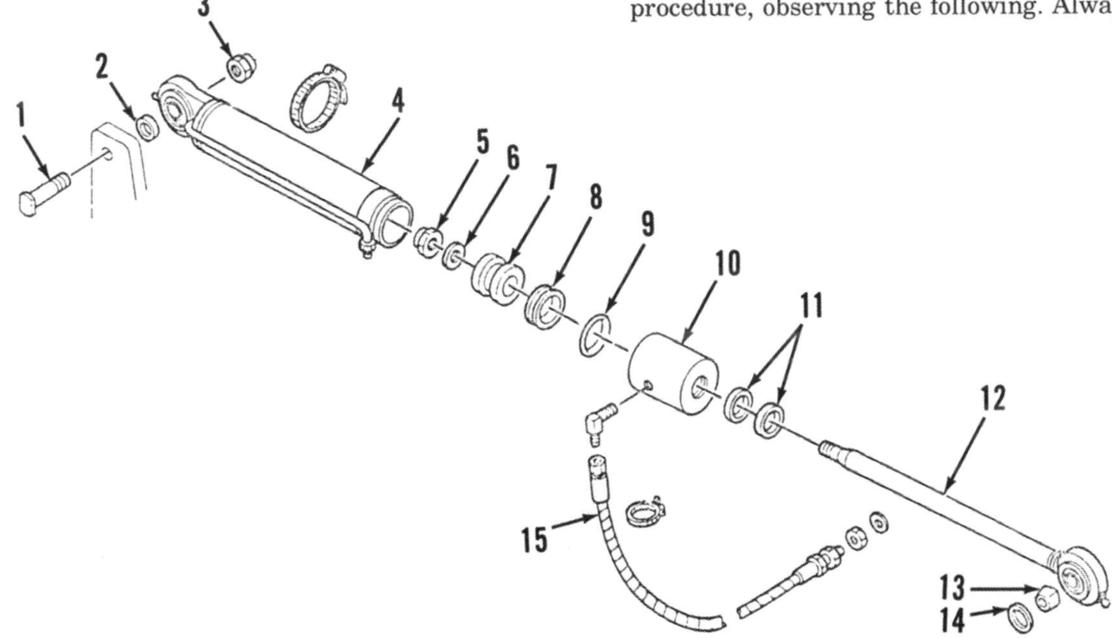

Fig. 44—Exploded view of separate hydrostatic steering cylinder used on narrow four-wheel-drive models. Standard width four-wheel-drive models with integral cylinder is shown in Fig. 45.

1. Special screws
2. Washer
3. Nut
4. Cylinder
5. Nut
6. Spacer
7. Piston
8. Back up ring
9. Seal
10. Bearing cap
11. Seals
12. Shaft
13. Bearing
14. Lock ring
15. Hose

"O" rings, scraper rings and seals when reassembling. Tighten cap screws (4) to 94 N•m (69 ft.-lbs.) torque.

Tighten tie rod joints (13) in piston rod (1) to 300 N•m (221 ft.-lbs.) torque. Tighten slotted nuts of tie rod ends attached to steering knuckles to 210 N•m (155 ft.-lbs.) torque. If necessary, adjust toe-in to 0-7.5 mm (0-0.3 inch). Tighten clamp screws (12) to 56 N•m (41 ft.-lbs.) torque.

STEERING VALVE

All Models With Hydrostatic Steering

38. REMOVE AND REINSTALL. Disconnect the pressure and return lines from the pump (U—Fig. 40) and drain fluid from the system. Reconnect the lines after draining to prevent the entrance of dirt. Remove the instrument console lower panels, then disconnect the four supply and return lines from the control valve. Cover openings to prevent the entrance of dirt. Remove the four bolts from the base of the steering column and remove the control valve.

Reinstall steering valve assembly by reversing removal procedure. Start engine and operate steering system lock to lock several times to purge air from system.

39. OVERHAUL. Remove steering valve as described in paragraph 38, then thoroughly clean exterior of unit. Unbolt and remove steering column, then place steering valve in a special holding fixture. Remove the seven cover retaining screws (10—Fig. 46). Remove cover (9), stator (7), rotor (6), spacer ring (2) and "O" rings (5 and 8). Remove distributor plate (4), drive shaft (1) and "O" ring (3). Suction valve parts (37 and 38) will fall from threaded holes. Hold steering valve vertically and turn valve spool and sleeve to align cross pin (20) parallel to flat side of housing. With cross pin in this position and housing in horizontal position, remove sleeve (22), spool (21), thrust bearing (18) and bearing races (17 and 19) from housing. Remove cross pin (20) from rotary valve and separate spool (21) from sleeve (22). Remove leaf springs (23 and 23S) from spool.

If necessary, the pressure relief valve (32 through 36) can be removed after unscrewing plugs (32 and 34); however, pressure must be checked and adjusted if plug (34) is turned. Steering relief pressure is adjusted by turning plug (34).

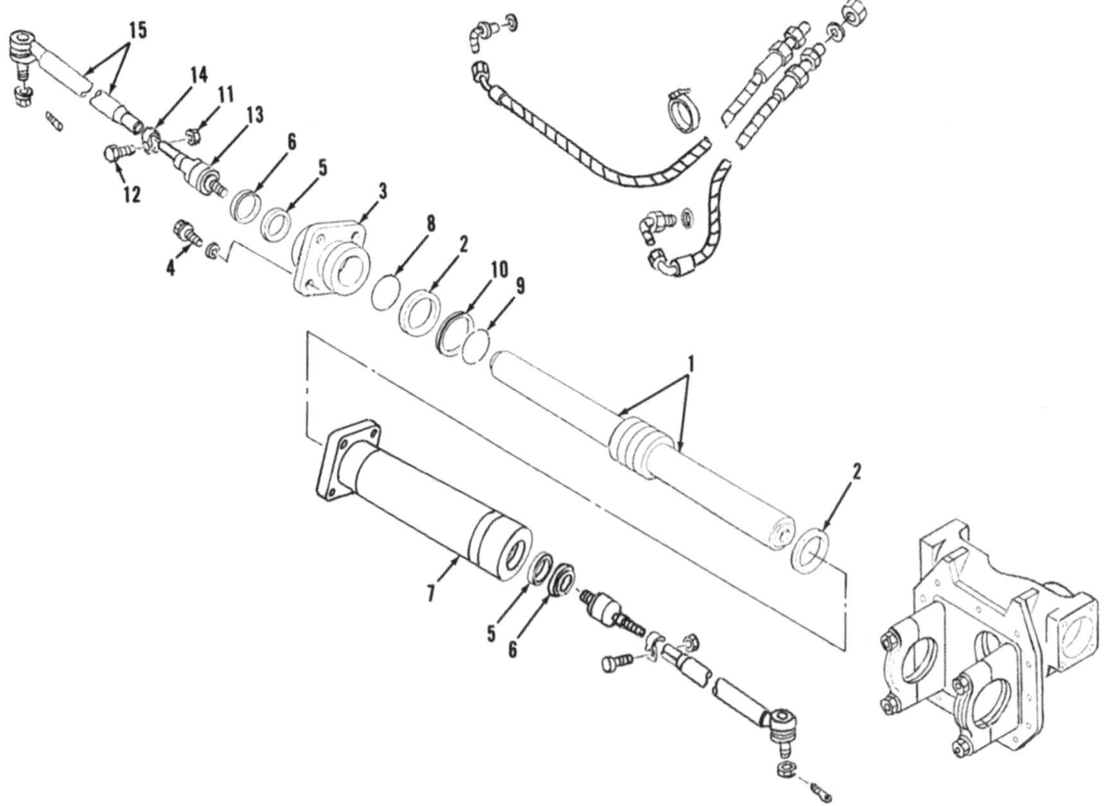

Fig. 45—Exploded view of double acting, integral steering cylinder used on four-wheel-drive models with standard tread width.

1. Piston & rod
2. Wear rings (2)
3. End plate
4. Cap screw (4)
5. Piston rod seals
6. Wiper seals
7. Cylinder
8. "O" ring
9. Expander ring
10. Piston seal
11. Nut
12. Screw
13. Tie rod joint
14. Clamp
15. Tie rod

Clean and inspect all parts for excessive wear or other damage and renew parts as necessary. Housing (15), spool (21) and sleeve (22) are available only as an assembly. Use all new "O" rings and seals when reassembling. Lubricate all interior parts with clean steering fluid.

Insert spool (21) into sleeve (22) aligning leaf spring slots and insert cross pin (20) into sleeve and spool. Insert the two flat springs (23S), then install the two arched springs (23) between the flat springs as shown in Fig. 47. Install retainer (24—Fig. 46) over leaf springs. Place bearing race (17), thrust bearing (18) and bearing race (19) on spool. Use Danfoss N. SJ.150-9000-11 or equivalent to install the "O" ring and back-up ring into position in outer spool, then insert complete valve spool assembly into housing (15). Lubricate "O" ring (3) and install in groove of housing, followed by valve plate (4). Make sure all the holes in valve plate and housing are aligned. Turn valve spool assembly so pin (20) is parallel with flat port face of housing (15) where hoses attach. Insert check valve ball (13) and plug (14) into hole indicated in Fig. 48. Insert the suction valve balls (37) and pins

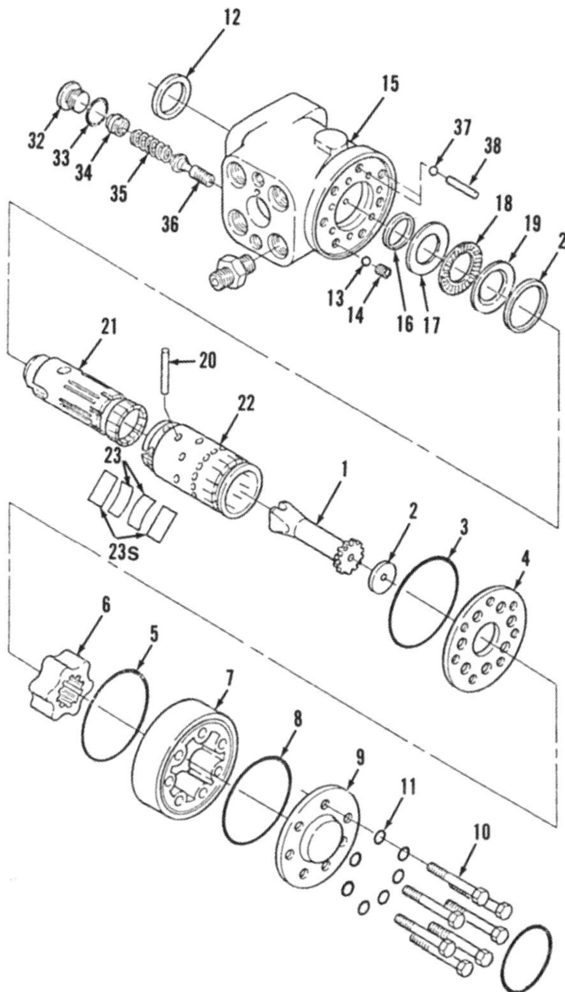

Fig. 46—Exploded view of hydrostatic steering valve with integral pressure relief valve (33 through 36). Notice that the two outer springs (23S) are straight.

1. Drive shaft
2. Spacer
3. "O" ring
4. Distributor plate
5. "O" ring
6. Rotor
7. Stator
8. "O" ring
9. End cover
10. Cap screw (7)
11. Seal rings
12. Seal
13. Check valve ball
14. Check valve plug
15. Housing
16. Seal
17. Bearing race
18. Thrust bearing
19. Bearing race
20. Cross pin
21. Valve spool
22. Valve sleeve
23. Leaf springs (2)
23S. Flat leaf springs (2)
24. Leaf spring retainer ring
32. Plug
33. "O" ring
34. Adjusting plug
35. Relief valve spring
36. Relief valve
37. Suction valve ball
38. Suction valve pin

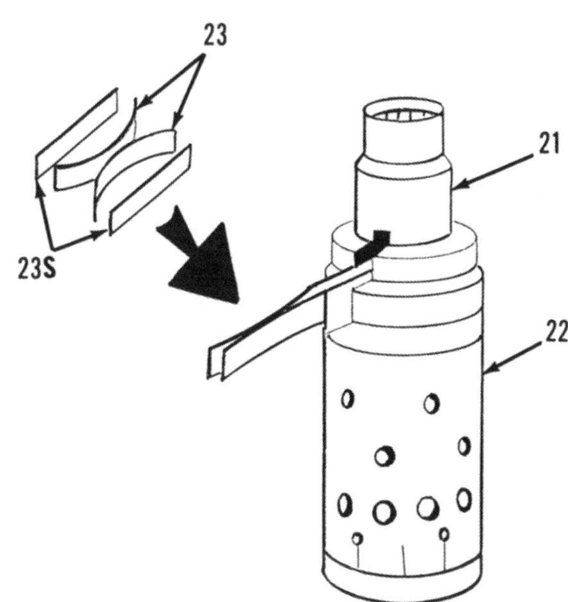

Fig. 47—Arch of the two center leaf springs (23) should be together and the flat springs (23S) should be on the outside as shown.

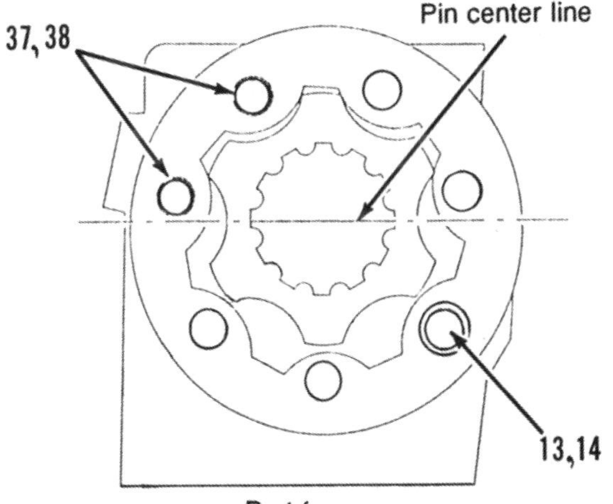

Fig. 48—The center line of pin (20—Fig. 46) should be aligned parallel with the port face of housing as shown when assembling. Refer to text.

(38) in the two holes indicated. Scribe a reference line across splined end of rotor shaft (1—Fig. 46) parallel with the groove at other end for pin. Be sure that valve spool pin is still aligned with the port face of the housing and install the rotor shaft as shown in Fig. 48. When correctly assembled, one lobe of the rotor will be positioned straight away from the port face (up in Fig. 48) when pin center line is located as shown. Install "O" ring (5—Fig. 46), pump body (7) and pump rotor (6). Install spacer (2), "O" ring (8) and end cover (9). Install the seven retaining screws (10). The one screw with the pin attached should be in location identified as "7" in Fig. 49. Tighten the screws in the order shown in Fig. 49, first to 10.8 N•m (8 ft.-lbs.), then to 28.4 N•m (21 ft.-lbs.) torque. Install relief valve assembly (32 through 36—Fig. 46) if removed. If setting of plug (34) has been disturbed, refer to paragraph 32 for checking and adjusting relief valve pressure.

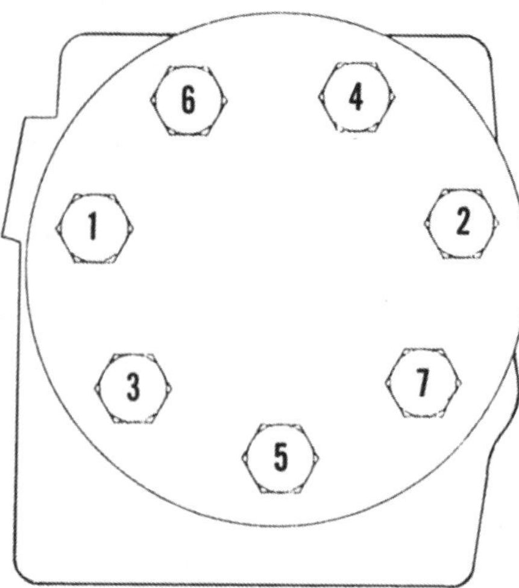

Fig. 49—The one screw with pin attached should be installed in position (7). Numbers indicate the order in which screws should be tightened.

ENGINE AND COMPONENTS

R&R ENGINE AND CLUTCH

All Models

40. Disconnect battery, remove vertical exhaust muffler (models so equipped), then unbolt and remove the complete hood. On models without cab, disconnect electrical connectors between front and rear wiring harnesses. For access to wiring on models with cab, remove steering column access panel. Disconnect electrical connectors between front wiring harness and safety start switch sender wire, windshield washer motor sender wire and rear wiring harness at multi-connector. Remove any connectors retaining front harness to engine.

Disconnect proofmeter drive cable from oil pump drive gear on left side of engine. Disconnect power steering or hydrostatic steering pressure and return tubes at the pump and reservoir unit and at hydrostatic steering cylinder. If so equipped, disconnect drag link from steering arm. Disconnect support struts from rear hood assembly. On models with cab, remove retaining bolts, then withdraw rear hood support. Remove bolts and nuts securing front end of fuel tank. On models without cab, remove nuts and bolts mounting the battery support bracket to rear hood panel. Disconnect wiring from starter motor, then unbolt and remove the starter motor.

On models with cab, close heater water control valves and disconnect and plug the hoses. On all models, disconnect the fuel shut-off cable and throttle control rod from fuel injection pump. Detach fuel leak-off tube from neck of fuel tank filler tube. Close fuel shut-off valve, then disconnect fuel line from shut-off valve at bottom of tank. If equipped with low exhaust, disconnect exhaust pipe from engine manifold. If equipped with engine oil cooler, disconnect and drain lines and position out of the way. If equipped with air conditioning, discharge refrigerant from the system using a refrigerant recovery/recycling machine. Then disconnect and plug interfering lines. Refer to paragraph 9 and remove the drive shaft, if equipped with four-wheel drive.

Use split support stands designed for this series of tractors, if available. Insert wedges between front axle and front support to prevent tipping. Place supports under the front end of transmission housing and support engine so that engine and front axle assembly can be moved forward away from transmission. Remove the flywheel inspection cover and engine to transmission housing cap screws, then carefully move engine and front axle assembly straight forward away from transmission and rear part of tractor.

Disconnect interfering hoses and remove necessary parts, then support the front axle and engine separately. Be sure that both engine and front axle are securely supported, then remove screws attaching engine to front axle and separate.

Reassemble tractor by reversing the disassembly procedure. Refer to Fig. 50 for recommended torque when tightening engine to transmission screws. Refill and bleed systems as outlined in specific sections. Evacuate and recharge air conditioner if so equipped.

CYLINDER HEAD

All Models

41. REMOVE AND REINSTALL. To remove the cylinder head, proceed as follows: Remove the vertical muffler, if so equipped, and remove the engine hood. Disconnect exhaust pipe from exhaust manifold of models with low exhaust. Remove the exhaust manifold, battery, battery tray and battery support brackets from all models. Drain cooling system and remove the upper radiator hose. Disconnect the air intake hose from intake manifold, shut fuel supply valve at tank and remove the complete fuel filter assembly. Remove fuel injectors as outlined in paragraph 86. Remove the intake manifold. Disconnect ventilation tube from rocker cover and remove the rocker cover. Remove rocker arm assembly and push rods. Remove the cylinder head retaining screws, then lift cylinder head from engine.

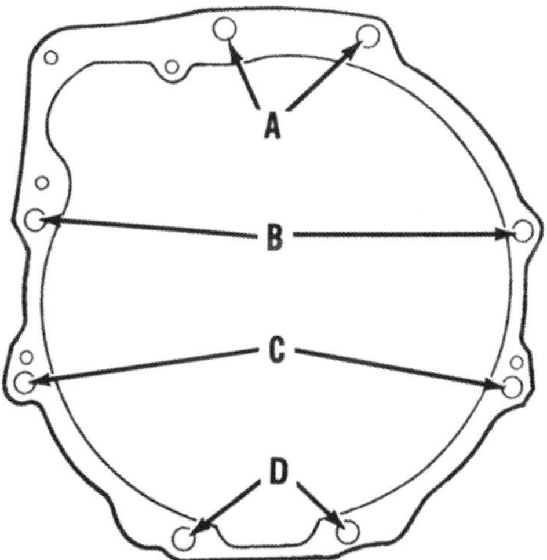

Fig. 50—Tighten bolts attaching engine to transmission housing to the following torque values.

A. 373-460 N•m (275-340 ft.-lbs.)
B. 190-230 N•m (140-170 ft.-lbs.)
C. 224-271 N•m (165-200 ft.-lbs.)
D. 57-76 N•m (42-56 ft.-lbs.)

NOTE: If cylinder head gasket has failed, check the mating surfaces of the cylinder head and block for flatness or evidence of erosion. Maximum allowable deviation from flatness is 0.15 mm (0.006 inch) overall or 0.08 mm (0.003 inch) in any 152 mm (6 inches). If cylinder head is not within specified flatness or is rough, the surface may be machined lightly, but lower surface of valve seat insert should never be less than 1.63 mm (0.064 inch) from machined gasket surface. Cylinder block may be machined lightly if not within specified flatness, but top of piston at TDC must not stand more than 0.3 mm (0.012 inch) above the gasket surface.

Do not use gasket sealer or compound when reassembling and be sure that gasket is correctly positioned on the two dowel pins. Tighten all of the cylinder head retaining screws in the sequence shown in Fig. 51 in three steps. First tighten screws to 122 N•m (90 ft.-lbs.); second tighten all screws to 135 N•m (100 ft.-lbs.); then tighten screws to 149 N•m (110 ft.-lbs.) final torque. Cylinder head retaining screws should only be tightened when engine is cold. Adjust valve clearance as described in paragraph 44. Reassemble by reversing disassembly procedure. Tighten inlet manifold retaining screws to 35 N•m (26 ft.-lbs.), exhaust manifold screws to 38 N•m (28 ft.-lbs.) and exhaust pipe to flange nuts to 31 N•m (23 ft.-lbs.) torque. Reinstall rocker arm cover retaining screws to 18 N•m (13 ft.-lbs.) torque. Bleed fuel injection system as outlined in paragraph 79.

VALVES, STEM SEALS AND SEATS

All Models

42. Exhaust valves are equipped with positive-type valve rotators and an "O" ring-type seal is used between the valve stem and the rotator body. Inlet valves stems are fitted with umbrella-type oil seals. Both inlet and exhaust valves seat on renewable-type valve seat inserts that are a shrink fit in cylinder head. Inserts are available in oversizes as well as standard sizes. Valves with 0.003 or 0.015 inch oversize stems are available for service.

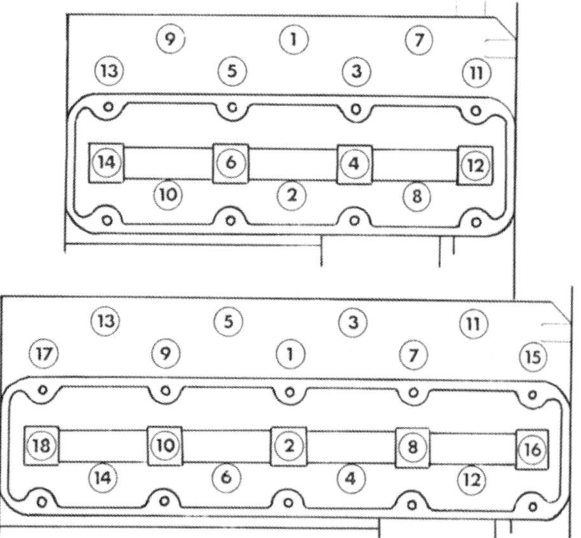

Fig. 51—Cylinder head retaining screws should be tightened in the sequence shown. Refer to text for torque value and procedure.

Correct face angle for both inlet and exhaust valves is within range of 44 degrees 15 minutes to 44 degrees 30 minutes (44.25-44.5 degrees). Valve seat angle should be 45 degrees to 45 degrees 30 minutes (45.0-45.5 degrees). Refer to the following valve, seat, stem and guide specifications:

Valve stem diameter—
Inlet. . . . . . 9.426-9.444 mm (0.3711-0.3718 inch)
Exhaust. . . 9.400-9.418 mm (0.3701-0.3708 inch)

Valve stem to guide clearance—
Inlet. . . . . . 0.025-0.069 mm (0.0010-0.0027 inch)
Exhaust. . . 0.051-0.094 mm (0.0020-0.0037 inch)

Valve seat width—
Inlet. . . . . 2.03-2.59 mm (0.080-0.102 inch)
Exhaust. . . . . . . 2.13-2.69 mm (0.084-0.106 inch)

Valve seat run out, Inlet and Exhaust—
Maximum . . . . 0.038 mm (0.0015 inch)

VALVE GUIDES AND SPRINGS

All Models

43. Inlet and exhaust valve guides are integral with the cylinder head and are not renewable. Valve stem to guide clearance should be 0.025-0.069 mm (0.0010-0.0027 inch) for the inlet valves; 0.051-0.094 mm (0.0020-0.0037 inch) for exhaust valves. Both inlet and exhaust valves with 0.003 or 0.015 inch oversize stems are available for service. Guides can be reamed oversize and valve with oversize stem fitted. Machine valve seat after reaming guide to make sure seat is concentric with guide.

Inlet and exhaust valve springs are interchangeable. Free length should be 54.6 mm (2.15 inches). Springs should be straight and square when removed. Renew spring if not square, shows evidence of rusting or is otherwise questionable. Springs should exert the following pressure when compressed to the following test heights:

Pressure at 44.20 mm . . . . 27.7-31.3 kg
(1.74 inches) . . . . (61-69 lbs.)

Pressure at 33.53 mm . . . . 57.8-63.1 kg
(1.32 inches) . . . . (125-139 lbs.)

VALVE CLEARANCE

All Models

44. CHECK AND ADJUST. Valve clearance (tappet gap) should be set with engine cold. Correct clearance is 0.36-0.46 mm (0.014-0.018 inch) for inlet; 0.43-0.48 mm (0.017-0.019 inch) for exhaust valves.

Valves can be adjusted statically using Fig. 52 and Fig. 53 as a guide with the crankshaft set in only two positions. Turn the crankshaft until the "0" degree mark on flywheel is aligned with the timing pointer as shown in Fig. 54, then check the two front (No. 1 cylinder) rocker arms. If rocker arms are loose, No. 1 cylinder is on compression stroke and the valves shown in Fig. 52 can be adjusted. If the front two rocker arms are tight, adjust the valves shown in Fig. 53. In either event, turn the crankshaft one complete revolution until "0" degree mark (Fig. 54) is again aligned and adjust remaining valves as indicated in other illustration. Reinstall rocker arm cover retaining screws to 18 N•m (13 ft.-lbs.) torque.

EXHAUST VALVE ROTATORS

All Models

45. The positive-type valve rotators (17—Fig. 55) need no maintenance, but each exhaust valve should rotate slightly when engine is running. Observe the valves with the engine running and install a new rotator on any exhaust valve that does not turn.

CAM FOLLOWERS

All Models

46. The semi-mushroom-type cam followers (tappets) can be removed after removing the engine cam-

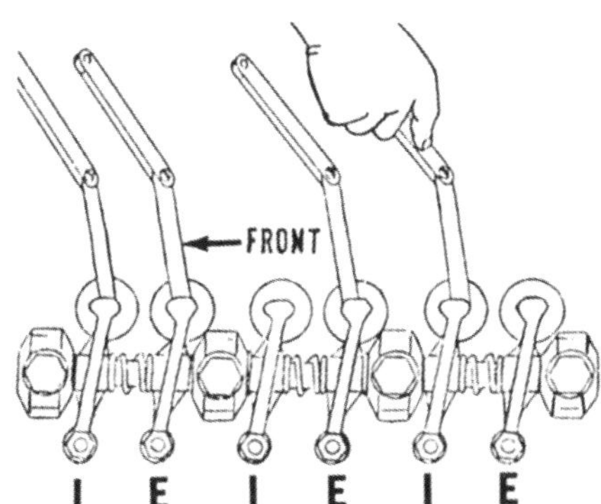

Three cylinder models

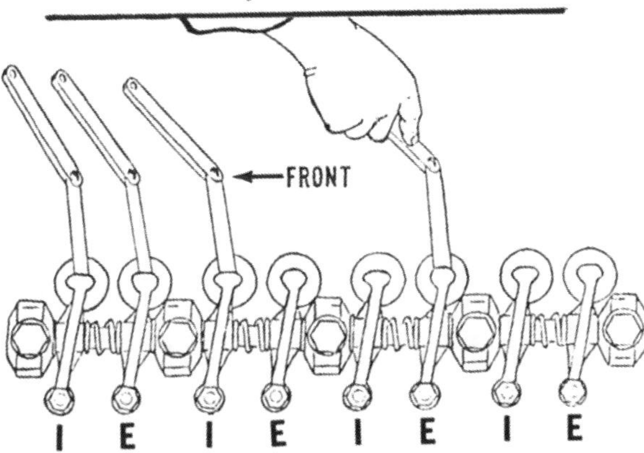

Four cylinder models

Fig. 52—With "0" degree (TDC) mark aligned and number one (front) cylinder on compression stroke, adjust the indicated valves. Turn the crankshaft one complete revolution until the "0" mark is again aligned and adjust the valves indicated in Fig. 53. Illustration for three-cylinder models is shown at top and four-cylinder models is shown at bottom.

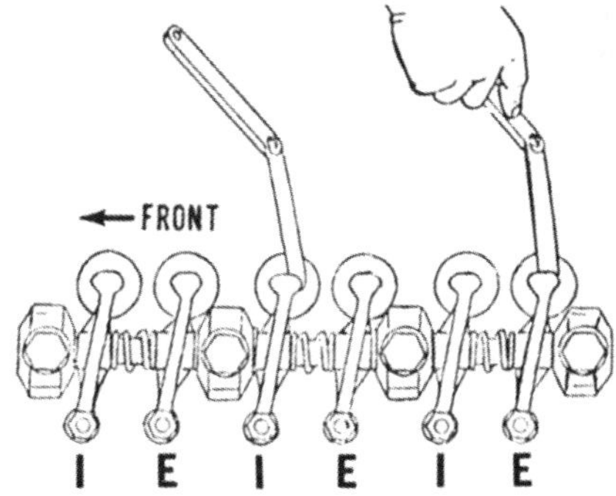

Three cylinder models

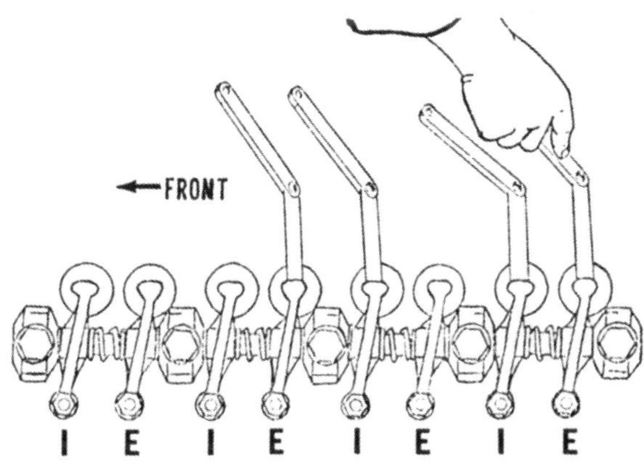

Four cylinder models

Fig. 53—With "0" degree (TDC) mark aligned and number one (front) cylinder on exhaust stroke, adjust the indicated valves. Turn the crankshaft one complete revolution until the "0" degree mark is again aligned and adjust the valves indicated in Fig. 52. Illustration for three-cylinder models is shown at top and four-cylinder models is shown at bottom.

shaft as outlined in paragraph 54. Cam followers are 25.118-25.130 mm (0.9889-0.9894 inch) diameter and operate in unbushed cylinder block bores. Desired clearance is 0.015-0.053 mm (0.0006-0.0021 inch) in the 25.15-25.17 mm (0.990-0.991 inch) diameter bores.

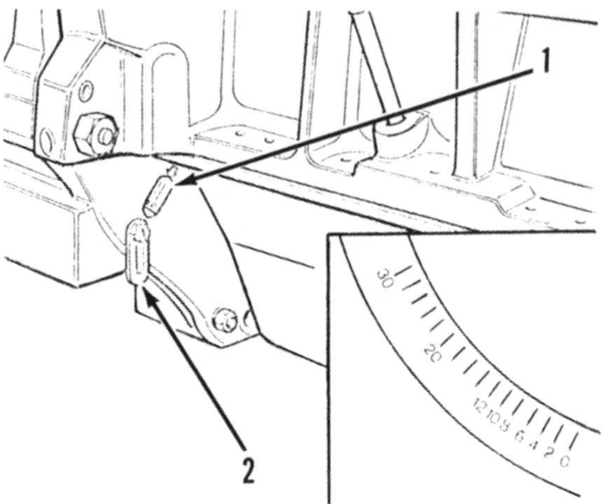

Fig. 54—Flywheel has degree marks as indicated so that crankshaft can be easily located with front (number 1) cylinder at Top Dead Center (0 degree).

1. Window
2. Cover

ROCKER ARMS

All Models

47. To remove rocker arms, lift hood, swing battery tray out and remove rocker arm cover. Loosen, but do not remove, the cylinder head screws that retain the rocker arm assembly (Fig. 55). Four screws are used on 3-cylinder models, five screws on 4-cylinder models. Lift the cylinder head bolts, shaft and rocker arms away from the engine as an assembly.

Disassemble the rocker arms and shaft, by removing the cylinder head and support retaining screws. Rocker arms and shaft should not show excessive wear. Inside diameter of new rocker arm is 25.48-25.50 mm (1.003-1.004 inches) and diameter of new shaft is 25.45-25.50 mm (1.000-1.001 inches). Inside diameter of supports should be 25.48-25.50 mm (1.003-1.004 inches).

When reassembling, be sure that notch (N—Fig. 55) is up and toward front end of engine so that rocker

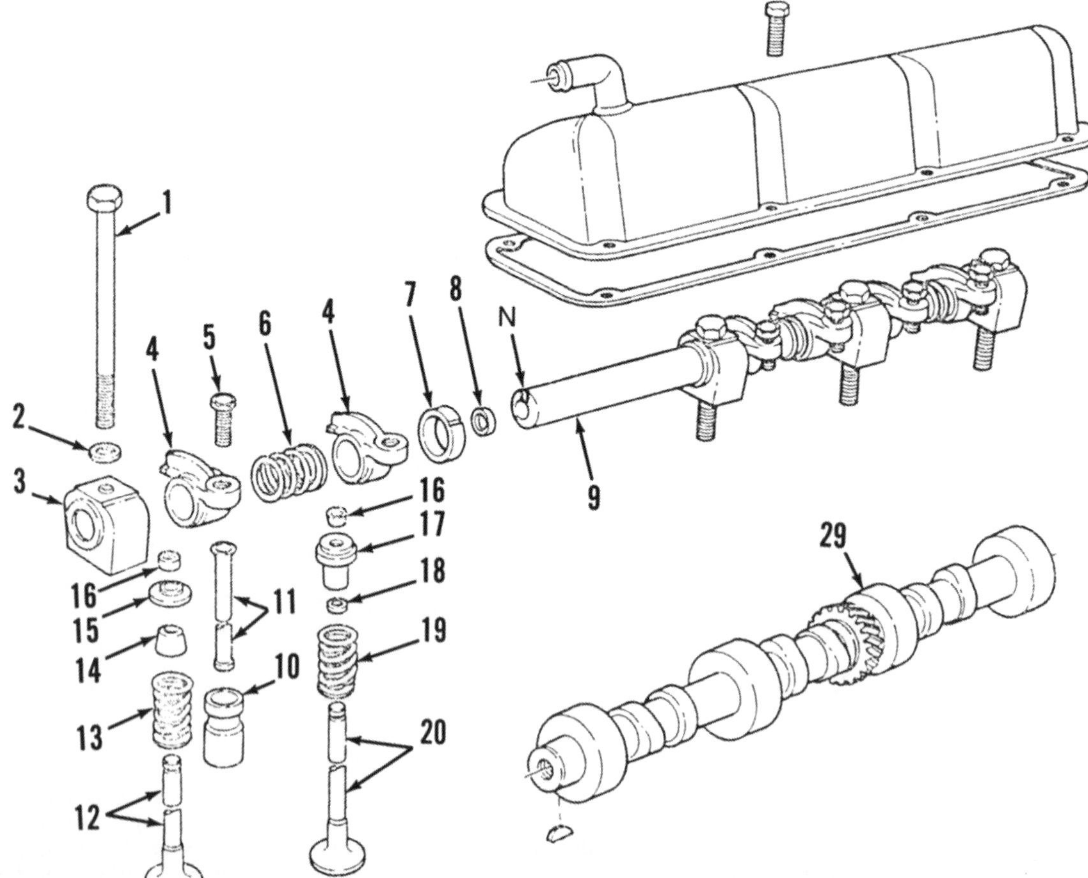

N. Notch
1. Cylinder head screws
2. Flat washers
3. Support
4. Rocker arms
5. Adjusting screws
6. Spring
7. Spacer
8. End plug
9. Rocker arm shaft
10. Cam follower
11. Push rod
12. Inlet valve
13. Spring
14. Umbrella type stem seal
15. Inlet retainer
16. Retainer key
17. Exhaust rotator
18. Exhaust stem seal
19. Spring
20. Exhaust valve
29. Camshaft

Fig. 55—Partially exploded view of rocker arm assembly. Cylinder head retaining screws (1) also retain the rocker arm shaft (9) and supports (3). Notch (N) in end of shaft must be up and toward front of engine.

arm oiling holes will be correctly positioned. Back each of the rocker arm adjusting screws out two turns before installing on engine. Tighten the retaining cylinder head screws evenly until seated against head, then tighten to the recommended torque as outlined in paragraph 41. Adjust valve clearance as described in paragraph 44. Reinstall rocker arm cover retaining screws to 18 N•m (13 ft.-lbs.) torque.

TIMING GEAR COVER

All Models

48. REMOVE AND REINSTALL. To remove the timing gear cover (11—Fig. 56), remove hood, drain cooling system and radiator hoses. Disconnect hose from the front mounted air cleaner. Remove the grille and disconnect lines from oil coolers, if so equipped. Disconnect battery ground cable and wires to the headlights. Unbolt radiator shell upper support bracket from the shell. Remove the front drive shaft and shield from models with front wheel drive. Disconnect steering linkage and steering hoses that would interfere with the removal of the front axle. Support front of tractor in such a way that it will not interfere with removal of the front support and axle. A hoist may be attached to front support or special stands can be attached to sides. Install wedges between the front support and the axle to prevent tipping, then support the axle with a suitable jack or special safety stand to prevent tipping while permitting the front support and axle to be moved safely away from the tractor. Unbolt front support from engine, then carefully roll axle and front support assembly forward.

After the front support, axle and radiator are separated from the front of the engine, remove the fan belt

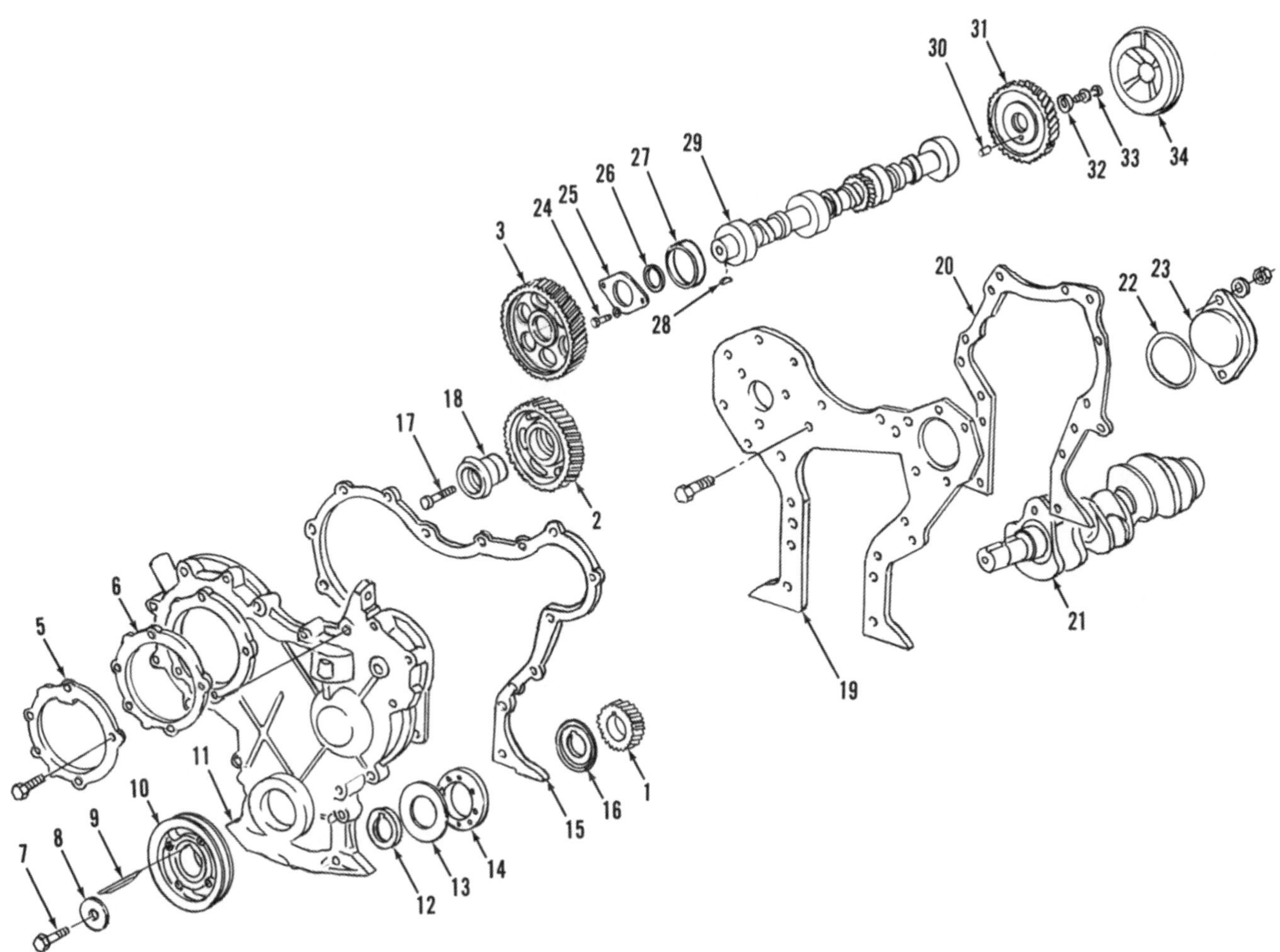

Fig. 56—Exploded view of timing gears and cover.

1. Crankshaft gear
2. Camshaft drive (idler) gear
3. Camshaft gear
5. Cover plate
6. Gasket
7. Pulley retaining screw
8. Washer
9. Key
10. Pulley
11. Front (timing gear) cover
12. Spacer
13. Dust seal
14. Oil seal
15. Gasket
16. Oil slinger
17. Screw
18. Adapter shaft
19. Engine front plate
20. Gasket
21. Crankshaft
22. "O" ring
23. Power steering pump cover
24. Screw
25. Thrust plate
26. Spacer
27. Bearings
28. Key
29. Camshaft
30. Pin
31. Hydraulic pump drive gear
32. Washer
33. Screw
34. Cover

and alternator front mounting bolt. Drain engine oil and remove the oil pan. Remove cap screw and washers from center of crankshaft pulley, then use a suitable puller to remove the crankshaft pulley. **Be careful not to damage the crankshaft or pulley when removing.** Remove the power steering or hydrostatic steering pump if so equipped. Remove the retaining screws, then remove front cover from dowel pins. The diesel fuel injection pump drive gear cover does not need to be removed from timing gear cover.

Oil slinger, front oil seal and dust seal can be renewed as described in paragraph 63 after timing gear cover is removed. Refer to paragraphs 49, 50, 51, 52 and 53 for servicing timing gears.

Clean gasket surfaces of engine front plate and timing gear cover, then reverse removal procedure to install timing gear cover. Tighten the front cover retaining screws to 22 N•m (16 ft.-lbs.) torque. Screws retaining stamped oil pan should be tightened to 30 N•m (22 ft.-lbs.) and screws retaining cast pan should be tightened to 38 N•m (28 ft.-lbs.) torque. Tighten crankshaft pulley retaining screw to 224 N•m (165 ft.-lbs.) torque.

Reattach front axle, radiator and front support to engine by reversing removal procedure. Tighten the retaining screws to 339-420 N•m (250-310 ft.-lbs.) for models with stamped oil pan; to 240-298 N•m (180-220 ft.-lbs.) torque for models with cast oil pan.

Fig. 57—View showing timing marks aligned on crankshaft gear (1), camshaft drive (idler) gear (2), camshaft gear (3) and injection pump drive gear (4). Fuel injection pump drive gear (4) is marked with both "3" and "4." The timing mark on the camshaft drive (idler) gear should be aligned with the correct mark (3 or 4) depending upon the number of engine cylinders.

1. Crankshaft gear
2. Idler gear
3. Camshaft gear
4. Injection pump gear

TIMING GEARS

All Models

49. Before removing any gears, first refer to paragraph 47 and remove the rocker arm assembly to avoid possible damage to the pistons or valve train. Damage could result if either camshaft or crankshaft is turned independently from the other unless the rocker arms are removed.

The timing gear train consists of crankshaft gear (1—Fig. 57), camshaft drive (idler) gear (2), camshaft gear (3) and fuel injection pump drive gear (4).

50. CAMSHAFT DRIVE (IDLER) GEAR AND SHAFT. To inspect or remove the camshaft drive (idler) gear and adapter, first remove the timing gear cover as outlined in paragraph 48. Check end play and backlash between idler gear and other timing gears before removing the gear. Backlash between the crankshaft gear (1—Fig. 57) and idler gear (2) should be 0.025-0.23 mm (0.001-0.009 inch). Backlash between the idler gear (2) and the camshaft gear (3) should be 0.025-0.23 mm (0.001-0.009 inch). Backlash between the idler gear (2) and fuel injection pump drive gear (4) should be 0.025-0.30 mm (0.001-0.012 inch). Correct end play on shaft (adapter) is 0.025-0.28 mm (0.001-0.011 inch).

Remove center screw (17—Fig. 56), then lift shaft (adapter) and gear (2) from face of cylinder block. Inside diameter of bushing inside gear (2) should be 50.813-50.838 mm (2.0005-2.0015 inches). Outside diameter of adapter shaft (18) should be 50.762-50.775 mm (1.9985-1.9990 inches). Renew shaft (adapter) and/or gear if bearing surfaces are worn excessively. Inspect gear teeth for wear or scoring.

To install the camshaft drive (idler) gear, position crankshaft with front (number 1) piston at top dead center and turn the camshaft gear and fuel injection pump drive gear so that timing marks point to the center of the camshaft drive (idler) gear retaining screw. Install the camshaft drive (idler) gear (2—Fig. 57) so that all the timing marks are aligned as shown, then install shaft (adapter) and retaining screw. Recheck timing marks after shaft (adapter) retaining screw is tightened to 140 N•m (103 ft.-lbs.) torque.

51. CAMSHAFT GEAR. To remove the camshaft gear (3—Fig. 56), first remove the timing gear cover as outlined in paragraph 48. Check camshaft end play and backlash between the camshaft gear (3) and camshaft drive (idler) gear (2) before removing the gear. Backlash should be 0.025-0.23 mm (0.001-0.009 inch). Camshaft end play can be measured with a feeler gage as shown in Fig. 58 with camshaft pried forward. Measure end play between rear face of gear (3—Fig. 56) and front face of thrust plate (25). Correct end play is 0.025-0.18 mm (0.001-0.007 inch).

Remove the rocker arm assembly as outlined in paragraph 47 and the camshaft drive (idler) gear as outlined in paragraph 50. Damage could result if either camshaft or crankshaft is turned independently from the other unless the rocker arms are removed. Remove the camshaft gear center retaining screw and washer, if so equipped, then pull the gear from the shaft. If equipped with center screw, the camshaft gear will be a hand push fit on shaft. If not equipped with center screw, gear will be much tighter fit and camshaft and gear should be removed as outlined in paragraph 54, then gear should be pressed from shaft.

Tighten gear to camshaft retaining screw to 58 N•m (43 ft.-lbs.) and screws retaining thrust plate (25) to cylinder block to 31 N•m (23 ft.-lbs.) torque.

52. CRANKSHAFT GEAR. The crankshaft gear (1—Fig. 56) should not be removed unless a new gear is to be installed. Backlash between the crankshaft gear (1—Fig. 57) and idler gear (2) should be 0.025-0.23 mm (0.001-0.009 inch).

To remove the crankshaft gear (1), first remove the timing gear cover as outlined in paragraph 48, rocker arm assembly as outlined in paragraph 47 and the camshaft drive (idler) gear as outlined in paragraph 50. Damage could result if either camshaft or crankshaft is turned independently from the other unless the rocker arms are removed.

Remove the spacer (12) and oil slinger (16), if not already off, then use appropriate puller to remove the gear from the crankshaft. Special Nuday tool number SW 501 with insert number SW.501-1 or V.L. Churchill tool numbers CPT 6040-B, CT 6069-A and CT 6069-1 are available to remove and install the tight fitting crankshaft gear.

Fig. 58—Measure end play of camshaft before removing camshaft or gear. Screwdriver (S) can be used to gently push the camshaft and gear forward while using a feeler gage (F) to measure the gap between thrust plate (25) and rear surface of gear (3).

53. INJECTION PUMP DRIVE GEAR. Refer to paragraph 90 for removal or other service procedures to the injection pump drive gear.

CAMSHAFT AND BEARINGS

All Models

54. To remove the camshaft, first remove the engine as outlined in paragraph 40 and the timing gear cover as outlined in paragraph 48. Remove the engine mounted hydraulic pump as outlined in paragraph 195 (from models so equipped) and the rocker arms as outlined in paragraph 47. Withdraw the push rods, then invert the engine assembly to allow the cam followers to fall away from the camshaft. Check camshaft end play and backlash between the camshaft gear (3—Fig. 56) and camshaft drive (idler) gear (2) before removing the camshaft or gear. Refer to paragraph 51 for checking procedures and for removal of the camshaft gear. Remove the clutch, flywheel and rear engine plate. If engine is not equipped with an engine driven hydraulic pump, remove the pump gear cover from the engine left side. On all models, remove the oil filter. Work through opening (H—Fig. 59) in block and push the camshaft rear cover plate (34) from the cylinder block bore using a punch (P) as shown. Remove the engine oil pump, drive shaft and gear (10—Fig. 66) and the hydraulic pump drive gear (31—Fig. 56).

Even with engine inverted, it may be necessary to push cam followers away from camshaft. Withdraw

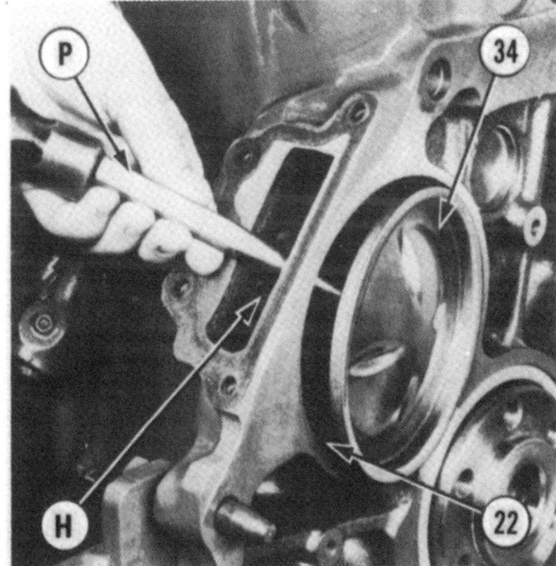

Fig. 59—Camshaft rear cover plate (34) can be removed with a punch (P) as shown. "O" ring is shown at (22).

camshaft carefully from the front of the cylinder block. The camshaft is supported in four journals of three-cylinder engines, five journals of four-cylinder engines. Journals are 60.693-60.719 mm (2.3895-2.3905 inches) diameter and should have 0.025-0.076 mm (0.001-0.003 inch) clearance in the installed bushings. Bushings can be removed and installed using a suitable piloted bearing driver.

NOTE: Be sure to align oil holes in bearings with the oil passages in the cylinder block.

New camshaft bushings are presized and should not require resizing if carefully installed using proper size tool.

Install camshaft by reversing the removal procedure. Refer to Fig. 57 and paragraph 51 for installing and timing camshaft drive gear. If removed, tighten screw retaining the hydraulic pump drive gear to 58 N•m (43 ft.-lbs.) torque. Tighten the screw retaining camshaft gear to front of camshaft to 58 N•m (43 ft.-lbs.) and screws retaining thrust plate (25) to cylinder block to 31 N•m (23 ft.-lbs.) torque.

CONNECTING ROD AND PISTON UNITS

All Models

55. The connecting rod and piston units can be removed from above after removing the cylinder head and oil pan. Be sure to remove the top ridge from the cylinder bores before attempting to withdraw the assemblies.

Connecting rod and bearing cap are numbered to correspond to their respective cylinder bores. When renewing the connecting rod, be sure to stamp the cylinder number on new rod and cap.

When assembling, it is important that identification or notch on top of piston is toward front end of engine. The connecting rod is symmetrical and can be installed either way without affecting durability or performance.

Be sure that notches in rod and cap and the tangs of the bearing liners are on the same side of the engine. Cylinder numbers are usually stamped on the same side as notches for the tangs. Tighten the connecting rod retaining nuts to 85 N•m (63 ft.-lbs.) torque.

PISTONS AND RINGS

All Models

56. Pistons may be fitted with three conventional compression rings and one oil control ring or may be fitted with one "L" shaped "head land" ring at the top of the piston, one conventional straight sided second ring and a conventional oil control ring.

Installation of the "L" shaped top ring is possible in only one position. The second compression ring for pistons with "head land" top ring should be installed with chamfered (beveled) inner diameter toward the top. Compression ring side clearance in groove should be 0.112-0.155 mm (0.0044-0.0061 inch) for top ring; 0.99-0.142 mm (0.0039-0.0056 inch) for second ring. Ring end gap should be 0.38-0.76 mm (0.015-0.030 inch) for top ring; 0.33-0.71 mm (0.013-0.028 inch) for the second ring. The oil control ring and expander can be installed with either side up.

If the conventional top ring has a beveled edge on the inside diameter, the bevel should be toward the top. The second ring may have a stepped edge on the inside diameter that should be toward top of the piston. The third compression ring may have either a beveled or stepped edge on the inside diameter that should be toward top of piston. Compression ring side clearance in groove should be 0.112-0.155 mm (0.0044-0.0061 inch) for the top ring; 0.99-0.142 mm (0.0039-0.0056 inch) for the second and third rings. Ring end gap should be 0.38-0.76 mm (0.015-0.030 inch) for the top ring; 0.33-0.71 mm (0.013-0.028 inch) for the second and third ring. The oil control ring and expander can be installed with either side up.

If rings have a dot or "TOP" mark on side of ring, the mark or "TOP" mark should be assembled toward top of the piston, regardless of type of ring. Piston ring sets are available in standard size and oversizes.

PISTONS AND CYLINDERS

All Models

57. Some 4630 models are equipped with pistons fitted with "L" shaped ("head land") top rings, a conventional second compression ring and one oil control ring. Some 4630 models and all other models are fitted with three conventional compression rings and one oil control ring. Oversized pistons are available for all models.

Three-cylinder models can be fitted with thin wall sleeves if reboring will exceed the largest available oversize piston. The thin-walled sleeves should be used only with standard size or 0.10 mm (0.004 inch) oversized pistons.

Standard cylinder bore diameter is 111.778-111.841 mm (4.4007-4.4032 inches) for all models. Cylinder bore out of round should be less than 0.03 mm (0.0015 inch) and should be corrected if more than 0.127 mm (0.005 inch). Cylinder bore taper should be less than 0.025 mm (0.001 inch) and should be corrected if taper exceeds 0.127 mm (0.005 inch). Cylinder head surface of block should be flat within 0.08 mm (0.003 inch) over any 152 mm (6 inches) surface. Cylinder block should have less than 0.15

mm (0.006 inch) difference in height in the overall length.

58. INSTALLING SLEEVES. The thin walled flanged sleeves should be selectively and carefully fitted to the block bore. After installation, the flange should be flush with the top surface of the block. Counterbore depth is critical and should be carefully done. To install sleeves, first measure sleeve outside diameter in at least four different locations to determine the average diameter. Bore the cylinder block until the bore diameter is 0.05 mm (0.002 inch) smaller than the sleeve average diameter, but no larger than the same size as the sleeve. Refer to Fig. 60 and machine the counterbore in block to the dimensions shown. Clean and dry the cylinder and block bore and chill the sleeves in liquid nitrogen or dry ice for 15 minutes. Push the sleeves into position as far as possible until sleeve flange is seated in block counterbore. The sleeve flange should be flush with the surface of the cylinder block. Block or sleeves may be surfaced lightly if necessary, but no more than 0.13 mm (0.005 inch) should be removed. Bore and hone the installed sleeve to the standard size of 111.778-111.841 mm (4.4007-4.4032 inches). Do not attempt to install pistons larger than 0.10 mm (0.004 inch) oversized when block is fitted with sleeves.

59. FITTING PISTONS. Original diameter of standard size pistons is 111.56-111.62 mm (4.3922-4.3947 inches), measured at right angles to piston pin at bottom of the skirt. Oversized pistons are available for all models. Piston skirt to cylinder clearance should be 0.069-0.094 mm (0.0027-0.0037 inch). The top of the piston should be 0.28-0.58 mm (0.011-0.023 inch) above the head surface of the cylinder block.

PISTON PINS

All Models

60. The 38.092-38.100 mm (1.4997-1.5000 inches) diameter floating-type piston pins are retained in piston bosses by snap rings and are available in standard size only. The piston pin should have 0.0076-0.0127 mm (0.0003-0.0005 inch) clearance in the pin bosses at 21° C (70° F). The piston pin should have 0.013-0.018 mm (0.0005-0.0007 inch) in rod bushing. Oil hole in piston pin bushing must be drilled to 6.4 mm (0.25 inch) after bushing is pressed into the connecting rod. Ream the bushing to 38.108-38.115 mm (1.5003-1.5006 inches) to provide the correct pin to bushing clearance.

The connecting rod is symmetrical and can be installed either way without affecting durability or performance; however, the crankshaft bearing tangs and cylinder numbers stamped on rod and cap are usually on the right side of the engine away from the camshaft. When installing the piston, piston pin and connecting rod assembly, it is important that identification or notch on top of piston is toward front end of engine.

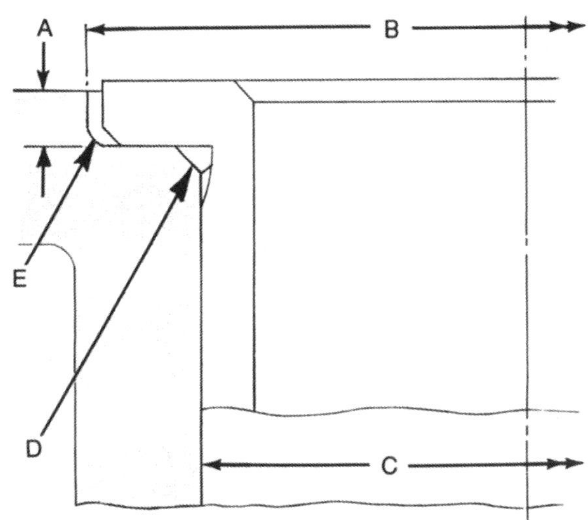

Fig. 60—Refer to the diagram above and the following dimensions when fitting the thin-walled lipped sleeves to cylinder.

A. 2.41-2.51 mm (0.095-0.099 inch)
B. 120.55-120.73 mm (4.746-4.753 inches)
C. Bore cylinder 0.000-0.002 mm (0.000-0.002 inch) smaller than average sleeve OD
D. 0.50-0.75 mm (0.020-0.030 inch) with 45 degree chamfer
E. 0.381 mm (0.015 inch) maximum radius

CONNECTING RODS AND BEARINGS

All Models

61. Connecting rod bearings are precision type, renewable from below after removing the oil pan and connecting rod bearing caps. When removing the bearing caps, notice which side of the engine the cylinder identification numbers are located. Normally the identification numbers are located on the side away from the camshaft. The connecting rods are symmetrical and can be installed either way; however, it is important that the pistons be installed in the cylinder bore with the identification notch toward the front of the engine.

Refer to paragraph 55 for removal of the complete rod and piston units and to paragraph 60 for installing new bushings for the piston pins. Refer to the following for connecting rod and crankshaft crankpin specifications.

Connecting Rod—
Piston pin bushing ID .... 38.108-38.115 mm (1.5003-1.5006 inches)
Rod twist, maximum .... 0.30 mm (0.012 inch)
Rod bend, maximum .... 0.10 mm (0.004 inch)

Crankshaft Crankpin—
Standard Diameter, Blue .... 69.840-69.850 mm (2.7496-2.7500 inches)
Standard Diameter, Red ..... 69.850-69.860 mm (2.7500-2.7504 inches)
Out of round, maximum .... 0.005 mm (0.0002 inch)

Crankshaft bearing liners can be of two materials—copper-lead or aluminum-tin alloy. The bearings will have an identification marking as follows:

Copper-Lead.......................... PV or G
Aluminum-Tin...................... G and AL

Standard size bearing liners of each material are available in two thicknesses and are color-coded to indicate as follows:

Copper-Lead Bearing Thickness—
Red .... 2.395-2.408 mm (0.0943-0.0948 inch)
Blue........................ 2.405-2.418 mm (0.0947-0.0952 inch)

Aluminum-Tin Alloy Bearing Thickness—
Red .... 2.390-2.403 mm (0.0941-0.0946 inch)
Blue........................ 2.400-2.413 mm (0.0945-0.0950 inch)

In production, crankshaft crankpins are color-coded to indicate crankpin diameters as follows:

Crankshaft Crankpin Diameter—
Red .... 69.850-69.860 mm (2.7500-2.7504 inches)
Blue...................... 69.840-69.850 mm (2.7496-2.7500 inches)

When installing a new crankshaft and color-code marks are visible on connecting rods and crankpin journals, the bearing liners can be fitted as follows: If color-code markings on both rod and crankshaft crankpin are red, install two bearing liners with red markings. If color-code markings on both rod and crankshaft crankpin are blue, install two bearing liners with blue markings. If color-code marks on connecting rod and crankshaft crankpin do not match (one red and the other blue), install one red and one blue bearing liner.

NOTE: Be sure both bearing liners are of the same material; that is, either both are copper-lead or both are aluminum-tin alloy.

If the color-code marks are not visible, check clearance with Plasti-gage for proper clearance according to the bearing material as follows:

Crankpin Journal-To-Bearing Liner Clearance—
Copper-Lead Bearing Material .... 0.043-0.096 mm (0.0017-0.0038 inch)
Aluminum-Tin Alloy Bearing Material .... 0.053-0.107 mm (0.0021-0.0042 inch)

Bearing liners are available in undersizes to permit machining the crankshaft crankpin. Crankshaft crankpin should be machined to one of the following undersizes to fit the available undersize liners.

Bearing Undersize	Crankpin Journal Diameter
0.051 mm (0.002 inch)	69.789-69.799 mm (2.7476-2.7480 inches)
0.254 mm (0.010 inch)	69.50-69.606 mm (2.7400-2.7404 inches)
0.508 mm (0.020 inch)	69.342-69.352 mm (2.7300-2.7304 inches)
0.762 mm (0.030 inch)	69.088-69.098 mm (2.7200-2.7204 inches)
1.016 mm (0.040 inch)	68.834-68.844 mm (2.7100-2.7104 inches)

NOTE: When regrinding crankpin journals, maintain a 3.05-3.56 mm (0.12-0.14 inch) fillet radius and chamfer oil hole after journal is ground to size.

Crankpin journal to bearing liner clearance should be 0.043-0.096 mm (0.0017-0.0038 inch) if equipped with copper-lead bearing material; 0.053-0.107 mm (0.0021-0.0042 inch) if equipped with liners of aluminum-tin alloy bearing material. Connecting rod side float on crankshaft crankpin should be 0.18-0.33 mm (0.007-0.013 inch). Tighten connecting rod nuts to 85 N•m (63 ft.-lbs.) torque.

CRANKSHAFT AND MAIN BEARINGS

All Models

62. Crankshaft of three-cylinder models is supported in four main bearings; in five main bearings for four-cylinder engines. Crankshaft end thrust is controlled by flanged main bearing liner located in second position from the front of three-cylinder models; in the center position of four-cylinder engines.

The main bearing liners can be removed and new liners rolled in without removing the crankshaft, but removal of the crankshaft requires removal of the engine, timing gear cover, clutch and flywheel. The balancer assembly must be removed from four-cylinder models as described in paragraph 65. Be sure that all main bearing caps have an identification number before removing the cap so that all main bearing caps will be reassembled in same location. Main bearing caps are machined with the block so the half of the main bearing bore contained in a cap is correct only when correctly installed in the proper location in the one block with which it was machined.

Crankshaft main bearing liners may be of two different materials—copper-lead or aluminum-tin alloy. The bearings will have an identification marking as follows: The copper-lead liners will have a PV or a G mark and the aluminum-tin alloy liners will show G and AL.

Standard size main bearing liners of each material are available in two different thicknesses and are color-coded to indicate as follows:

Copper-Lead Bearing Thickness—

Red	3.162-3.175 mm (0.1245-0.1250 inch)
Blue	3.172-3.185 mm (0.1249-0.1254 inch)

Aluminum-Tin Alloy Bearing Thickness—

Red	3.162-3.175 mm (0.1245-0.1250 inch)
Blue	3.172-3.185 mm (0.1249-0.1254 inch)

In production, crankshaft main bearing journals are color-coded to indicate diameters as follows:

Crankshaft Main Journal Diameter—

Red	85.644-85.656 mm (3.3718-3.3723 inches)
Blue	85.631-85.644 mm (3.3713-3.3718 inches)

When installing a new crankshaft and color-code marks are visible on main bearing journals, the bearing liners can be fitted as follows: If color-code markings on both main bearing journal and main bearing bore in block are red, install two bearing liners with red markings. If color-code markings on both main bearing journal and main bearing bore in block are blue, install two bearing liners with blue markings. If color-code marks on crankshaft journal and bore in block do not match (one red and the other blue), install one red and one blue bearing liner.

NOTE: Be sure that both bearing liners are of the same material; that is, either both are copper-lead or both are aluminum-tin alloy.

If the color-code marks are not visible, check clearance with Plastigage for proper clearance according to the bearing material as follows:

Main Journal-To-Bearing Liner Clearance—

Copper-Lead Bearing Material	0.056-0.114 mm (0.0022-0.0045 inch)
Aluminum-Tin Alloy Bearing Material	0.056-0.114 mm (0.0022-0.0045 inch)

Length of the main journal machined for flanged bearing liners should be 37.06-37.11 mm (1.459-1.461 inches) and the distance between the thrust surfaces of the bearing liner is 36.91-36.96 mm (1.453-1.455 inches). The diameter of the crankshaft at the rear oil seal journal should be 122.12-122.28 mm (4.808-4.814 inches).

Main bearing liners are available in undersizes to permit machining the crankshaft journals. Crankshaft main journal should be machined to one of the following undersizes to fit the available undersize liners.

Bearing Undersize	Crankpin Journal Diameter
0.051 mm	85.580-85.593 mm
(0.002 inch)	(3.3693-3.3698 inches)
0.254 mm	85.390-85.402 mm
(0.010 inch)	(3.3618-3.3623 inches)
0.508 mm	85.136-85.148 mm
(0.020 inch)	(3.3518-3.3523 inches)
0.762 mm	84.882-84.894 mm
(0.030 inch)	(3.3418-3.3423 inches)
1.016 mm	84.628-84.640 mm
(0.040 inch)	(3.3318-3.3323 inches)

NOTE: When regrinding crankshaft main journals, maintain a 3.05-3.56 mm (0.12-0.14 inch) fillet radius and chamfer oil hole after journal is ground to size.

Crankshaft main journal to bearing liner clearance should be 0.056-0.114 mm (0.0022-0.0045 inch) for both types of bearing liner material. Crankshaft end play should be 0.10-0.20 mm (0.004-0.008 inch). Tighten main bearing journal cap retaining screws to 163 N•m (120 ft.-lbs.) torque. Be sure that main bearing caps are installed in correct location and that liner tangs in both block and cap are on the same side. Refer to paragraph 61 for specification for connecting rods and connecting rod journals of crankshaft. Refer to Fig. 57 for alignment of timing gear marks. The rear main bearing cap must be sealed to the block and contains the crankshaft rear seal. When installing the rear main cap, remove all traces of oil and apply a light coating of sealing compound to the mating surfaces of the cap and block (2—Fig. 61). Install new side seals and trim slightly higher than face of block and cap. Application of penetrating oil on side seals will cause them to swell and provide a tighter seal. Refer to paragraph 64 for installing the rear oil seal and paragraph 63 for installing the front oil seal. Refer to paragraph 65 for installation and timing of the balancer assembly on four cylinder engines.

Fig. 61—Mating surface (2) of the rear main bearing cap and block should be coated with a light coat of sealer.

CRANKSHAFT OIL SEALS

All Models

63. FRONT OIL SEAL. The crankshaft front oil seal is located in the timing gear cover and the cover must be removed to install a new seal. Refer to paragraph 48 and remove the timing gear cover. Drive out both the dust seal and the oil seal toward the inside of cover. Install dust seal in cover, then use a piloted seal driver of suitable size to install the new seal in cover with spring-loaded lip toward inside of cover.

64. REAR OIL SEAL. The crankshaft rear oil seal can be renewed after removing the clutch assembly as outlined in paragraph 103, the engine flywheel as outlined in paragraph 66 and the oil seal retainer plate (3-cylinder models) or engine rear plate (4-cylinder models). Check to be sure the leakage is from the rear seal and not side seals of rear main bearing cap, camshaft rear cover or some other area. Refer to paragraph 62 if rear main bearing cap must be resealed.

Old rear oil seal can be pried from the bore in block and cap. Thoroughly clean the bore and journal. Diameter of the crankshaft rear oil seal journal should be 122.12-122.28 mm (4.808-4.814 inches). Apply a light coat of high temperature grease to the seal bore, to journal and to lip of seal. Push the new seal squarely into bore using a special tool (such as Ford No. 1301) and screws threaded into crankshaft flywheel mounting holes. Seal is installed flush with rear surface of block during original production, but should be 1.5 mm (0.06 inch) below flush on subsequent installations. Seal must be square with crankshaft to seal properly. Attach a dial indicator to the crankshaft and check seal runout. Seal is not square if runout exceeds 0.38 mm (0.015 inch). Use a new gasket and tighten seal retainer plate (3-cylinder

der models) retaining screws to 19 N•m (14 ft.-lbs.) torque. Tighten screws retaining 4-cylinder engine rear plate to 31 N•m (23 ft.-lbs.) torque.

ENGINE BALANCER

Four-Cylinder Models

65. A Lanchester-type engine balancer is used on all 4-cylinder engines. The balancer is driven at twice crankshaft speed by a gear machined on the crankshaft immediately in front of the center main bearing.

To remove the balancer, first remove the oil pan as outlined in paragraph 67, then unbolt and remove the balancer unit from the lower side of the crankcase. Drive out roll pins (26—Fig. 62) located at front of balancer casting and remove shafts and gears. Renew shafts and/or bushings in gears if worn excessively. Desired shaft to bushing clearance is 0.005-0.020 mm (0.0002-0.0008 inch). Shaft diameter is 25.133-25.400 mm (0.9895-1.000 inch). End float between balancer gears and support should be 0.20-0.51 mm (0.008-0.020 inch).

To reassemble, place gears in housing with the timing marks aligned as shown in Fig. 63 and marks toward front (roll pin) end of housing. Place new thrust washer at each side of balance gear and insert shafts. Check backlash between balancer gears. If backlash between balance gears is 0.05-0.25 mm (0.002-0.010 inch), drive roll pins into place. If backlash is excessive, renew gears. Install gasket (27—Fig. 62) and assemble the balancer to the bottom of crankcase, aligning timing marks (B and C—Fig. 64). The timing mark (C) will be visible when number 2 piston is at top dead center. Tighten screws attaching balancer to the bottom of crankcase to 88 N•m (65 ft.-lbs.) torque. Mount a dial indicator as shown in Fig. 65 and measure backlash between balancer gear and crankshaft gear (C). Backlash should be 0.05-

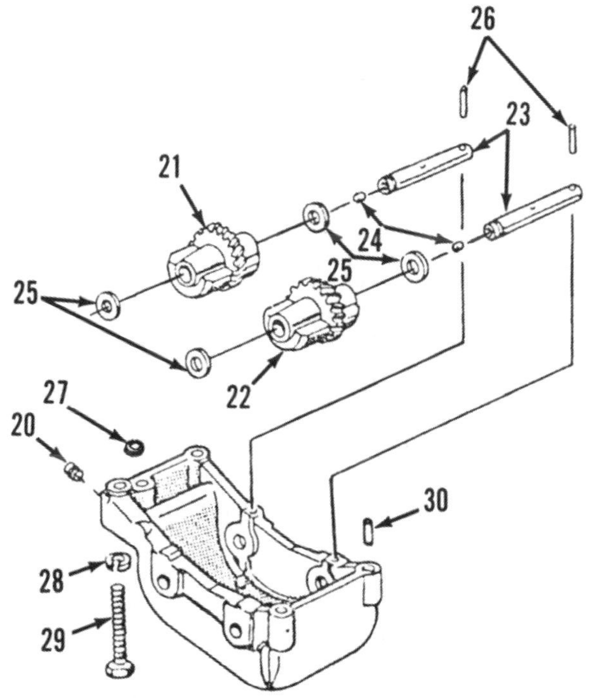

Fig. 62—Exploded view of dynamic balancer used on 4-cylinder engines.

20. Pipe plug
21. Drive gear & bushings
22. Driven gear & bushings
23. Shafts
24. Tapered plugs
25. Thrust washers
26. Spring pins
27. Gasket
28. Lock washers
29. Screws
30. Dowel pins

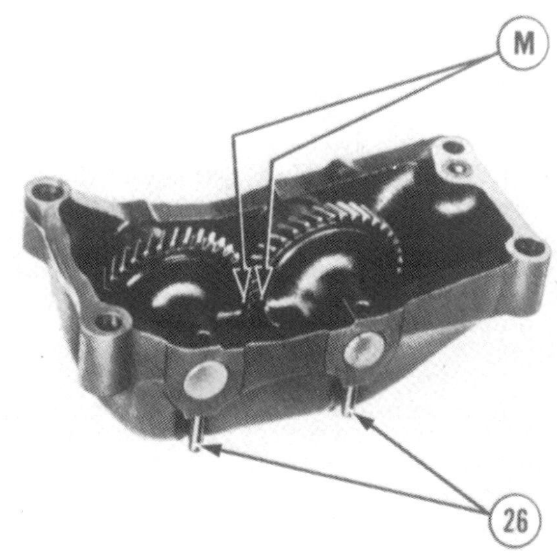

Fig. 63—View of 4-cylinder balancer assembly with balancer gear timing marks aligned. Refer to text.

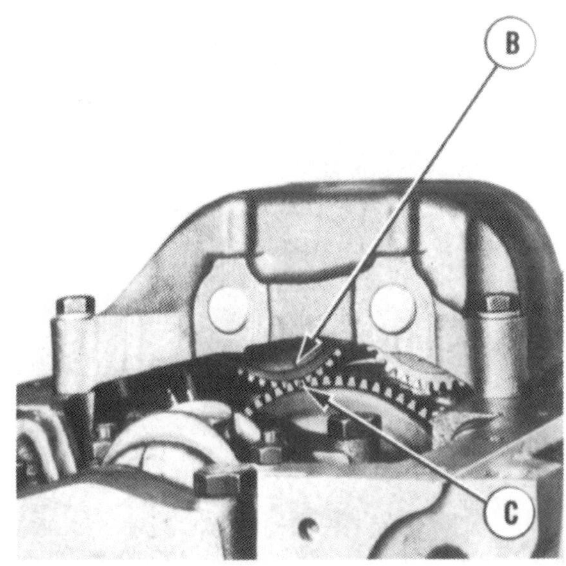

Fig. 64—View of 4-cylinder balancer assembly correctly timed to the engine crankshaft. Refer to text.

0.20 mm (0.002-0.008 inch). Check to be sure that timing marks (Fig. 63 and Fig. 64) are aligned before installing oil pan.

FLYWHEEL

All Models

66. The flywheel can be removed after splitting the tractor between the engine and transmission as outlined in paragraph 102 and removing the clutch as outlined in paragraph 103. The flywheel can be installed in one position only. The flywheel retaining screws also retain the clutch shaft pilot bearing retainer. When reinstalling flywheel, tighten retaining cap screws to 217 N•m (160 ft.-lbs.) torque.

The starter ring gear is installed from the front face of the flywheel, so flywheel must be removed before new ring gear can be installed. Heat old gear before removing. Heat new gear evenly until gear expands enough to slip onto flywheel. Tap the gear all the way around to be sure that it is properly seated, then allow gear to cool.

NOTE: Be sure to heat gear evenly. If any portion is heated to a temperature higher than 260° C (500° F), gear will wear rapidly.

OIL PAN

All Models

67. To remove stamped oil pan, drain engine oil and remove the dipstick. Remove the front drive shaft and shield from models with front-wheel drive. The stamped oil pan can be unbolted and removed from all models.

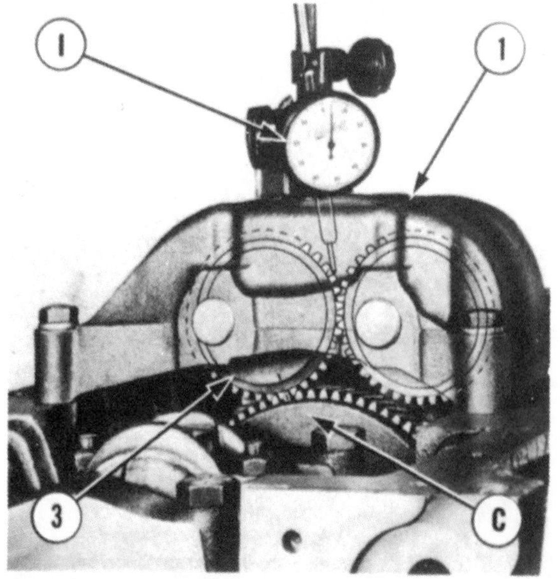

Fig. 65—Mount a dial indicator as shown to measure backlash between balance gear and drive gear on crankshaft. Refer to text.

Cast pans are heavy and should be supported while removing. The front support is also attached to the front of cast oil pans. To remove the cast oil pan, drain oil and remove the dipstick. Remove the front drive shaft from front-wheel-drive models. Remove the hood panels from all models. Disconnect power steering or hydrostatic steering hoses from steering cylinders and cover openings to prevent the entrance of dirt. Disconnect steering cooler lines (U—Fig. 40). Support tractor under the front of transmission with a jack and support front end with a hoist. Remove the front support to cylinder block screws one at a time and replace with screws 200 mm (8 inches) long. Remove the nuts from the studs passing through the front lip of oil pan. Move the front axle forward carefully approximately 38 mm (1½ inches). Support oil pan with a jack, unbolt oil pan and lower away from the engine.

On all models, oil pan retaining screws should be tightened beginning in the center and working toward ends. Tighten screws retaining stamped pan to 30 N•m (22 ft.-lbs.) torque. Tighten screws retaining cast pan to 38 N•m (28 ft.-lbs.) torque. Tighten the four screws and two stud nuts attaching front support to engine and front of cast oil pan to 240-298 N•m (180-220 ft.-lbs.) torque. Remainder of the installation is the reverse of removal procedure.

OIL PUMP AND RELIEF VALVE

All Models

68. To remove the oil pump, first remove the oil pan as outlined in paragraph 67. then remove the two retaining cap screws and remove pump from the cylinder block. Refer to Fig. 66 for exploded view of oil pump and drive gear assembly. The floating shaft (9) will usually be removed with the pump.

To disassemble the pump, remove clip (1) and oil screen (2), then remove the screws retaining screen cover (3) and pump cover (4) to pump body (11). Remove pump covers and rotor set (5), noting which direction outer rotor was placed in pump body. Remove retainer plug (6), spring (7) and relief valve (8).

Check pump for unusual wear or damage. Measure clearance between pump body (11) and outer rotor as shown in Fig. 67. Body to rotor clearance should be 0.15-0.28 mm (0.006-0.011 inch). Measure clearance between tips of the inner rotor and the arcs of the outer rotor. Rotor to rotor clearance should be 0.025-0.150 mm (0.001-0.006 inch). Measure end play of rotors as shown in Fig. 68. Rotor end play should be 0.025-0.089 mm (0.0010-0.0035 inch). The oil valve relief spring (7—Fig. 66) should exert 4.85-5.4 kg. (10.7-11.9 lbs.) when compressed to a height of 27.2

mm (1.07 inches). Engine oil pressure should be maintained at 414-483 kPa (60-70 psi) at 2000 rpm.

Assemble pump by reversing disassembly procedure. Tighten screws securing screen cover (3) and pump cover (4) to pump body (11) to 11 N•m (96 in.-lbs.) torque. Screws securing oil pump to the crankcase should be tightened to 49 N•m (36 ft.-lbs.) torque. Prime oil pump by immersing pump in clean oil and turning rotor shaft prior to installing pump in engine.

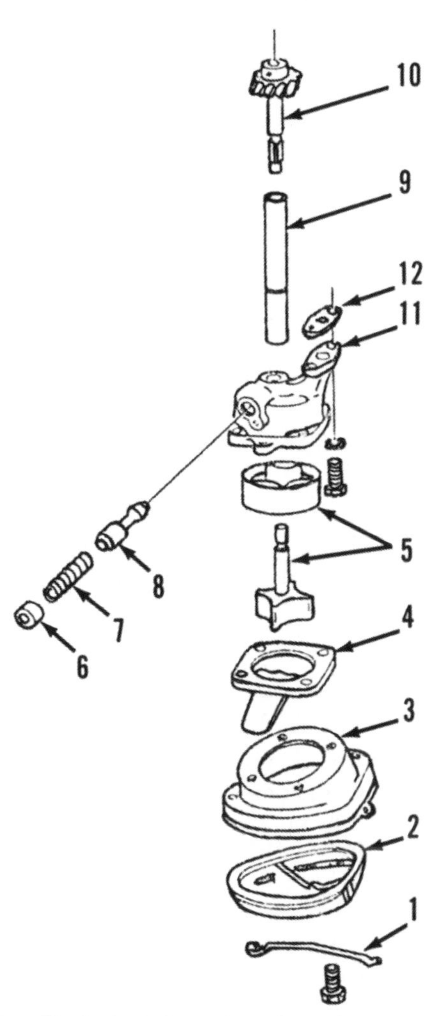

Fig. 66—Exploded view of engine oil pump.

1. Retainer clip
2. Screen
3. Screen cover
4. Pump cover
5. Rotor set
6. Plug
7. Relief valve spring
8. Relief valve
9. Floating shaft
10. Drive shaft & gear
11. Pump body
12. Gasket

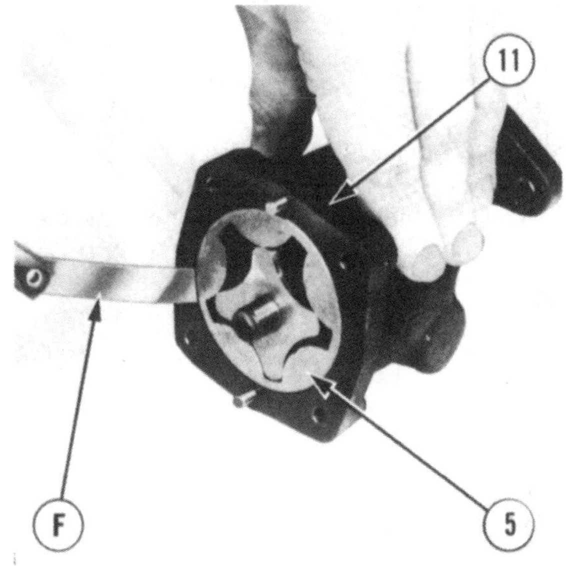

Fig. 67—Measure clearance between oil pump body (11) and outer rotor (5) with a feeler gage (F) as shown.

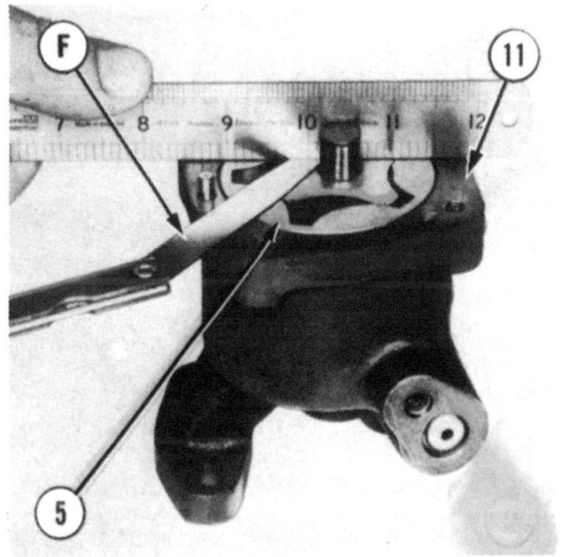

Fig. 68—Measure end play of rotors in pump body with a feeler gage (F) as shown.

DIESEL FUEL SYSTEM

69. The diesel fuel system consists of three basic components: The fuel filters, injection pump and injection nozzles. When servicing any unit associated with the fuel system, the maintenance of absolute cleanliness is of utmost importance. Of equal importance is the avoidance of nicks or burrs on any working parts.

Probably the most important precaution that service personnel can impart to owners of diesel powered tractors is to urge them to use an approved fuel that is absolutely clean and free of foreign material. Extra precaution should be taken to make certain that no water enters the fuel storage tanks.

TROUBLESHOOTING

All Models

70. If the engine will not start, or does not run properly after starting, refer to the following paragraphs for possible causes of trouble.

71. FUEL NOT REACHING INJECTION PUMP. If fuel will not run from the supply line when disconnected from the pump, check the following:

Be sure the supply valve is open.
Check the fuel filters for being clogged (include the filter screen at the fuel supply valve).
Bleed the fuel filters as outlined in paragraph 79.
Check lines and connectors for damage.

72. FUEL REACHING NOZZLES BUT ENGINE WILL NOT START. Fuel should flow from injection lines when loosened and engine is cranked with the starter. If fuel reaches the nozzles, but engine will not start, check the following:

Check the cranking speed. If too slow, engine may not start.
Check throttle control rod adjustment as in paragraph 93.
Check pump timing as outlined in paragraph 91.
Check fuel lines and connections for pressure leakage.
Check engine compression.

73. ENGINE HARD TO START. If the engine is hard to start, check the following:

Check the cranking speed. If too slow, engine may not start.
Bleed the fuel filters as outlined in paragraph 79.
Check the fuel filters for being clogged (include the filter screen at the fuel supply valve).
Check for water in fuel or improper fuel.
Check for air leaks at suction side of transfer pump.
Check engine compression.

74. ENGINE STARTS, THEN STOPS. If the engine will start, but then stops, check the following:

Check fuel filters for clogs or restrictions.
Check for water in the fuel.
Check for restrictions in the air intake.
Check engine for overheating.
Check for air leaks at suction side of transfer pump.

75. ENGINE SURGES, MISFIRES OR POOR GOVERNOR REGULATION. Make the following checks:

Bleed the fuel filters as outlined in paragraph 79.
Check for clogged or restricted fuel lines or clogged fuel filters.
Check for water in the fuel.
Check pump timing as outlined in paragraph 91.
Check injector lines and connections for leakage.
Check for faulty or sticking injector nozzles.
Check for faulty, leaking or sticking engine valves.

76. LOSS OF POWER. If the engine does not develop full power or speed, check the following:

Check throttle control rod adjustment as in paragraph 93.
Check maximum no-load speed adjustment as in paragraph 92.
Check for clogged or restricted fuel lines or clogged fuel filters.
Check for air leaks at suction side of transfer pump.
Check pump timing as outlined in paragraph 91.
Check engine compression.
Check for improper engine valve clearance adjustment as outlined in paragraph 44.
Check for faulty, leaking or sticking engine valves.

77. EXCESSIVE BLACK SMOKE AT EXHAUST. If the engine emits excessive black smoke from the exhaust, check the following:

Check for restrictions in the air intake, such as a clogged air filter.
Check pump timing as outlined in paragraph 91.
Check for faulty or sticking injector nozzles.
Check engine compression.

FILTERS AND BLEEDING

All Models

78. MAINTENANCE. The lower part of the sediment separator is glass and it should be possible to see any sediment or water that has accumulated in the bowl. Small amounts of contaminants should be drained by loosening the drain plugs (1 and 2—Fig. 69) located in the bottom of the sediment bowl and filter housing. It is important to keep the glass sediment bowl clean so contaminants can be clearly seen.

After every 600 hours of operation, the fuel filter should be renewed. Turn fuel off at tank shut-off valve, then remove screws (3 and 4) and disassemble filter and sediment bowl assemblies (Fig. 70).

Clean all parts that are to be reinstalled and install new filter element (9) and all gaskets. Be sure the glass sediment bowl (15) is clean and clear before reassembling. Turn the fuel supply on at the tank and bleed the fuel system as outlined in paragraph 79 after reassembling.

Many fuel system contaminants have either originated or accumulated in the tractor's fuel tank. A filter (4—Fig. 71) located in the tank may become so restricted with solid contaminants that fuel does not enter the fuel filter assembly. The filter can be removed from the tank and cleaned or renewed, but do not overlook accumulation of similar contaminants in the fuel tank.

79. BLEEDING. Turn the fuel supply valve on at fuel tank and open the bleed screw (5 or 6—Fig. 72). Close the bleed screw when a steady stream of fuel without bubbles flows from the opened bleed point. Open the bleed screw (17—Fig. 73) on fuel injection pump and crank engine with starter until a steady stream of fuel without bubbles flows from the bleed point, then close the bleed screw. Press the primer plunger (7—Fig. 72) several times until resistance is felt, indicating that system is free of air. Loosen the fuel injection pressure connection at each fuel injection nozzle and crank the engine with starter until fuel flows at the loosened connections. Tighten fuel line connections and start engine.

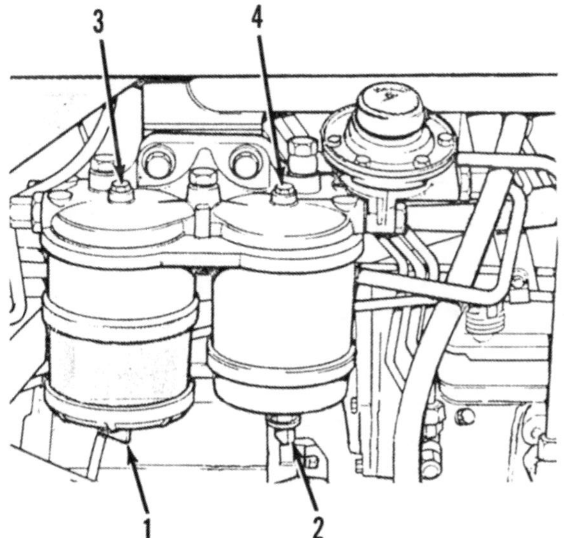

Fig. 69—Drain plugs are located at the bottom of the fuel system sediment bowl (1) and at the bottom of the fuel filter (2).

FUEL INJECTORS

CAUTION: Fuel leaves the injection nozzle with sufficient force to penetrate the skin. When testing, stay clear of the nozzle spray.

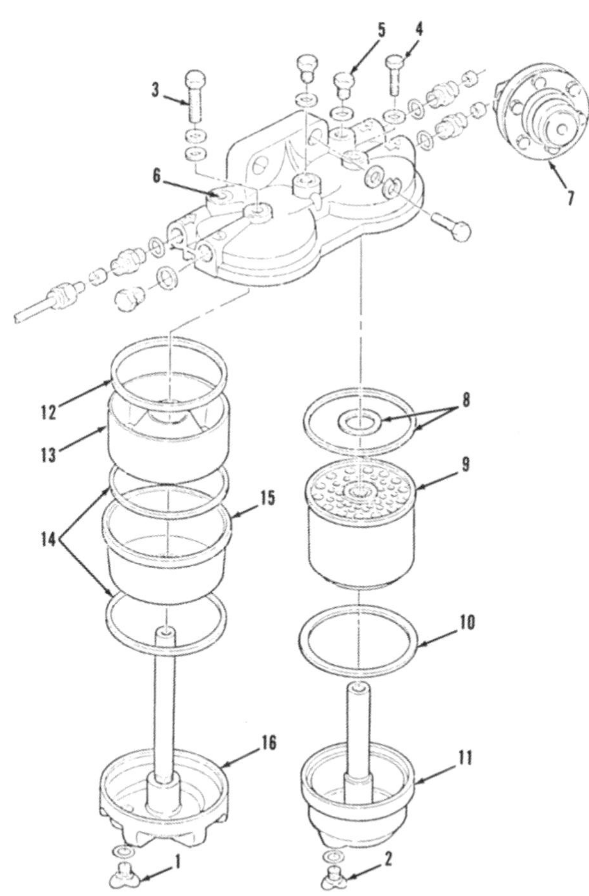

Fig. 70—Exploded view of the fuel filter assembly typical of all models. Slight differences may be noted.

1. Drain plug
2. Drain plug
3. Screw attaching sediment bowl
4. Screw attaching fuel filter
5. Bleed screw (except 4830 models)
6. Location of bleed screw on 4830 models
7. Fuel primer
8. Gaskets
9. Fuel filter
10. Gasket
11. Filter base
12. Gasket
13. Top element
14. Gaskets
15. Sediment bowl
16. Sediment bowl base

All Models

80. LOCATING A FAULTY INJECTOR. If the engine does not run properly and a faulty injector is suspected, locate the faulty injector as follows: With the engine running, loosen the high-pressure connection at each injector in turn, thereby allowing fuel to escape at the union rather than enter the injector. As in checking for malfunctioning spark plugs in a spark ignition engine, the faulty injector is the one that least affects the running of the engine when its line is loosened.

81. INJECTOR TESTING. A complete job of testing and adjusting fuel injectors requires the use of a special tester, such as shown in Fig. 74. The injector should be removed and tested for opening pressure, spray pattern, seat leakage and leak-back.

Operate the tester until oil flows, then connect the injector to the tester. Close tester valve to shut-off the passage to the gage and operate the tester lever to be sure that injector is in operating condition and that nozzle is not plugged. If oil does not spray from all four holes in the nozzle, if test lever is hard to operate or if other obvious defects are noted, remove the injector from the tester and service as outlined in paragraph 87. If the injector operates without undue pressure on the lever and fuel is sprayed from all the holes in nozzle, proceed with the following tests.

82. OPENING PRESSURE. Open the shut-off valve to the gage and operate the tester lever slowly. Note the pressure indicated on the gage at which the nozzle spray occurs. This pressure should be approximately 18,380-19,910 kPa (2666-2887 psi).

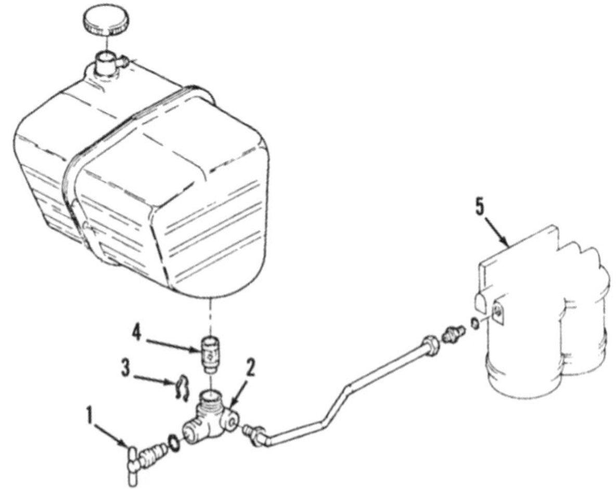

Fig. 71—View of fuel shut-off valve (1 & 2), showing relative position of tank filter (4) and fuel filters (5).

1. Shut-off valve
2. Valve housing
3. Retainer clip
4. Tank filter screen
5. Fuel filters

If the gage pressure is not within limits, remove the cap nut and turn the adjusting screw (3—Fig. 76) as

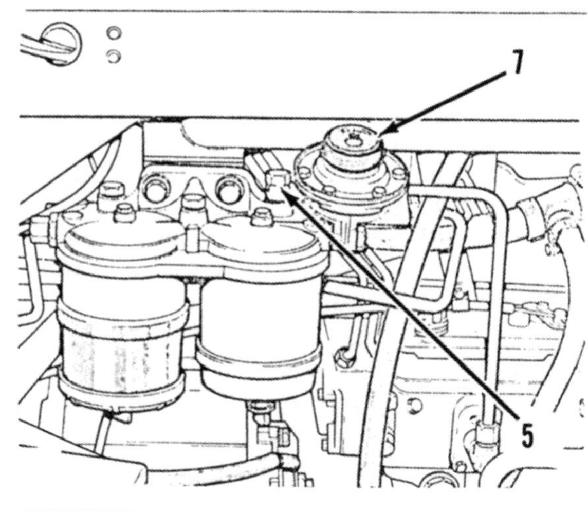

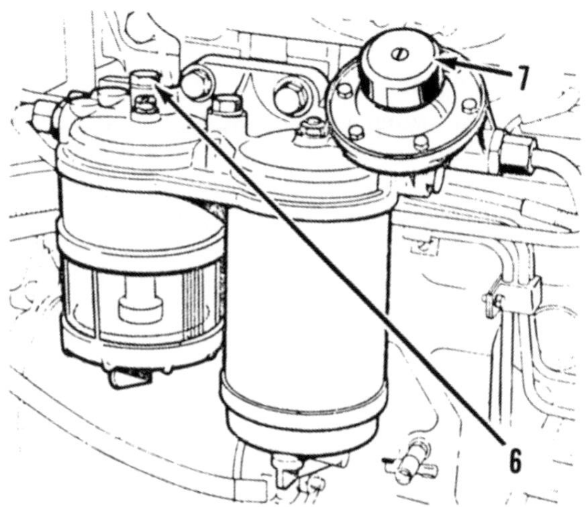

Fig. 72—View of fuel filter assemblies used showing location of bleed screws. Screw (5) is location for all models except 4830; screw (6) is location for 4830 models. Primer pump is shown at (7) for all models.

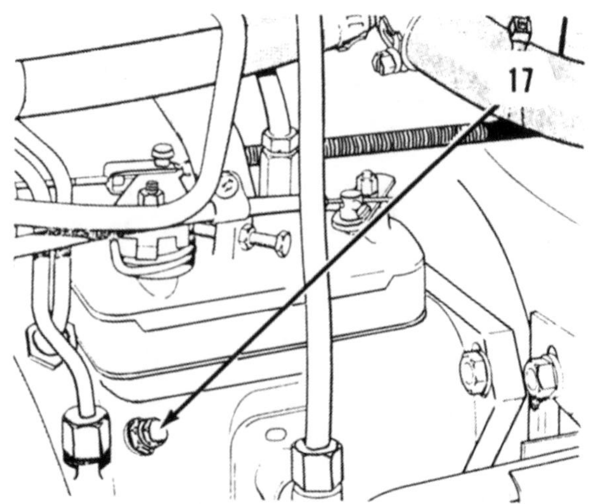

Fig. 73—View showing location of fuel bleed screw (17) on the injection pump.

required to bring the opening pressure within specified limits. If opening pressure is erratic or can not be properly adjusted, remove the injector from the tester and overhaul injector as outlined in paragraph 87. If the opening pressure is within limits, check the spray pattern as outlined in the following paragraph.

83. SPRAY PATTERN. Operate the tester lever slowly and observe the nozzle spray pattern. All four of the sprays must be similar and spaced at approximate intervals of 110, 90, 70 and 90 degrees in nearly a horizontal plane. Each spray must be well atomized and should spread to a 76 mm (3 inches) diameter cone approximately 9.5 mm (3/8 inch) from the nozzle tip. If the spray pattern does not meet these conditions, remove the injector from the tester and overhaul as outlined in paragraph 87. If the spray pattern is satisfactory, proceed with seat leakage test as outlined in the following paragraph.

84. SEAT LEAKAGE. Close the valve to shut-off pressure to the tester gage and operate the tester lever quickly for several strokes. Wipe the nozzle tip dry with clean blotting paper, open the valve to the tester gage and press the lever down slowly to bring the gage pressure to 1035 kPa (150 psi) below the nozzle opening pressure and hold this pressure for one minute. Touch a piece of blotting paper to the nozzle tip as shown in Fig. 74. The resulting oil blot should not be larger than 12.7 mm (1/2 inch) in diameter. If the nozzle tip drips oil or if blot is excessively large, remove the injector from the tester and overhaul as outlined in paragraph 87. If nozzle seat leakage is not excessive, proceed with nozzle leak-back test as outlined in the following paragraph.

85. NOZZLE LEAK-BACK. Operate the tester lever to bring gage pressure to approximately 15,860 kPa (2300 psi), release the lever and note the time that it takes for the gage to drop from 15,170 kPa (2200 psi) to 10,345 kPa (1500 psi). If the time is less than 5 seconds, the nozzle is worn or there are dirt particles lodged in the nozzle. If the time to drop the pressure is too long (40 seconds or more), the needle may be too tight in the bore. Refer to paragraph 87 for disassembly, cleaning and overhaul information.

NOTE: A leaking tester connection, a leaking check valve in tester or leaking test gage will also indicate excessively fast leak-back. If injectors consistently show excessively fast leak-back, the tester should be suspected as faulty rather than all injectors.

86. REMOVE AND REINSTALL. Carefully clean all dirt and other foreign material from lines, connectors, injectors and cylinder head area around injectors before removing the injectors. Disconnect injector leak-off line at each injector and at fuel return line. Disconnect the injector line at pump and at injector. Cover all openings and lines to prevent the entrance of dirt. Remove the two nuts retaining the injector and carefully remove the injector from the cylinder head.

Make sure injector seats in cylinder head are clean and free from any carbon deposit before reinstalling injectors. Install a new copper washer in seat and a new dust sealing washer around the body of injector. Insert injector in cylinder head bore, install retaining washers and nuts and tighten the nuts evenly and alternately to 23 N•m (17 ft.-lbs.) torque. Install a new gasket below and above each leak-off banjo fitting, then tighten the banjo bolt to 8 N•m (72 in.-lbs.) torque. Reconnect the leak-off line to the return line. Check fuel injector connections to be sure that they

Fig. 74—A fuel injector tester, such as the one shown, is necessary for checking and adjusting the fuel injector assemblies.

Fig. 75—View of nozzle attached to the tester for checking the opening pressure.

are clean and properly aligned with fittings at injector and at pump. Reconnect injector pressure lines, tightening to 27 N•m (20 ft.-lbs.) torque.

It is suggested that the fuel filter be serviced as described in paragraph 78 each time injectors are serviced. Refer to paragraph 79 for bleeding the fuel system. Since lines to injectors are opened, it will be necessary to crank engine with fitting at injectors loosened to fill the lines. Start and run engine to be sure that injectors are properly sealed and that injector pressure line and leak-off line connections are not leaking.

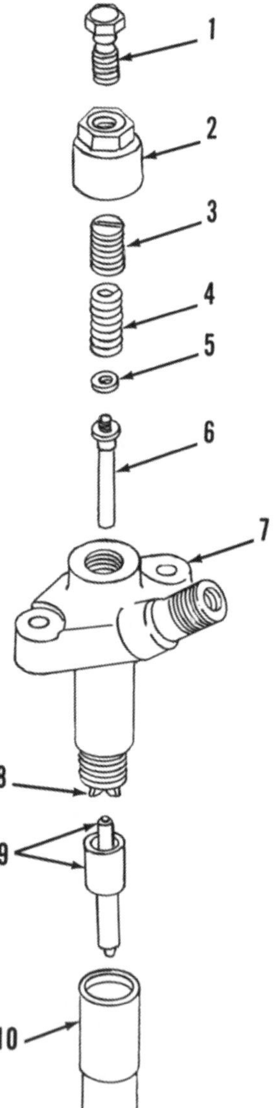

Fig. 76—Exploded view of injection nozzle typical of all models.

1. Leak-off line connector
2. Cap nut
3. Adjusting screw
4. Spring
5. Washer
6. Spindle
7. Holder
8. Alignment pins
9. Nozzle body & valve needle
10. Nozzle retaining nut

87. OVERHAUL. Do not attempt to overhaul diesel injectors unless complete and proper equipment is available.

Secure the injector holding fixture in a vise and locate the injector in the holding fixture. Never clamp the injector body in a vise. Remove the cap nut (2—Fig. 76) and back-off the adjusting screw. Lift the upper spring disc, injector spring and lower spring seat from injector. Remove the nozzle retaining nut using an appropriate special socket, then remove nozzle and valve. The nozzle and valve are a lapped fit to each other and parts from one must never be interchanged with parts from another matched unit. Place all parts in clean fuel oil or calibrating fluid as they are disassembled. Clean the injector unit as follows: Soften hard carbon deposits by soaking in a suitable carbon solvent, then use a soft wire (brass) brush to remove carbon from the exterior. Rinse the nozzle and needle immediately after cleaning to prevent carbon solvent from corroding the highly finished surfaces. The pressure chamber of nozzle can be cleaned with a special reamer. Clean spray holes with proper size of wire probe held in a pin vise as shown in Fig. 77. To reduce the chance of breakage, wire should protrude from vise only far enough to pass through pin holes. Rotate pin vise, but do not apply undue pressure.

Valve seat in nozzle is cleaned by inserting a special seat scraper into nozzle, then rotating the scraper. Refer to Fig. 78. The annular groove can be cleaned with scraper shown in Fig. 79.

After cleaning, back-flush the nozzle using a back-flush adapter attached to nozzle tester as shown in Fig. 80. Rotate the nozzle while operating the lever to flush nozzle.

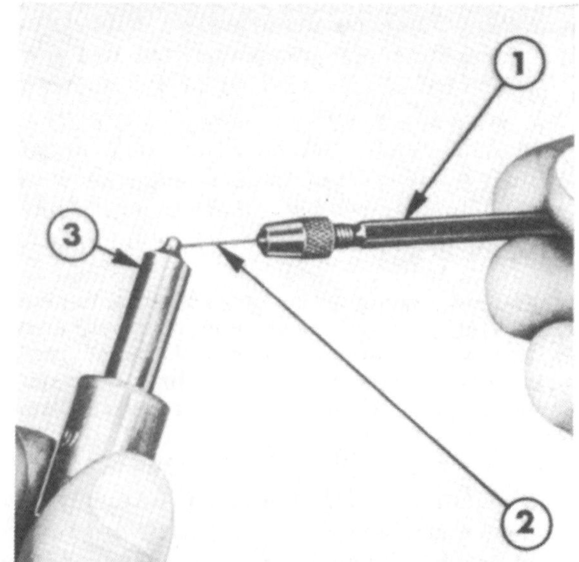

Fig. 77—Clean nozzle spray holes with wire probe held in a pin vise.

Seat in nozzle can be polished using a small amount of tallow on the end of a polishing stick. Be sure that carbon is thoroughly flushed from nozzle before polishing seat.

If the leak-back test was longer than 40 seconds as tested in paragraph 85, or if needle is sticking in bore of nozzle, the needle should be lapped to the nozzle. Use special polishing lapping compound such as Bacharach 66-0655. Place the small diameter of nozzle in a drill chuck and rotate at not more than 450 rpm. Apply a small amount of polishing compound on the barrel of valve needle taking care not to allow any compound on tip or beveled portion. Insert the valve needle into the slowly rotating nozzle. Work the needle in and out several times taking care not to put any pressure against seat. Withdraw needle, remove nozzle and back-flush the nozzle and needle as shown in Fig. 80.

NOTE: Do not lap valve for longer than 5 seconds at a time and allow parts to cool between lappings.

Assemble nozzle and needle while still wet from cleaning. Valve needle should slide easily in nozzle. If any sticking is noted, be sure that parts are clean and temperature has normalized.

Rinse all parts in clean fuel oil or calibrating fluid before reassembling, then assemble while still wet. Position the nozzle and needle valve (9—Fig. 76) on injector body (7) and be sure that dowel pins (8) in body are correctly located in nozzle. Install the nozzle

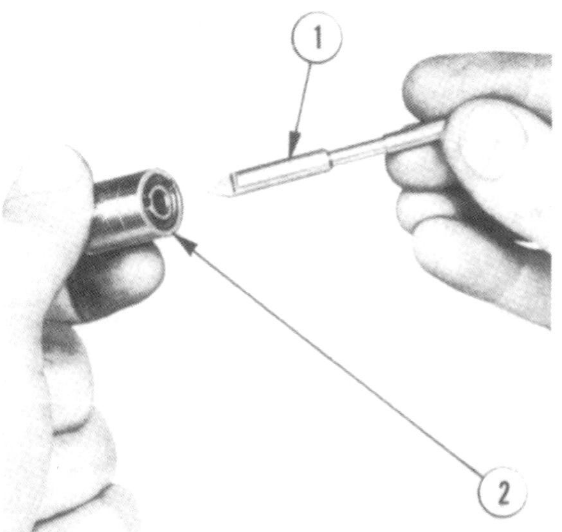

Fig. 78—Use a special scraper to clean carbon from valve seat in nozzle body.

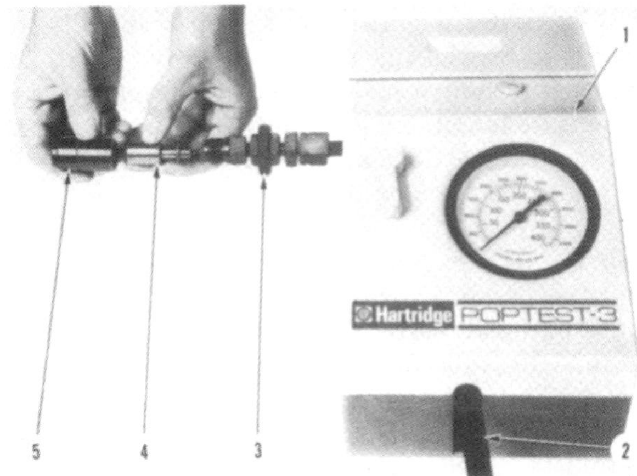

Fig. 80—A back-flush attachment can be installed on nozzle tester to clean the nozzle by reversing the flow of test fluid.

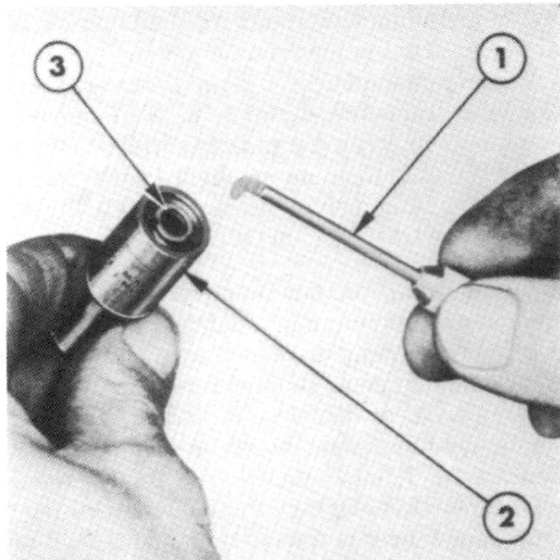

Fig. 79—Scraper can be used to clean the annular groove of nozzle.

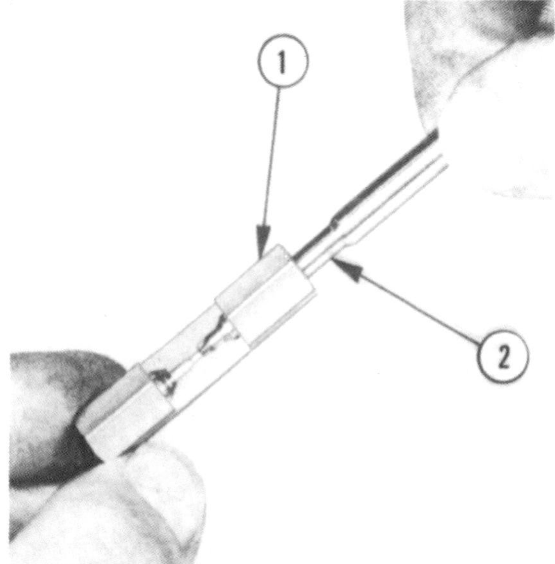

Fig. 81—The tip of the needle valve can be cleaned using special scraper shown.

retaining nut (10) and tighten nut to 70 N•m (50 ft.-lbs.) torque. Install spindle (6), washer (5), spring (4) and pressure adjusting screw (3). Connect the injector to tester (Fig. 75) and adjust opening pressure as outlined in paragraph 82. Use a new copper washer and install cap nut (2—Fig. 76). Recheck opening pressure after tightening cap nut (2) to 70 N•m (50 ft.-lbs.) torque to be sure that opening pressure has not changed.

Retest injector as outlined in paragraphs 81 through 85. Install new nozzle and needle assembly or complete injector assembly if still faulty. Nozzles should be thoroughly flushed with calibrating fluid prior to storage.

FUEL INJECTION PUMP

All Models

88. TIMING. Refer to paragraph 90 for installation and timing of the injection pump drive gear. The pump is internally timed and except for adjusting to compensate for timing gear backlash, timing can be considered correct if mark on pump is aligned with the "0" degree mark on the engine front plate.

89. REMOVE AND REINSTALL. Thoroughly clean the pump, lines and connections in the area around pump. Remove the lines from the pump to the injectors, disconnect the fuel inlet and outlet (return) lines and immediately cover all openings to prevent the entrance of dirt. Disconnect the throttle control rod and the fuel shut-off cable. Remove the cover plate from the timing gear cover and remove the three

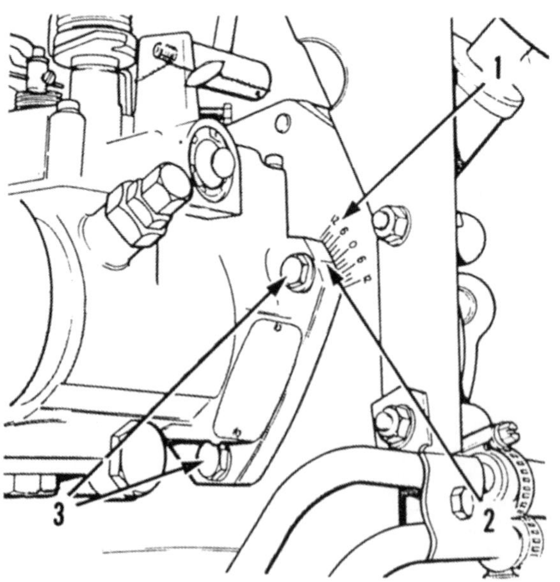

Fig. 82—View of timing marks on pump mounting ear and engine front plate. Timing is correct when mark on pump is aligned with "0" degree mark on plate. If front cover or pump is not marked, stamp mark before removing pump.

screws attaching the injection pump to the pump hub. Unbolt the injection pump from the engine front plate and withdraw pump. It may be necessary to use a puller attached to the three threaded holes in gear to push the pump drive shaft from the gear. The pump drive gear will remain in timing gear housing and will not change the gear timing. Gear cannot be removed unless cover is removed; however, engine crankshaft should not be turned with pump removed.

To reinstall the fuel injection pump, reverse the removal procedures. Align the scribed mark on pump body with the "0" degree mark on engine front plate as shown in Fig. 82 and tighten the pump retaining nuts to 24 N•m (18 ft.-lbs.) torque. Tighten the screws attaching the injection pump gear to the injection pump shaft to 38 N•m (28 ft.-lbs.) torque. Connect fuel lines, throttle control rod and fuel shut-off cable, then bleed the fuel system as outlined in paragraph 79.

INJECTION PUMP DRIVE GEAR

All Models

90. To remove injection pump drive gear, first remove timing gear cover as outlined in paragraph 48, rocker arm assembly as outlined in paragraph 47 and camshaft drive (idler) gear as outlined in paragraph 50. Damage could result if either camshaft or crankshaft is turned independently from the other unless the rocker arms are removed. Remove the three screws attaching gear to pump shaft, then remove the gear. It may be necessary to use a puller attached to the three threaded holes in gear to pull the gear from pump drive shaft.

Before installing the gear, turn the engine crankshaft so that timing marks on crankshaft gear (1—Fig. 83) and camshaft gear (3) point toward the center of the idler (camshaft drive) gear (2). Remove the self-locking cap screw retaining the idler gear and reinstall the gear with marks aligned with marks on crankshaft and camshaft gears. Tighten the idler gear cap screws to 140 N•m (103 ft.-lbs.) torque.

Turn the pump so that pump drive gear can be installed on dowel pin in pump drive and the correct timing mark on pump drive gear can be aligned with the mark on idler gear. The fuel injection pump drive gear (4) is marked with both "3" and "4." The timing mark on the idler gear should be aligned with the correct mark ("3" or "4") depending upon the number of engine cylinders. Tighten the screws attaching the injection pump gear to the injection pump shaft to 38 N•m (28 ft.-lbs.) torque. Refer to paragraph 48 and install the timing gear cover and to paragraph 47 and install the rocker arm assembly.

ADJUSTMENTS

All Models

91. TIMING. Refer to paragraph 90 for installation and timing of the injection pump drive gear. The pump is internally timed and except for adjusting to compensate for timing gear backlash, timing can be considered correct if mark on pump is aligned with the "0" degree mark on the engine front plate.

Internal timing should be accomplished only by trained and experienced injection pump servicing stations that have the necessary special tools and equipment.

Fig. 83—View showing timing marks aligned on crankshaft gear (1), camshaft drive (idler) gear (2), camshaft gear (3) and injection pump drive gear (4). Fuel injection pump drive gear (4) is marked with both "3" and "4." The timing mark on the idler gear should be aligned with the correct mark (3 or 4) depending upon the number of engine cylinders.

1. Crankshaft gear
2. Idler gear
3. Camshaft gear
4. Injection pump gear

92. SPEED. To check idle speed, first start the engine and bring to a normal operating temperature. Disconnect the throttle linkage from the governor arm of the fuel injection pump. Hold the governor arm against the slow idle speed screw (1—Fig. 84 or Fig. 85) and note the engine rpm. Idle speed should be 600-850 rpm for all models. If idle speed is incorrect, loosen lock nut and turn stop screw (1) as required, then tighten the lock nut.

The maximum high (no-load) speed should be as follows:

Model	High Idle Speed
3230	2175 rpm
3430 Without Cab	2175 rpm
3430 With Cab	2155 rpm
3930 Without Cab	2175 rpm
3930 With Cab	2155 rpm
4630 Without Cab	2200 rpm
4630 With Cab	2375 rpm
4830 Without Cab	2200 rpm
4830 With Cab	2375 rpm

The high speed stop screw is adjusted in a similar manner to slow idle speed; however, the adjustment screw is located under a cover (2—Fig. 84 or Fig. 85) that is sealed to prevent tampering. Do not adjust the

Fig. 84—View of DPA fuel injection pump showing the idle speed adjustment screw (1). The high speed stop screw is located under sealed cover (2). The DSP pump used on some models is shown in Fig. 85.

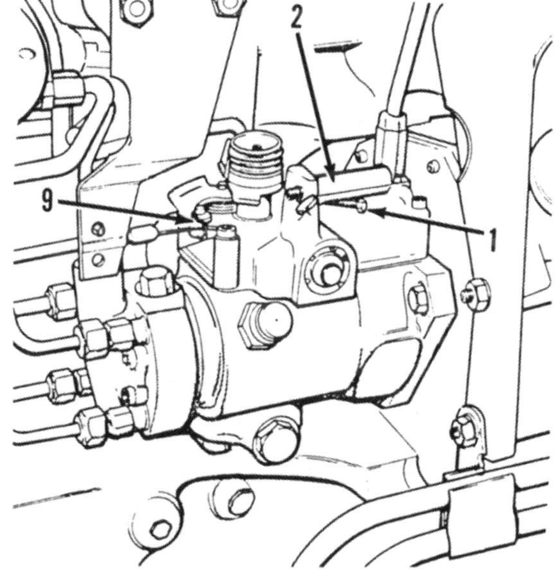

Fig. 85—View of DSP fuel injection pump showing the idle speed adjustment screw (1). The high speed stop screw is located under sealed cover (2). The DPA pump used on some models is shown in Fig. 84.

high speed screw to permit engine speeds higher than manufacturer's recommended speed.

93. THROTTLE LINKAGE. Refer to exploded views of typical throttle linkage in Fig. 86 and Fig. 87. To check and adjust the throttle linkage, first disconnect the throttle link (11—Fig. 86) or cable swivel (9—Fig. 87) from the injection pump governor control arm and proceed as follows. Refer to paragraph 92 and check engine governed speeds and adjust if necessary. Move the speed control hand lever to the rear against the slow speed stop, hold the injection pump governor arm against the low idle stop screw, then attempt to enter end of link (11—Fig. 86) or swivel (9—Fig. 87) into hole of governor arm. Adjust the length of link (11—Fig. 86) on models without cab by loosening lock nut (10), then turning rod end (9) as required. Tighten lock nut (10) when length of rod is correct. Models with cab are equipped with cable (13—Fig. 87). To adjust position of swivel (9), loosen screw (10), move swivel as required, then tighten screw (10). On all models, be sure that linkage operates freely throughout the entire range of travel and does not limit movement of governor lever on pump. Friction disc (5—Fig. 86 or Fig. 87) should provide sufficient tension so that hand lever will remain in any desired position, yet move without excessive binding. Tighten or loosen adjustment nut (8) as required. Recheck governed speeds as described in paragraph 92 with linkage connected to be sure that linkage does not limit operation.

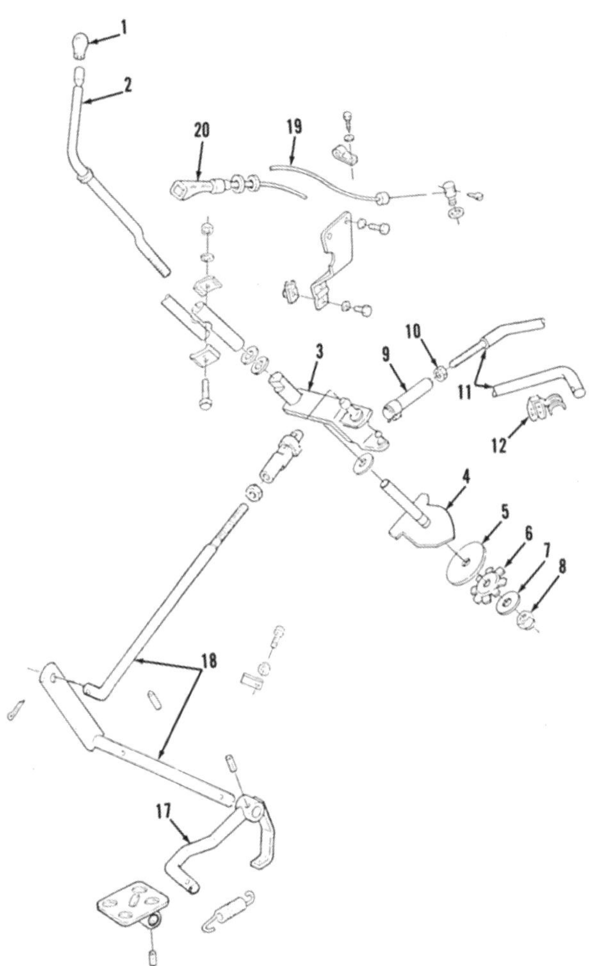

Fig. 86—Exploded view of throttle linkage typical of all models without cab.

1. Knob
2. Hand lever
3. Throttle arm lever
4. Plate
5. Friction disc
6. Spring
7. Washer
8. Nut
9. Rod end
10. Locknut
11. Throttle link
12. Retainer
17. Foot throttle
18. Linkage
19. Stop cable
20. Fuel shut-off control

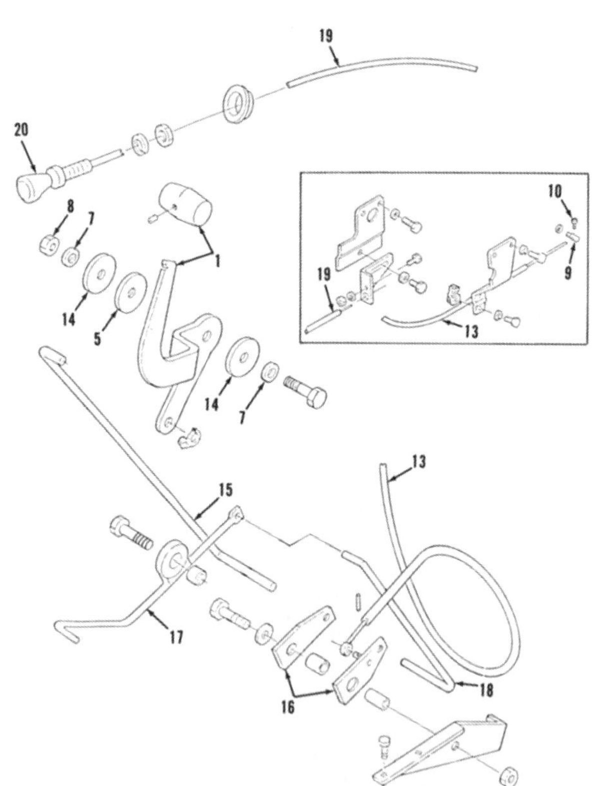

Fig. 87—Exploded view of throttle linkage typical of models with cab.

1. Hand lever
5. Friction disc
7. Washers
8. Nut
9. Swivel
10. Set screw
13. Throttle cable
14. Washers
15. Linkage
16. Arm assembly
17. Foot throttle
18. Linkage
19. Stop cable
20. Fuel shut-off control

COOLING SYSTEM

RADIATOR PRESSURE CAP AND THERMOSTAT

All Models

94. All models are equipped with a 69 kPa (10 psi) radiator pressure cap. On three-cylinder models, the thermostat is located in front of cylinder head under the thermostat housing/water outlet. On four-cylinder models, a separate thermostat housing is attached to the front of cylinder head and thermostat is located under the water outlet housing. The thermostat can be removed from all models after draining coolant and removing the water outlet housing. The thermostat should begin to open at 82° C (180° F) and be fully open at 95° C (203° F).

Be sure thermostat is correctly positioned before tightening the outlet housing retaining screws to 19 N•m (14 ft.-lbs.) torque.

RADIATOR

All Models

95. REMOVE AND REINSTALL. Remove hood, drain cooling system and disconnect radiator hoses. Disconnect the air inlet hose from the front-mounted air cleaner. Remove the grille and disconnect lines from oil coolers, if so equipped. Disconnect battery ground cable and wires to the headlights. Unbolt radiator shell upper support bracket from the shell. Unbolt and remove radiator and fan shroud assembly from tractor.

To reinstall, reverse removal procedure.

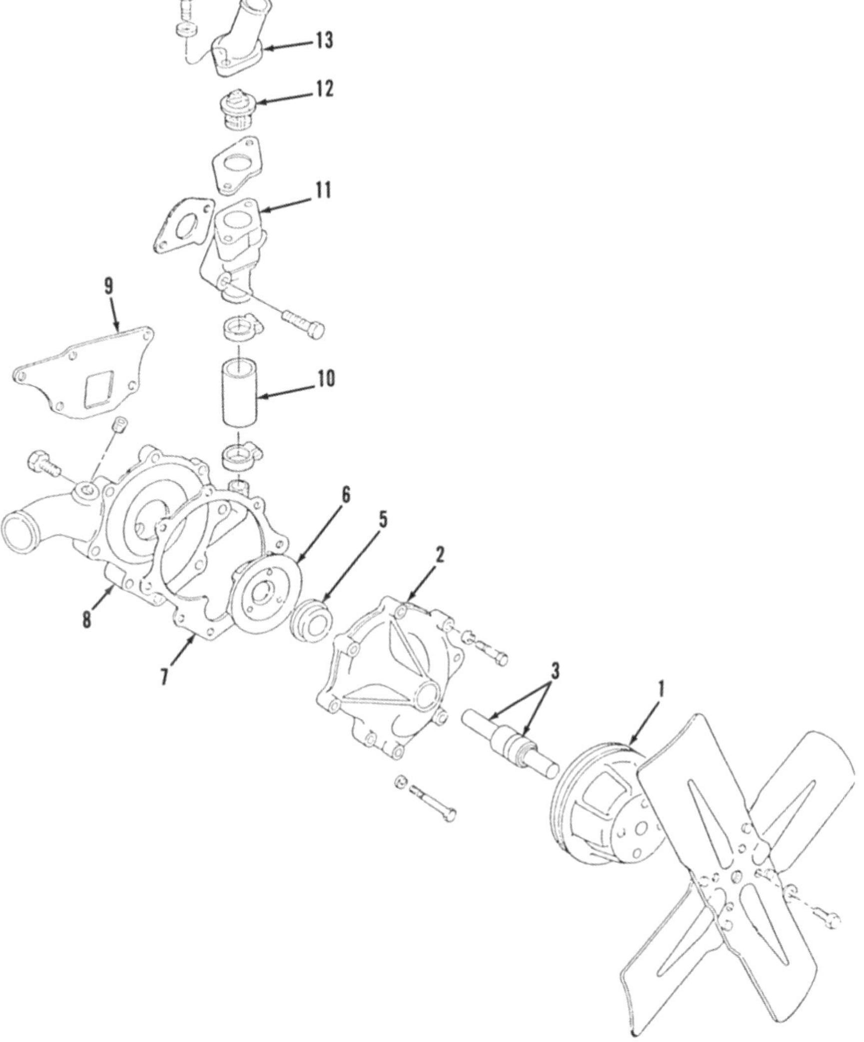

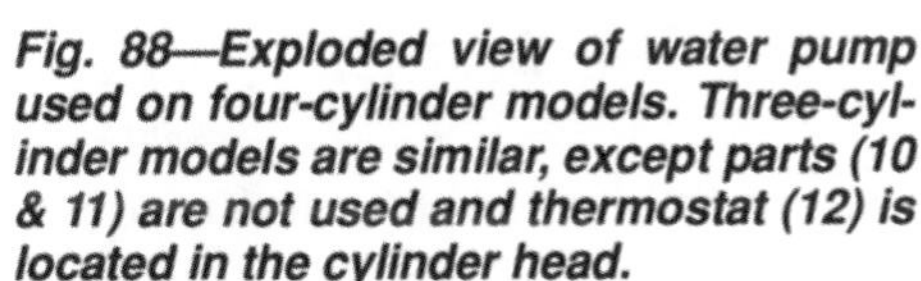
Fig. 88—Exploded view of water pump used on four-cylinder models. Three-cylinder models are similar, except parts (10 & 11) are not used and thermostat (12) is located in the cylinder head.

1. Pulley
2. Housing
3. Shaft & bearing assy.
5. Seal assy.
6. Impeller
7. Gasket
8. Rear cover
9. Gasket
10. By-pass hose
11. Thermostat housing
12. Thermostat
13. Outlet housing

WATER PUMP

All Models

96. The water pump can be removed after removing the radiator as outlined in paragraph 95. Refer to Fig. 88 and disassemble pump as follows:

Unbolt and remove the fan, then use a puller to remove pulley (1) from shaft (3). Remove rear cover (8), then press shaft and bearing (3) forward out of impeller (6) and housing (2). Drive seal (5) out of housing toward rear.

Press new shaft and seal assembly (3) into housing (2) until the bearing is flush with the front face of the housing. Coat outer diameter of seal flange lightly with thread sealer, then press seal (5) into housing until flange is flush with top of housing. Support shaft and press impeller (6) onto shaft until flush with rear face of the housing as shown in Fig. 89. Support rear of shaft and press pulley onto front as shown in Fig. 90. Press pulley onto shaft until distance (A—Fig. 90) from rear of housing (2) to the front pulley flange is 63 mm (2.48 inches) as shown. Tighten screws attaching rear cover (8—Fig. 88) to housing (2) to 27 N•m (20 ft.-lbs.) torque. Tighten screws securing water pump to cylinder block to 35 N•m (26 ft.-lbs.) torque. The "V" belt should be tightened until deflection midway between pulleys is 12-19 mm (1/2 - 3/4 inch).

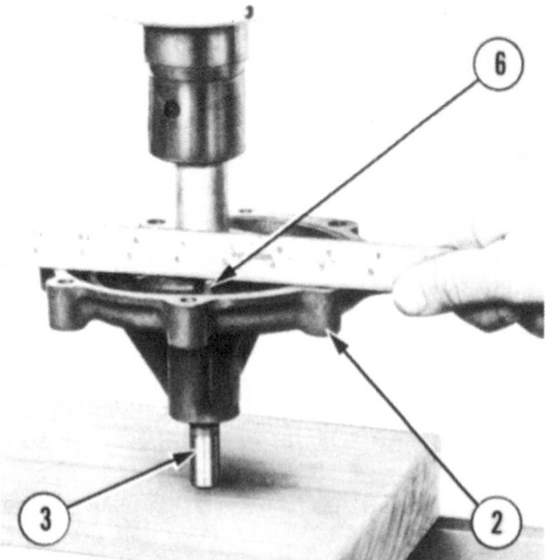

Fig. 89—Press impeller (6) onto shaft (3) until flush with rear surface of housing (2) as shown.

Fig. 90—Pulley should be pressed onto shaft until distance (A) is 63 mm (2.48 inches) as shown.

ELECTRICAL SYSTEM

ALTERNATOR AND REGULATOR

All Models

97. A large number of electrical components can be damaged very quickly using improper procedures, and certain alternator components can be damaged by procedures that would not affect a DC generator. Observe the following to prevent unnecessary damage.

1. Be sure negative post of all batteries is grounded. This precaution should also be observed when connecting booster battery and charger.

2. Never short across any alternator terminal.

3. Never attempt to "polarize" alternator.

4. Always disconnect all battery ground connections before removing or replacing any electrical unit.

5. Never operate an alternator with an open circuit. Be sure all leads are connected and tightened before starting engine. Also, be sure battery cables are not corroded or otherwise damaged leaving an open connection when engine is running.

98. A 55-amp alternator is used on models without cab and a 70-amp alternator is installed where the tractor is originally equipped with a cab. Regulator controlled voltage should be 13.6-14.4 volts. New brushes are 17 mm (0.67 inch) and minimum length is 5 mm (0.20 inch). Brush spring tension should be 1.3-2.7 N (4.7-9.8 oz.). Refer to Fig. 91 and Fig. 92, then tighten alternator screws, nuts and bolts to the following torques:

Through-bolts . . . . 5.5 N·m (48 in.-lbs.)
Shaft nut . . . . 37.5 N·m (27.5 ft.-lbs.)
Rectifier (7) attaching nuts . . . . 3.5 N·m (31 in.-lbs.)
Regulator & brushes assy. (10) attaching screws . . . . 2.5 N·m (24 in.-lbs.)
Main output terminal nut . . . . 4 N·m (35 in.-lbs.)

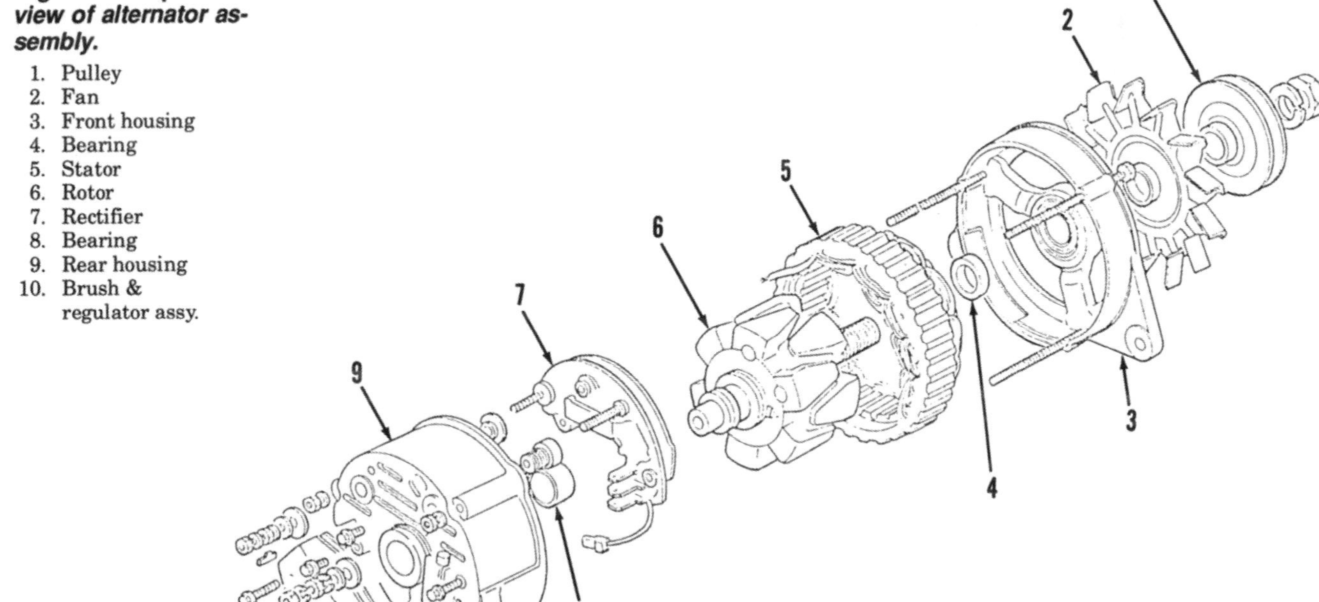

Fig. 91—Exploded view of alternator assembly.

1. Pulley
2. Fan
3. Front housing
4. Bearing
5. Stator
6. Rotor
7. Rectifier
8. Bearing
9. Rear housing
10. Brush & regulator assy.

STARTING MOTOR

All Models

99. The engine is originally equipped with a D8NN-11000-CE starting motor. The lever pivot (25—Fig. 93) is eccentric and can be rotated to adjust the drive pinion closer to the nose of end housing (33). To check clearance, engage the solenoid with 6-volt power source, then check clearance with feeler gage. If clearance is not 0.25-0.51 mm (0.010-0.020 inch), loosen the lock nut and turn the pivot screw to obtain proper clearance, then tighten lock nut. Recheck clearance after tightening the lock nut. If clearance cannot be adjusted to within limits, lever (24) may be worn excessively. Refer to the following specifications:

Brush spring tension—
Minimum with new brushes . . . . 1190 gms. (42 oz.)
Brush minimum length. . . . . 7.9 mm (0.31 inch)
Commutator diameter. . . . . 38.9 mm (1.53 inches)
Armature shaft end play, max.. . . . . 0.51 mm (0.020 inch)
Armature shaft run-out, max.. . . . . 0.13 mm (0.005 inch)

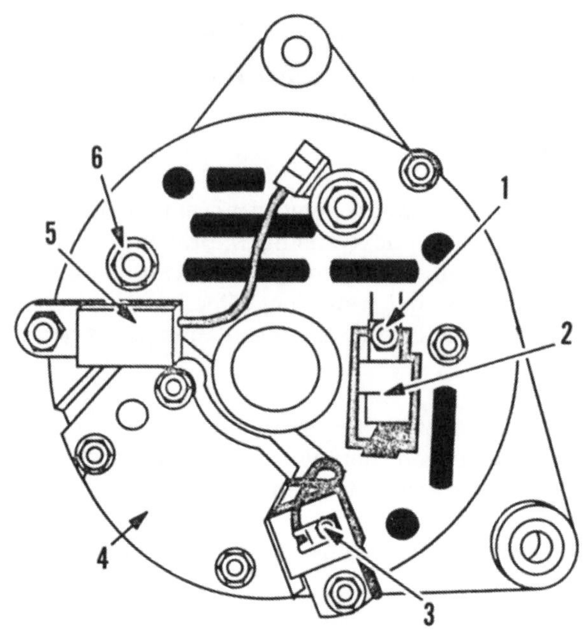

Fig. 92—View of alternator rear housing showing terminals for identification.

1. Indicator terminal
2. Main output terminals
3. Battery temperature sensing terminal
4. Brushes and regulator
5. Radio interference suppressor
6. Tachometer/overspeed indicator terminal

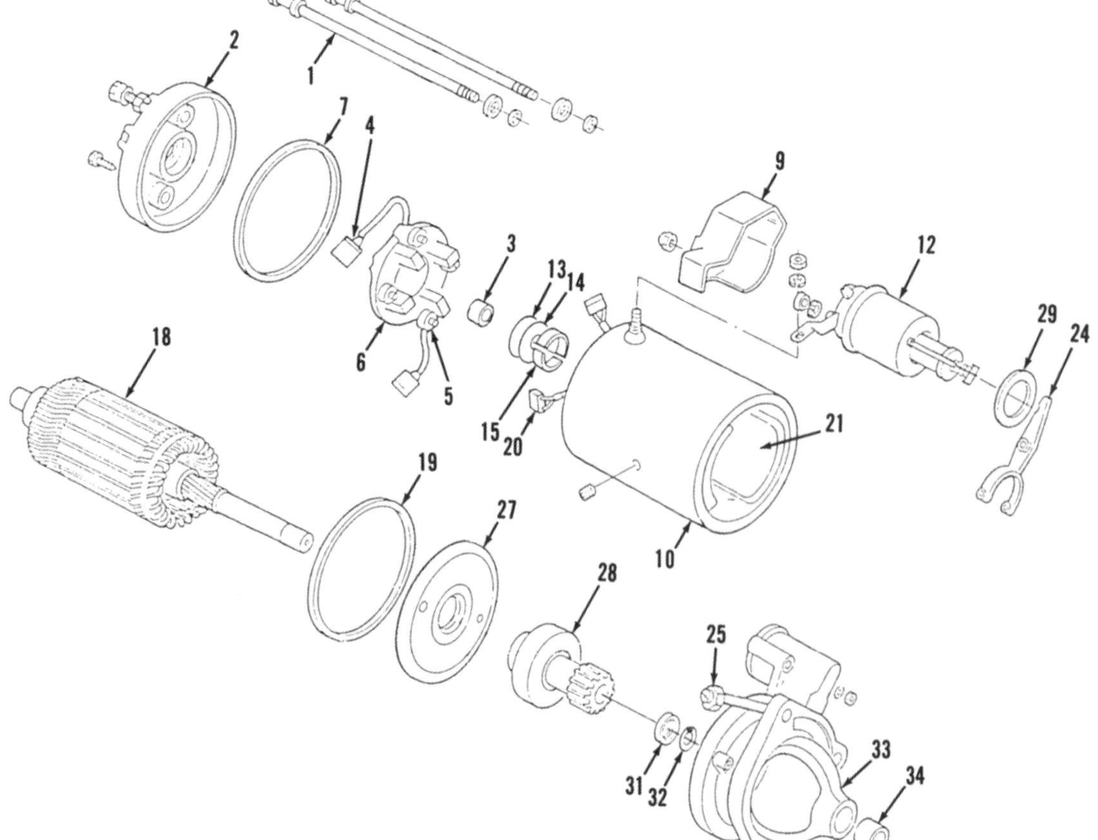

Fig. 93—Exploded view of starter assembly.

1. Through-bolts
2. End cover
3. Bushing
4. Brushes
5. Brush springs
6. Brush holder
7. Seal
9. Insulator
10. Shell
12. Solenoid
13. Thrust washer
14. Thrust washer
15. Brake assembly
18. Armature
19. Seal
20. Field brushes
21. Field coils
24. Lever
25. Lever pivot
27. Plate
28. Starter drive
29. Solenoid seal
31. Retainer
32. Snap ring
33. Drive end housing
34. Bushing

Test volts 12
No-load test—
Amps 100
Rpm 5500-7500
Loaded test warm—
Amps 250-300
Rpm 150-200

CLUTCH

100. A 330 mm (13 inches) diameter, single plate, dry clutch is used on all models. The clutch pressure plate used on tractors without a cab has a diaphragm-type spring and is available only as a complete assembly. Tractors with cabs have a clutch pressure plate that has 16 coil springs. The independent pto is driven by a splined hub in the clutch cover.

LINKAGE ADJUSTMENT

All Models

101. Clutch pedal free play should be 28-41 mm (1.1-1.6 inches). Pedal free play should be measured at the pedal pad. Tractors without cab use linkage rods; models with cabs use cable linkage.

To adjust clutch pedal free play on models without cab, refer to Fig. 94 and proceed as follows: Loosen lock nut (1), remove cotter pin, then remove clevis pin (3). Turn clevis (2) to lengthen or shorten operating rod as required to correct pedal free travel. Reconnect clevis, install clevis pin and cotter pin, then tighten lock nut.

To adjust clutch pedal free play on models with cab, refer to Fig. 95 and proceed as follows: Loosen adjuster nut (1 or 2), then turn nuts as required to change pedal free travel. Nuts must be tightened against bracket to maintain adjustment after pedal free play is correct.

TRACTOR SPLIT

All Models

102. Disconnect battery, remove vertical exhaust muffler (models so equipped), then unbolt and remove the complete hood. On models without cab, disconnect electrical connectors between front and rear wiring harnesses. For access to wiring on models with cab, remove steering column access panel. Disconnect electrical connectors between front wiring harness and safety start switch sender wire, windshield washer motor sender wire and rear wiring harness at multi-connector. Remove any connectors retaining front harness to engine.

Disconnect proofmeter drive cable from oil pump drive gear on left side of engine. Disconnect power steering or hydrostatic steering pressure and return tubes at the pump and reservoir unit and at hydrostatic steering cylinder. If so equipped, disconnect drag link from steering arm. Disconnect support struts from rear hood assembly. On models with cab, remove retaining bolts, then withdraw rear hood

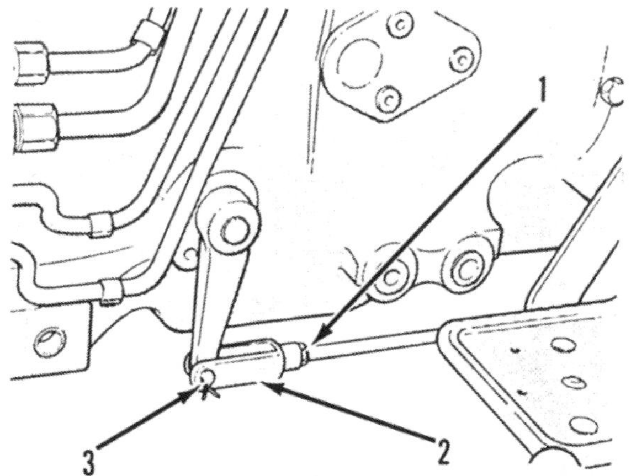

Fig. 94—Clutch is operated through rod linkage of models without cab. Pedal free play is adjusted at clevis (2).

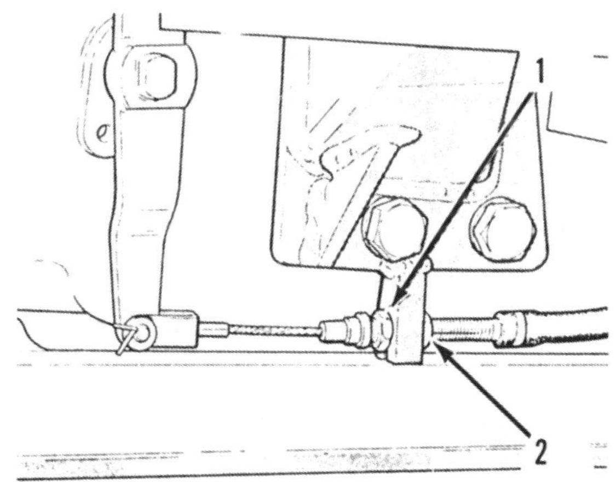

Fig. 95—Clutch is operated through cable on models with cab. Pedal free play is adjusted by nuts (1 & 2) which clamp cable housing to bracket.

support. Remove bolts and nuts securing front end of fuel tank. On models without cab, remove nuts and bolts mounting the battery support bracket to rear hood panel. Disconnect wiring from starter motor, then unbolt and remove the starter motor.

On models with cab, close heater water control valves and disconnect and plug the hoses. On all models, disconnect the fuel shut-off cable and throttle control rod from fuel injection pump. Detach fuel leak-off tube from neck of fuel tank filler tube. Close fuel shut-off valve, then disconnect fuel line from shut-off valve at bottom of tank. If equipped with low exhaust, disconnect exhaust pipe from engine manifold. If equipped with engine oil cooler, disconnect and drain lines and position out of the way. If equipped with air conditioning, discharge refrigerant from the system using a refrigerant recovery/recycling machine. Then disconnect and plug interfering lines. Refer to paragraph 9 and remove the drive shaft, if equipped with four-wheel drive.

Use split support stands designed for this series of tractors, if available. Insert wedges between front axle and front support to prevent tipping. Place supports under the front end of transmission housing and support engine so that engine and front axle assembly can be moved forward away from transmission. Remove the flywheel inspection cover and engine to transmission housing cap screws, then carefully move engine and front axle assembly straight forward away from transmission and rear part of tractor.

Reassemble tractor by reversing the disassembly procedure. Refer to Fig. 96 for recommended torque when tightening engine to transmission screws. Refill and bleed systems as outlined in specific sections. Evacuate and recharge air conditioner if so equipped.

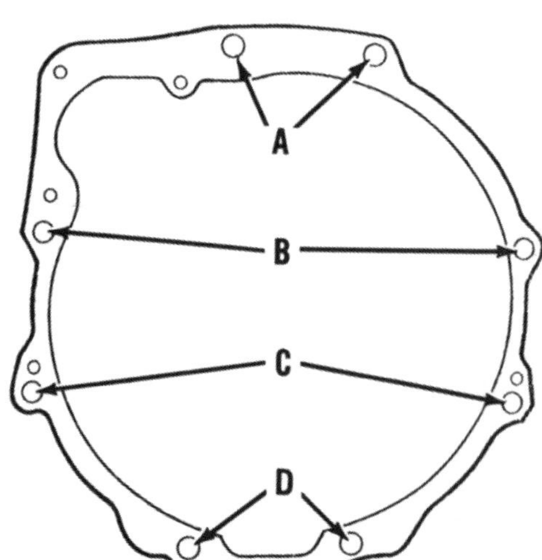

Fig. 96—Tighten bolts attaching engine to transmission housing to the following torque value.

A. 373-460 N•m (275-340 ft.-lbs.)
B. 190-230 N•m (140-170 ft.-lbs.)
C. 224-271 N•m (165-200 ft.-lbs.)
D. 57-76 N•m (42-56 ft.-lbs.)

R&R AND OVERHAUL

All Models

103. The clutch can be unbolted from the flywheel and removed after tractor is split as outlined in paragraph 102. Do not lose the spacers (8—Fig. 98), if installed between cover (9) and flywheel.

A diaphragm spring-type clutch pressure plate (Fig. 97) is used on models without cab and individual parts are not available. Models with cab are equipped with a pressure plate assembly that has 16 coil springs (6—Fig. 98). Overhaul and adjustment of the unit with coil springs can be accomplished as described in paragraph 104.

104. Standard clutch disassembly tools and a press are required to disassemble the clutch shown in Fig. 98. Place clutch cover assembly on bed of press or clutch disassembly tool, remove nuts (12), release lever plate (5) and the four pressure plate screws (14). Slowly release spring pressure, remove the unit from the press and complete disassembly as required. Renew any springs that are rusted, distorted or discolored by heat. Springs should have free length of 76.2 mm (3 inches) and should test 556 N (125 lbs.) when compressed to a height of 42.8 mm (1 11/16 inches). Renew any parts that are worn excessively, bent or otherwise questionable.

Clutch release lever height should be adjusted when reassembling using special tools available from Nuday Tool Co. Refer to Fig. 99 for cross section of tools and clutch assembly. Tools shown are part of kit number SW12B. Tool finger (C) should just pass freely over each of the clutch release levers (17). Turn nut (12) as required until adjustment is correct. Be sure to recheck adjustment of other release levers as another is adjusted. Lock position of adjuster nut (12) by staking after all are correctly adjusted. Remove the clutch assembly from test fixture.

The prelubricated, ball-type, pilot bearing, (1—Fig. 97 or Fig. 98) located in retainer (2), can be removed if renewal is required. Pack flywheel space in front of bearing with high-melting-point grease when installing new bearing. Tighten flywheel and bearing retainer screws to 136-149 N•m (100-110 ft.-lbs.) torque.

Most clutch friction discs (3—Fig. 97 and Fig. 98) are marked "FRONT" indicating which side is to be installed next to flywheel. Use proper alignment tool to center the friction disc and install clutch cover and pressure plate assembly. Tighten retaining nuts to 31-39 N•m (23-29 ft.-lbs.) torque.

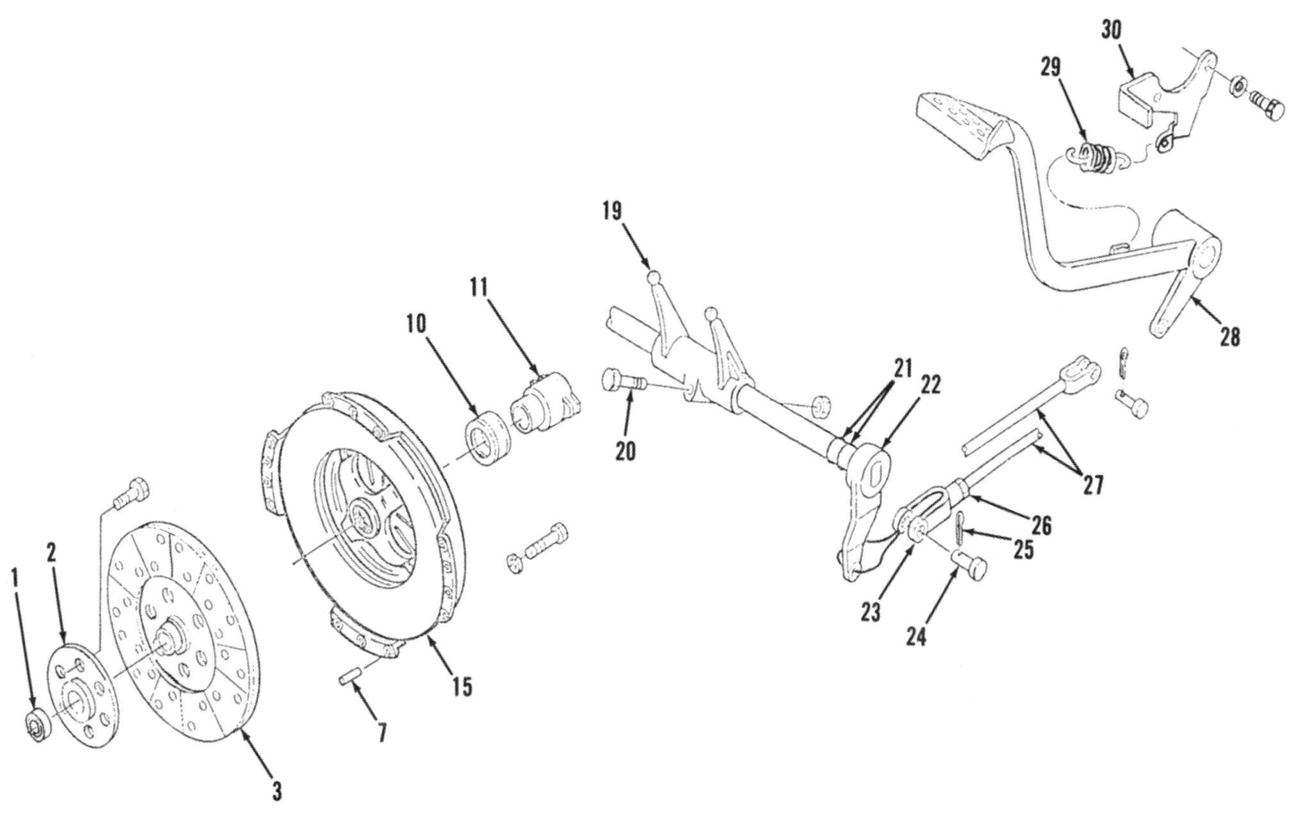

Fig. 97—Partially exploded view of clutch and linkage used on models without cabs.

1. Pilot bearing
2. Bearing retainer
3. Friction disc
7. Dowel
10. Release bearing
11. Hub
15. Clutch cover and pressure plate assembly
19. Release lever fork
20. Clamp screw
21. "O" rings (4)
22. Shaft & lever
23. Clevis
24. Clevis pin
25. Cotter pin
26. Locknut
27. Link rod
28. Pedal
29. Return spring
30. Pedal bracket

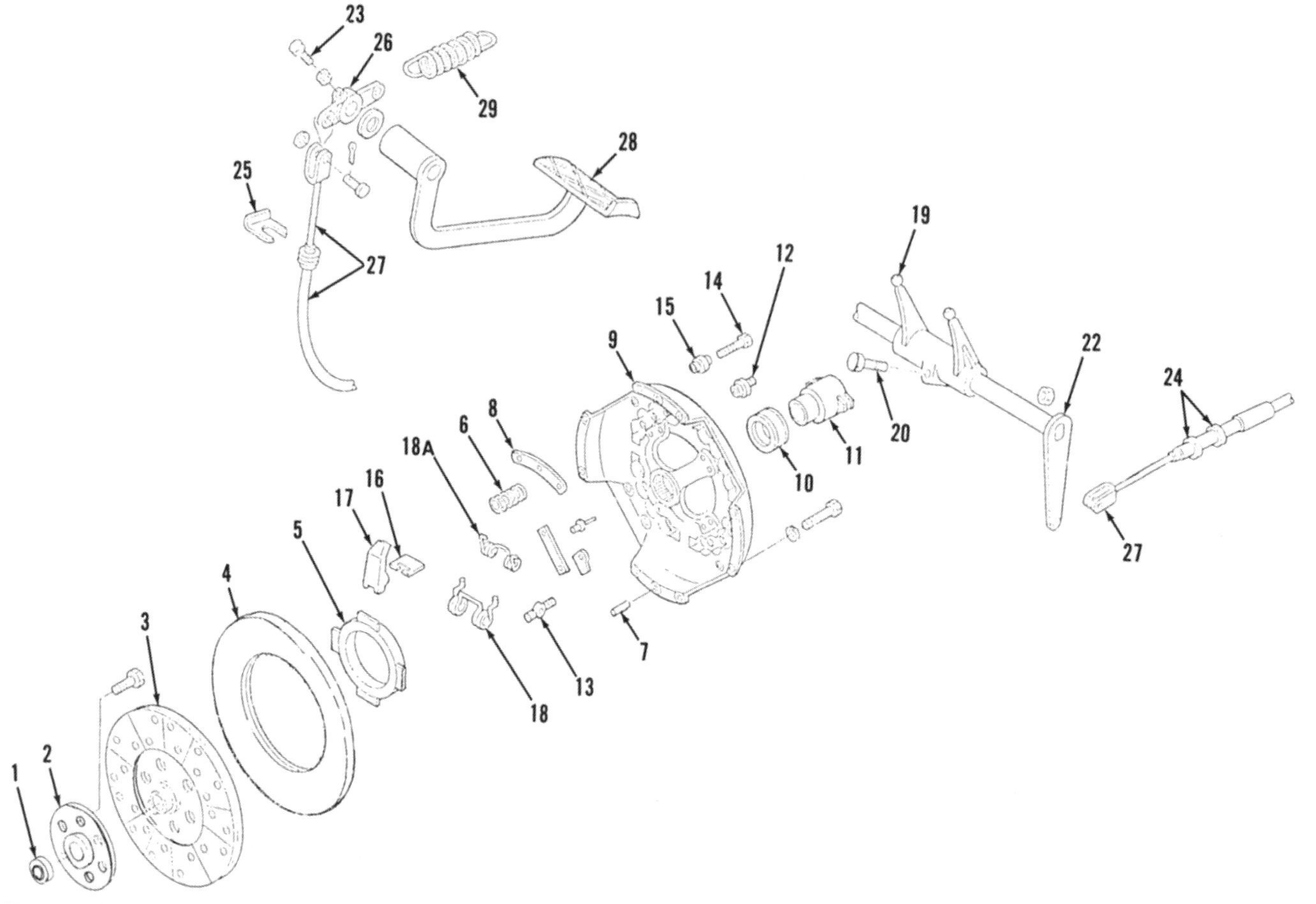

Fig. 98—Exploded view of clutch and operating mechanism used on models with cab.

1. Pilot bearing
2. Bearing retainer
3. Friction disc
4. Pressure plate
5. Release lever plate
6. Clutch springs (16)
7. Dowel
8. Spacer (4)
9. Clutch cover
10. Release bearing
11. Hub
12. Release eye bolt nut (4)
13. Eye bolt (4)
14. Pressure plate screw (4)
15. Ferrule (4)
16. Strut (4)
17. Release lever (4)
18. Release lever spring (4)
18A. Plate retainer spring (4)
19. Release lever fork
20. Clamp screw
22. Shaft & lever
23. Screw
24. Cable adjustment nuts
25. Clip
26. Lever
27. Cable and housing
28. Pedal
29. Return spring

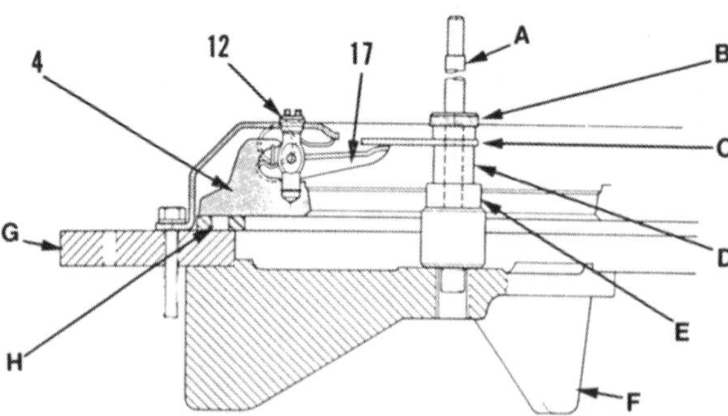

Fig. 99—Cross section of clutch and Nuday special tools required to set release lever height as suggested by the manufacturer.

4. Pressure plate
12. Release eye bolt nut
17. Release lever

A. Pillar spacer (Nuday SW12B/1)
B. Locknut (Nuday SW12B/12)
C. Gage finger (Nuday SW12B/13)
D. Gage finger body (Nuday SW12B/11)
E. Center spacer (Nuday SW12B/6)
F. Bridge spider (Nuday SW12B/14)
G. Base plate (Nuday SW12B/15)
H. Disc spacer (Nuday SW12B/10)

CONSTANT MESH TRANSMISSION (8 FORWARD - 2 REVERSE)

105. The constant mesh transmission provides 8 gear reduction ratios forward and 2 reverse ratios. Gear selection is accomplished by positioning the two gear shift levers. Shift pattern is shown in Fig. 100. A safety starting switch is provided and gear selector levers must be in neutral to start engine. An optional creeper transmission is available providing an additional four slower forward gears and one slower reverse gear. Refer to paragraph 116 for service to reduction (creeper) gearbox.

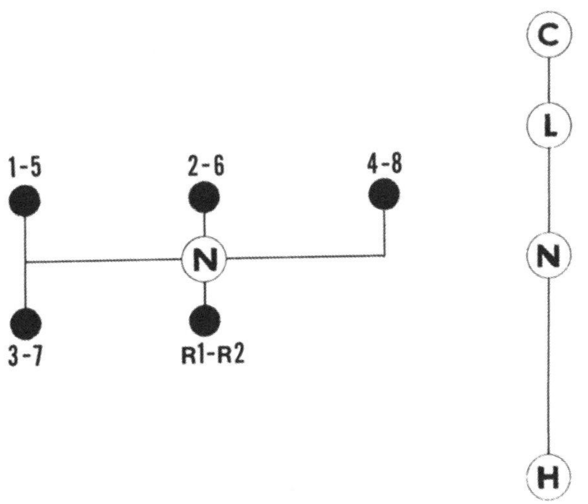

Fig. 100—Gear shifting patterns are shown. Numbers indicate the speed selection. The lower number is with range transmission in low (L) and the higher number is with range selector in high (H). Neutral position is identified by (N). Creeper gears (C) are only on models equipped with reduction (creeper) gearbox.

LUBRICATION

All Models With Constant Mesh Transmission

106. Transmission lubricant capacity is 12 L (12.7 quarts) for all models without creeper gears installed. Reduction unit (creeper transmission) lubrication is common with constant mesh transmission and the capacity is increased to 15 L (15.8 quarts). Maintain oil level to full mark on dipstick (D—Fig. 101) of models without cab. For models with cab, remove the floor cover, remove the transmission access panel and plug (L—Fig. 102) located on right side of transmission case. The recommended lubricant is Ford M2C134-D/C or equivalent for all models with constant mesh transmission. Transmission oil should be drained and transmission should be refilled with new oil every 1200 hours of operation.

REMOVE AND REINSTALL

Models Without Cab

107. To remove the constant mesh transmission, first split tractor between the engine and transmission housing as outlined in paragraph 102. Remove the steering unit and fuel tank as a unit from top of transmission housing. Support transmission housing from an overhead hoist. Support the rear axle center housing separately, then refer to paragraph 139 or 140 to separate transmission from center housing.

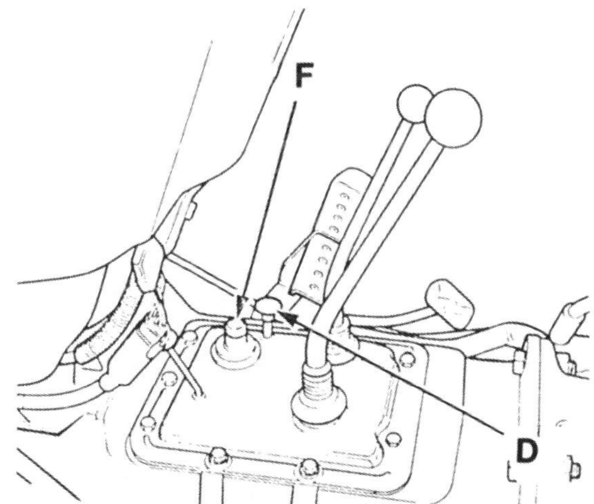

Fig. 101—View of constant mesh transmission top for models without cab. Dipstick is shown at (D) and filler plug at (F).

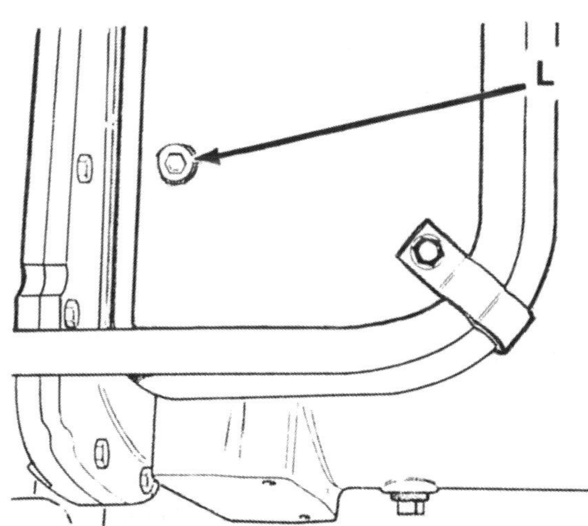

Fig. 102—View of constant mesh transmission oil level and fill plug (L) located on right side of transmission housing of models with cab.

Reinstall the transmission by reversing removal procedure. Refer to Fig. 154 for tightening torque of transmission to rear axle bolts and to Fig. 96 for tightening torques of engine to transmission bolts.

Models With Cab

108. To remove the constant mesh transmission, first refer to paragraph 212 and remove the tractor cab. Separate the engine from transmission as outlined in paragraph 102. Remove the steering unit and fuel tank as a unit. Support the rear axle center housing and the transmission housing separately. Unbolt the center housing from the transmission housing, then roll rear axle assembly away.

Reverse removal procedure to install transmission. Refer to Fig. 154 for tightening torque of transmission to rear axle bolts and to Fig. 96 for tightening torques of engine to transmission bolts.

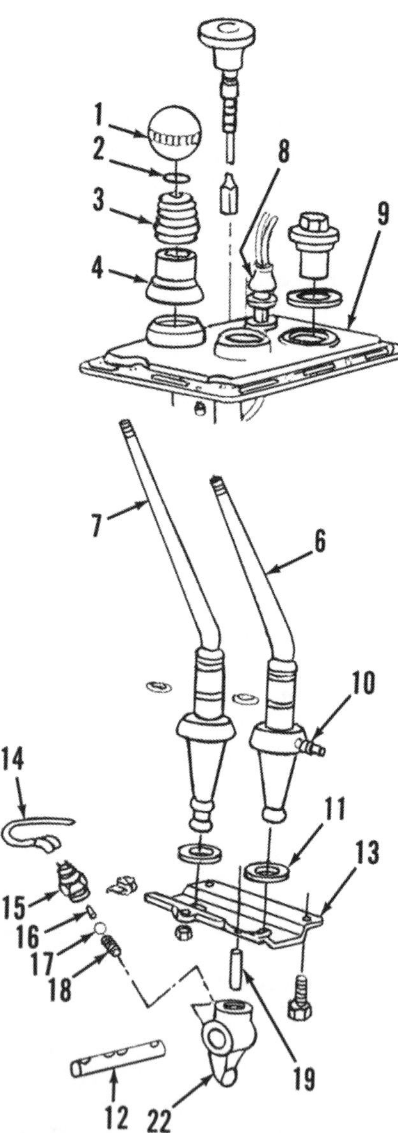

Fig. 103—Exploded view of 8-speed constant mesh transmission shift cover assembly.

1. Shift knob
2. Snap ring
3. Spring
4. Retainer
6. Range shift lever
7. Main shift lever
8. Grommet for wires
9. Cover
10. Pin
11. Washer
12. High/low shift rail
13. Stop plate
14. Wire to safety switch
15. Safety switch
16. Plunger
17. Ball
18. Spring
19. Dowel
22. Switch retainer

OVERHAUL

Models With Constant Mesh Transmission

109. SHIFT COVER. On models with cab, remove floor mat, gear shift boot and access panel. On all models, place shift levers in NEUTRAL and disconnect safety start switch wires. Remove screws attaching shift cover (9—Fig. 103), then raise cover enough to disconnect safety switch wires (14) from switch unit (15). Withdraw wires from cover, then lift cover from transmission.

The shift levers and safety switch (15) can be serviced with cover removed. The safety switch is actuated by spring (18), ball (17) and plunger (16). The plunger (16) is located inside bore in high/low shift rail (12); spring (18) and ball (17) are located in bore of retainer (22) below rail. The switch is closed only when the high/low lever is in neutral. Switch retainer (22) is positioned by locating pin (19) in shift cover that enters the vertical hole in top of retainer.

To disassemble the shift levers, unbolt and remove stop plate (13). Remove knobs (1). Depress lever spring (3) slightly, unseat snap ring (2), then withdraw lever downward, out of the cover.

Be sure to reconnect wires (14) to safety switch and be sure pin (19) enters bore of retainer (22) before installing cover. Tighten screws retaining cover to transmission to 36 N•m (26 ft.-lbs.) torque. Check to be sure safety switch operates properly.

110. SHIFT RAILS AND FORKS. To remove the shift rails and forks, remove the transmission as outlined in paragraph 107 or 108 and remove the top cover as outlined in paragraph 109. Remove the input shaft and front support plate as outlined in paragraph 111, pto drive shaft, output shaft assembly and secondary countershaft as outlined in paragraph 112.

Refer to exploded view of shift mechanism in Fig. 104 and to assembled view shown in Fig. 105.

Remove the four detent plungers (15—Fig. 104) and springs (14) from bores in housing. Remove a detent ball as each corresponding shift rail is removed. Move the high/low shift rail rearward until sliding coupling (44—Fig. 107) can be removed from the main shaft and high/low shift fork. Refer to Fig. 103 and remove starter safety switch (15) and retainer (22) as high/low shift rail (12) is being removed. Turn rail so that locknut can be loosened and set screw can be removed from the high/low shift fork (20—Fig. 104), then remove shift fork out the rear opening. Turn rail (12) so that locknut can be loosened and set screw can be removed from gate (19) at front of shaft. Remove shaft from rear of housing and catch gate (19) and detent ball (13). Loosen locknuts and remove set screws from remaining shift forks and arms, then withdraw rails one at a time from rear of housing. Remove shift forks, arms and detent balls as rail is withdrawn. Remove plug (8) and withdraw the two interlock plungers (7).

NOTE: The 4th-8th shift fork (6—Fig. 104) is located on the reverse idler shaft. Refer to paragraph 114 for shaft and fork removal.

To reassemble, first insert the two interlock plungers (7) into bore and install plug (8). Insert the 4th-8th

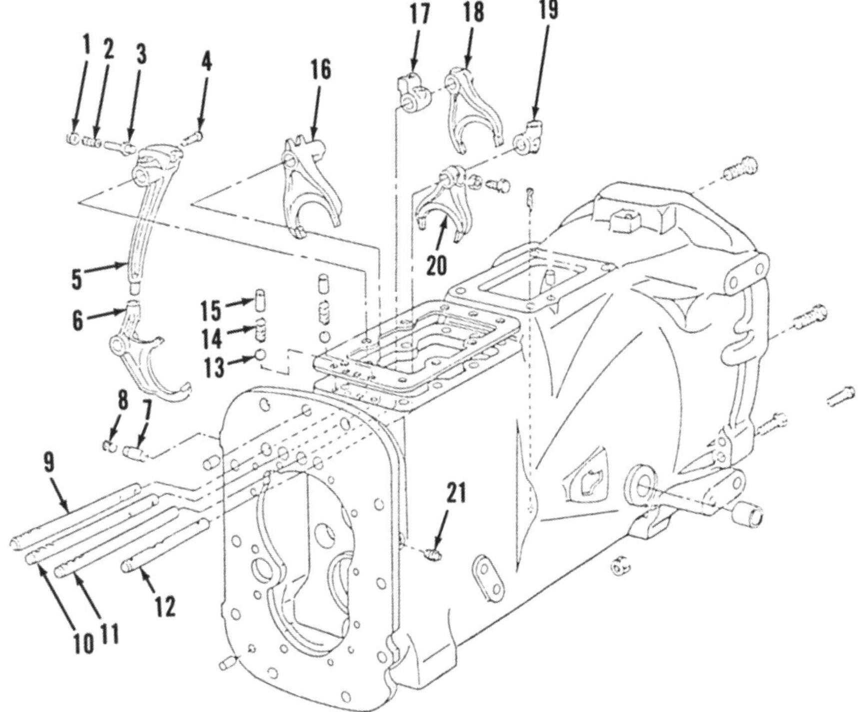

Fig. 104—Exploded view of constant mesh transmission shift mechanism. Refer to Fig. 105 for assembled drawing.

1. Retainer
2. Spring
3. Plunger
4. Pin
5. Selector arm
6. 4th-8th shift fork
7. Interlock plunger
8. Plug
9. 4th-8th shift rail
10. 2nd-6th and Reverse shift rail
11. 3rd-7th & 1st-5th shift rail
12. High/low shift rail
13. Detent balls (4)
14. Springs (4)
15. Plungers (4)
16. 2nd-6th & Reverse shift fork
17. Connector
18. 3rd-7th & 1st-5th shift fork
19. Gate
20. High/low shift fork
21. Oil level plug

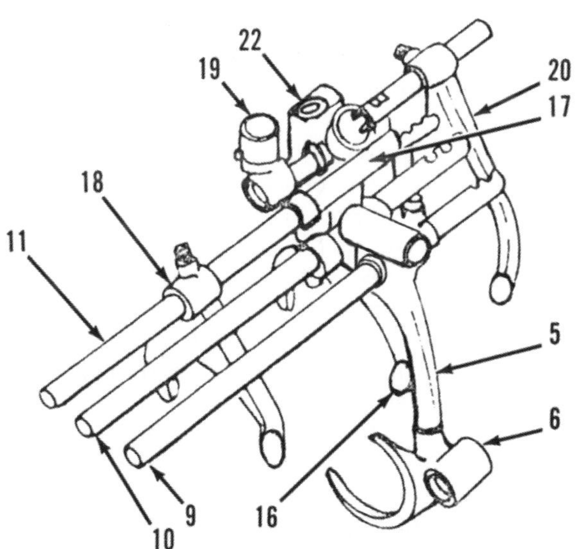

Fig. 105—Assembled view of shift mechanism for 8-speed constant mesh transmission. Refer to Fig. 104 for legend.

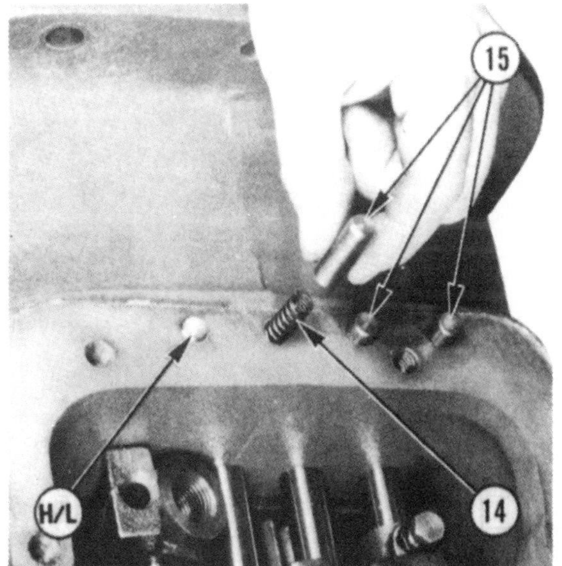

Fig. 106—View showing location of shift rail detent plungers, springs and balls. Plungers are retained by the transmission shift cover.

shift arm (5) in socket of shift fork (6) and install the upper 4th-8th shift rail (9) through arm so that set screw holes are aligned and the single detent is toward front. Install set screw and locknut, tightening both to 31 N•m (23 ft.-lbs.) torque. Move rail to neutral position.

Insert 2nd-6th-Reverse shift rail (10) into bore with three detents toward rear. Install shift rail through fork (16), install set screw and locknut, tightening both to 31 N•m (23 ft.-lbs.) torque. Move rail to neutral position.

Insert the 3rd-7th and 1st-5th shift rail (11) into bore with the single detent toward front. Install connector (17) and shift fork (18), install set screw and locknut, tightening both to 31 N•m (23 ft.-lbs.) torque. Move rail to neutral position.

Insert high/low shift rail (12) into bore and install shift fork (20) and gate (19). Install set screw and locknut, tightening both to 31 N•m (23 ft.-lbs.) torque. Position the sliding coupling on shift fork, then position coupling (44—Fig. 107) on dog-teeth at rear of main shaft.

Refer to paragraph 112 and install the output shaft and to paragraph 111 and install the input shaft front support plate. Be sure the transmission shifts properly. Insert detent balls (13—Fig. 104), springs (14) and plungers (15) as shown in Fig. 106. Install the transmission top cover as outlined in paragraph 109, then check to make sure that transmission still shifts properly. Refer to paragraph 107 or 108 and install the transmission assembly.

111. INPUT SHAFT, CLUTCH RELEASE BEARING SUPPORT AND FRONT SUPPORT PLATE. Depending on the work to be performed, it may be necessary to separate only the transmission from the engine as outlined in paragraph 102. However, to remove the input shaft (16—Fig. 108) and bearing assembly (14 and 15) from models with independent pto, it is necessary to first remove the transmission as outlined in paragraph 107 or 108.

Remove release bearing (10—Fig. 97 or Fig. 98) and hub (11) from support. Remove screw (20) from release fork (19) and slide shaft (22) from fork and housing. Unbolt and remove bearing support (1—Fig. 108) from front support plate (3) if desired. Unbolt front support plate (3) and remove with shafts and gears. Bearing (32) should remain on shaft (62), so front support plate (3) must be removed squarely.

Use a suitable puller and remove bearing (32), then remove snap ring (33) and thrust washer (34). Remove snap ring (72) and gear (70) from rear of shaft (62), then remove snap ring (69). Drive shaft (62) to the rear and withdraw from housing. Lift pto countershaft gear (35) from housing, then unbolt retainer (15). Remove retainer (15), input shaft (16) and bearing (14). Bearing (17) can be removed from bore in rear of shaft (16).

To reassemble, reverse removal procedures. Tighten screws attaching retainer (15) to 67 N•m (49 ft.-lbs.) torque. Screws attaching front support (3) and clutch release bearing support (1) should be tightened to 44 N•m (32 ft.-lbs.) torque. Clutch release fork clamp screw (20—Fig. 97 or Fig. 98) should be tightened to 47 N•m (35 ft.-lbs.) torque.

Refer to paragraph 102 to reattach engine to transmission housing if the transmission was not removed. Refer to paragraph 107 or 108 if transmission was removed.

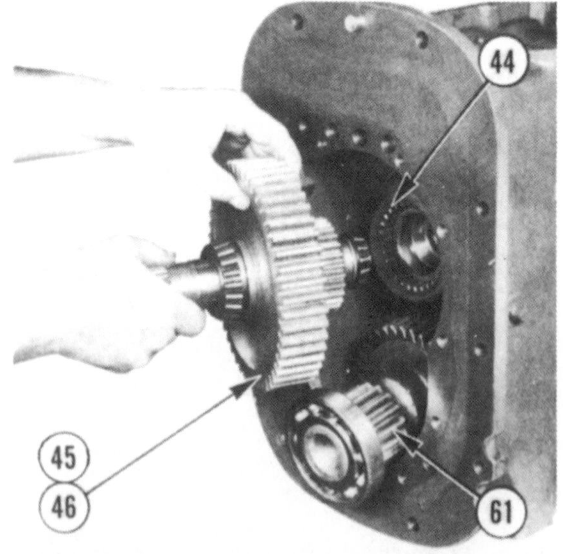

Fig. 107—Transmission output shaft and gear (45 & 46) should be removed and installed with secondary countershaft (61) laying on bottom of housing. Coupling (44) should be positioned on teeth of main shaft and should be correctly engaging the shift fork.

112. REAR SUPPORT PLATE, PTO DRIVE SHAFT, OUTPUT SHAFT AND SECONDARY COUNTERSHAFT. To remove the rear support plate (53—Fig. 108), pto countershaft (62) or the secondary countershaft (61), it is necessary to first remove the transmission as outlined in paragraph 107 or 108.

Refer to paragraph 111 and remove the clutch release shaft, bearing support (1), front support plate (3) and pto driven gear (35) from housing. Unbolt and remove the output shaft retainer (56), then remove screws attaching rear support plate (53) to transmission housing. Pry the support plate from the housing. Withdraw secondary countershaft (61) until free, then lower to bottom of housing. Withdraw output shaft and gear assembly (45 and 46—Fig. 107), then lift secondary countershaft (61) from housing.

Procedure for further disassembly and renewal of seals and bearings will be obvious. Refer to Fig. 108 and Fig. 109. Lips of all seals should be toward inside.

When assembling, be sure high/low coupling (44—Fig. 107) is correctly engaging shift fork and is positioned on teeth of main shaft. Lay the secondary countershaft (61) on bottom of housing, then install the output shaft and gear (45 and 46) as shown. Align the countershaft front bearing with bore in housing and bump shaft and bearing assembly forward until the front bearing seats against the snap ring. Install rear support (53—Fig. 108) with new gasket (52) and tighten the retaining screws to 44 N•m (32 ft.-lbs.) torque. Insert pto shaft (62) with bearing (68) into rear of transmission and pto gear (35) in front compartment of housing. Slide the pto shaft through gear and install thrust washer (34) and snap ring (33) on front of shaft and snap ring (69) in groove of rear support. Install bearing (32), front support plate (3) bearing support (1) and clutch release shaft as outlined in paragraph 111.

Install the transmission output shaft bearing retainer (56) with at least 1.524 mm (0.060 inch) thickness of shims (55) and tighten retaining screws to 44 N•m (32 ft.-lbs.) torque. Measure output shaft end play with a dial indicator. Remove the bearing retainer (56) and shims (55) equal to the shaft end play plus 0-0.050 mm (0-0.002 inch). Reinstall bearing retainer with the reduced thickness of shims and tighten screws to 44 N•m (32 ft.-lbs.) torque. Shaft should have slight preload in bearings. Measure the rolling torque of the output shaft by wrapping a string around the output shaft splines. Attach a spring scale to the string and measure rolling torque by pulling scale. Correct bearing preload will require 6-10 kg. (13-22 lbs.) of pull to keep shaft rolling. If rolling torque is incorrect, remove bearing retainer (56) and change thickness of shims (55) as required.

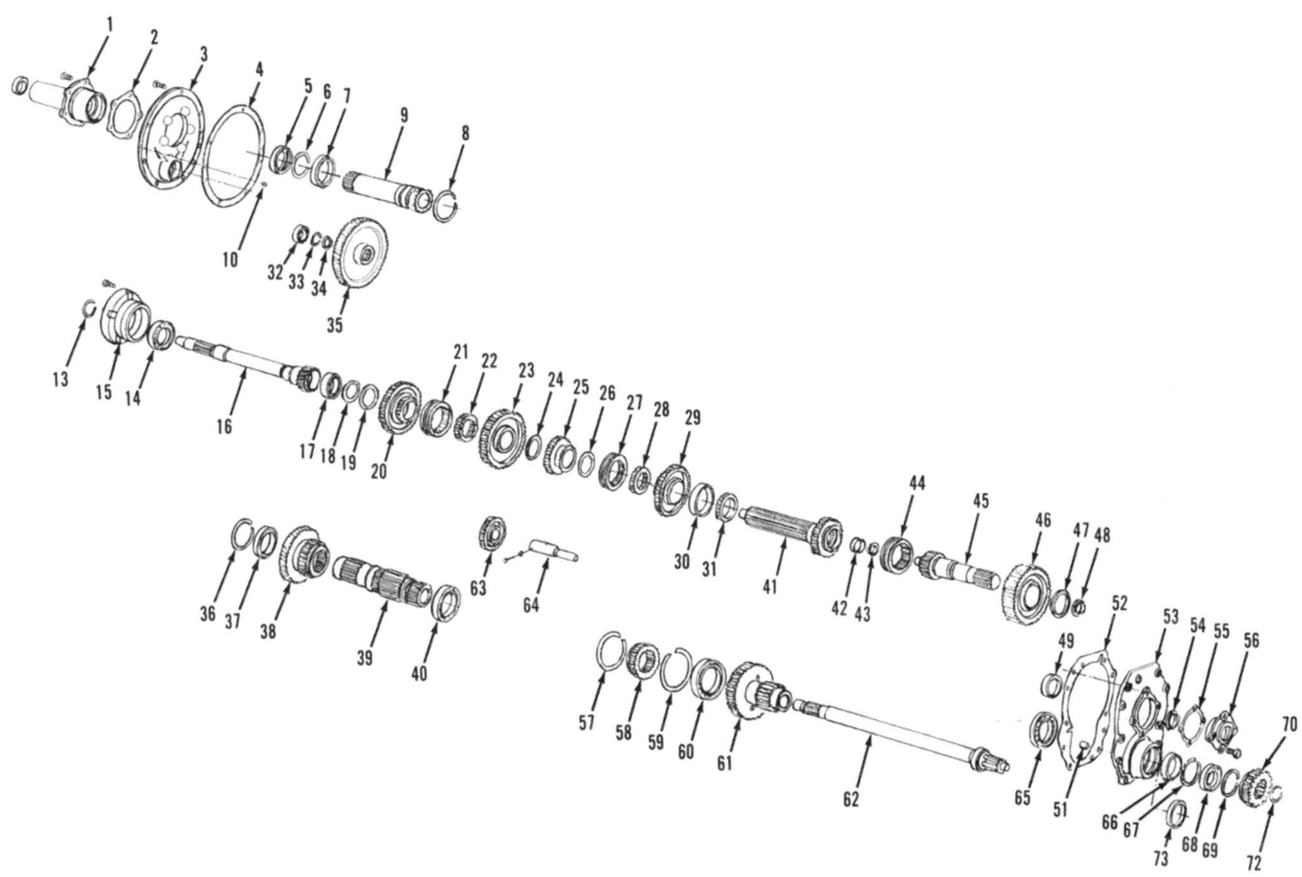

Fig. 108—Exploded view of 8-speed, constant mesh transmission gears, shafts and related parts. Refer to Fig. 109 for diagram showing location of gears.

1. Clutch release bearing support
2. Gasket
3. Front support plate
4. Gasket
5. Oil seal
6. Snap ring
7. Ball bearing
8. Snap ring
9. Pto drive shaft
10. Dowel pins
13. Snap ring
14. Ball bearing
15. Retainer
16. Transmission input (clutch) shaft
17. Ball bearing
18. Snap ring
19. Thrust washer
20. 3rd gear
21. Shift collar
22. Connector (hub)
23. 1st gear
24. Thrust washer
25. Reverse gear
26. Thrust washer
27. Shift collar
28. Connector (hub)
29. 2nd gear
30. Bearing cup
31. Bearing cone & rollers
32. Ball bearing
33. Snap ring
34. Thrust washer
35. Pto counter-shaft gear
36. Snap ring
37. Ball bearing
38. Countershaft main gear
39. Countershaft
40. Ball bearing
41. Main shaft
42. Bearing cup
43. Bearing cone & rollers
44. Shift collar
45. Output shaft
46. Output shaft gear
47. Thrust washer
48. Bearing cone & rollers
49. Bearing cup
51. Dowel pins
52. Gasket
53. Rear support plate
54. Oil seal
55. Shims
56. Bearing support
57. Snap ring
58. Shift collar
59. Snap ring
60. Ball bearing
61. Secondary countershaft
62. Pto countershaft
63. Reverse idler
64. Idler shaft
65. Ball bearing
66. Oil seal
67. Snap ring
68. Ball bearing
69. Snap ring
70. Hydraulic pump drive gear
72. Snap ring
73. Plug (w/o pto)

Remainder of assembly is the reverse of disassembly. Refer to paragraph 107 or 108 for installing transmission.

113. MAIN SHAFT ASSEMBLY. To remove the transmission main shaft (41—Fig. 108), it is necessary to first remove the transmission from the tractor as outlined in paragraph 107 or 108, and the shift cover as outlined in paragraph 109. Remove the clutch release shaft, transmission input shaft and front support plate as outlined in paragraph 111. Refer to paragraph 112 and remove the rear support plate, pto drive shaft, secondary countershaft and transmission output shaft, then remove the shift rails and forks as outlined in paragraph 110.

Remove snap ring (18) and thrust washer (19) from front of main shaft. Pull main shaft (41) from rear of transmission as gears, washers, connectors and shift collars (20 through 29) are lifted from housing.

Carefully inspect all parts and renew any that are excessively worn or damaged. Bushing inside gears are not serviced separately and gear/bushing assembly must be renewed if bushing is worn.

Reinstall main shaft by reversing removal procedure. Refer to Fig. 108 and Fig. 109 while assembling.

114. REVERSE IDLER. Removal of the reverse idler gear and shaft requires removal of the transmission main shaft as outlined in paragraph 113.

Refer to Fig. 110 and remove the cap screw (S) that retains the idler shaft. Push idler shaft forward and remove the 4th-8th shift fork (6—Fig. 104). The 4th-8th sliding gear (58—Fig. 108) can now be removed from the rear of main countershaft (39). Push idler shaft (64) rearward out of housing and remove the idler gear (63).

To reinstall the reverse idler, position the reverse idler gear (63) in housing with long hub toward front, then slide the shaft (64) through housing and gear with smaller diameter toward rear. Position the 4th-8th sliding gear (58) on main countershaft (39) with the shifter fork groove toward front. Slide idler shaft (64) forward far enough to install the shift fork (6—Fig. 104), then slide the shaft back through the fork and install retaining screw (S—Fig. 110). Always install a new rubber coated sealing washer under screw (S) and tighten to 24 N•m (17 ft.-lbs.) torque.

115. MAIN COUNTERSHAFT. To remove the main countershaft (39—Fig. 108), first refer to paragraph 113 and remove the transmission main shaft

Fig. 110—View showing the plug (8) that closes the bore for interlock plungers and screw (S) that retains the reverse idler shaft in housing.

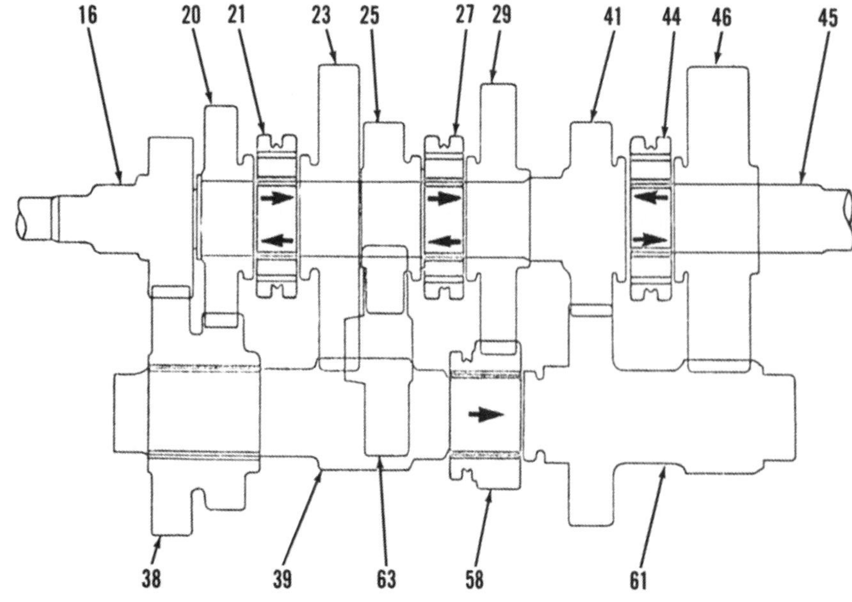

Fig. 109—Cross section drawing showing location of parts in 8-speed, constant mesh transmission. Refer to Fig. 108 for legend.

and to paragraph 114 and remove the reverse idler and shaft.

Remove snap ring (59) from secondary countershaft front bearing bore and snap ring (57) from bore behind main countershaft rear bearing. Use a suitable drift and drive main countershaft (39) toward rear until both front and rear bearings (37 and 40) are free from their bores and the assembly is resting in bottom of transmission housing. Place suitable wood or soft metal blocks between the countershaft main (cluster) gear (38) and rear supporting wall of main compartment, then continue to drive countershaft rearward forcing the gear and front bearing from shaft. Remove shaft (39) and bearing (40) from rear. Remove main gear (38) and bearing (37) from main compartment.

To reinstall the main countershaft, reverse the removal procedure. Refer to paragraphs 114 and 113 to install the reverse idler and main shaft assemblies.

REDUCTION GEARBOX

Models So Equipped

116. A reduction gearbox (creeper gearbox) is available as a dealer-installed option on models with constant mesh transmission. The gearbox is mounted at the rear of transmission in place of output shaft (45—Fig. 108), coupling (44), gear (46), rear cover (53) and related parts. Major components of reduction gearbox consist of coupling (8—Fig. 111 and Fig. 112), output shaft (3) and planetary assembly (1, 4, 6, 7, 13 and 20). Creep range ratios of either 5.7:1 or 10:1 below the low range can be installed, depending upon planetary gears (1).

Coupling (8—Fig. 111) is moved by the selector fork (10) and slides on the gear shift rail to provide one of three reduction ratios:

High—With coupling (8) forward, coupling engages the output shaft (3) with the transmission (41) main shaft in the same way as without reduction gearbox.

Neutral—Neutral detent position is shown in Fig. 111 and coupling (8) does not engage any driving member.

Low—Moving coupling slightly to the rear from neutral position, engages coupling (8) with the reduction carrier (7) to provide a standard low range. The number of gear teeth on the carrier (7) are the same as the standard output shaft gear (46—Fig. 108).

Creeper—Creeper reduction range is engaged by moving the coupling (8—Fig. 111 and Fig. 112) further to the rear, so that coupling (8) engages the intermediate ring gear (6). Power transmission is from the main shaft (41—Fig. 111), to secondary

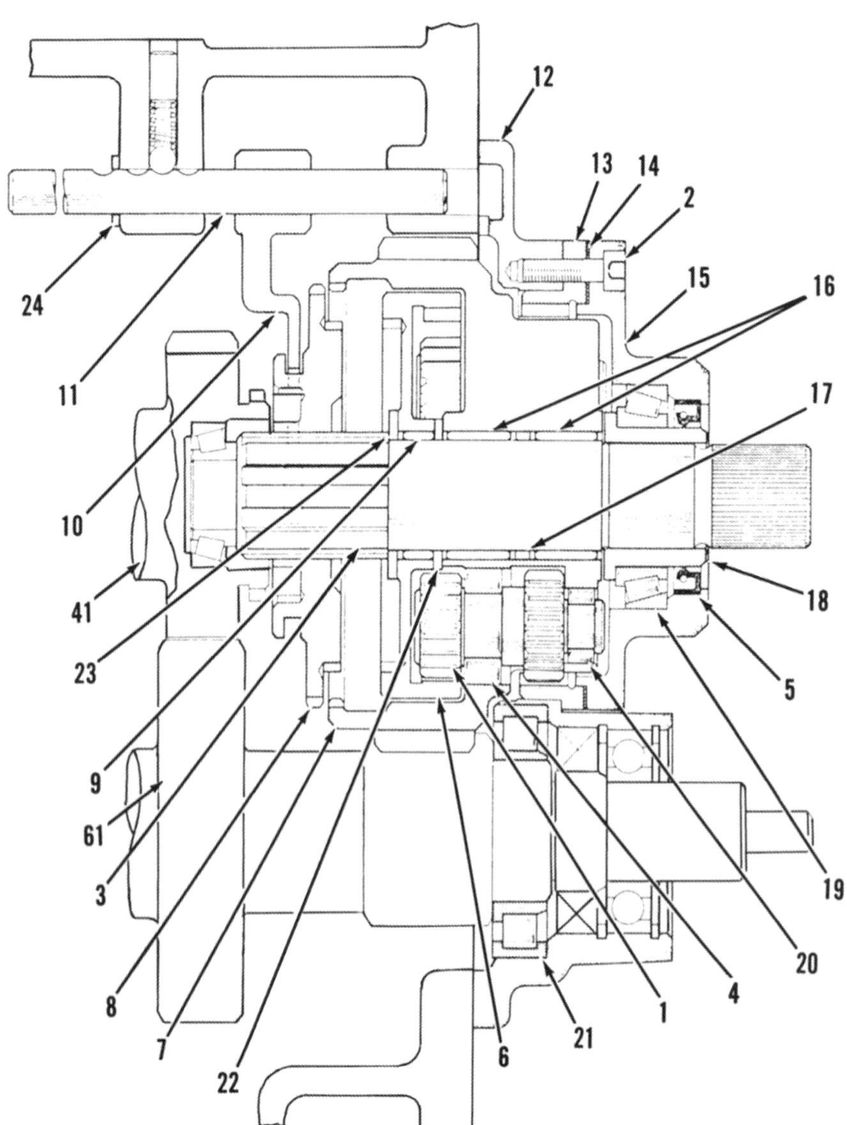

Fig. 111—Cross section of reduction gearbox available as a dealer-installed option on models with constant mesh transmission. Refer to Fig. 112 for legend except for transmission main shaft (41) and secondary countershaft (61).

countershaft (61), to carrier (7), to planetary gears (1), that are permitted to rotate around outer ring gear (13), to the intermediate ring gear (6), that is engaged with the coupling (8), which is splined to the output shaft (3).

117. OVERHAUL. Drain transmission, then separate transmission from the rear axle as outlined in paragraph 139 or 140. Remove shift cover as outlined in paragraph 109. On models with transmission pto, remove snap ring retaining pto coupling and slide coupling from pto countershaft. On models with independent pto, remove screws retaining the hydraulic pump drive gear shroud and remove shroud. Remove snap ring retaining pump drive gear and withdraw gear. Remove cap screws securing rear support plate (12—Fig. 112) to the transmission housing. Use suitable tools to pry support plate from transmission housing. Pto countershaft rear bearing will be withdrawn with the support plate.

NOTE: If needed, use a soft-faced mallet to tap pto countershaft as support plate (12) is pried rearward.

Withdraw reduction gear set from output shaft; ensure needle roller bearings (9 and 16) and intermediate ring gear (6) do not fall out. Disengage selector fork (10) from coupling (8). Withdraw output shaft (3) and coupling (8) assembly with thrust washer (23). Washer (23) may stick to intermediate ring gear (6).

Remove screws securing output shaft retainer (15), then separate retainer (15) and outer ring gear (13) from rear support plate (12). Note shims (14) located between retainer (15) and outer ring gear (13).

Lay reduction gear set with intermediate ring gear facing up. Lift intermediate ring gear (6) with thrust washer (23), if still in place, and needle roller bearing (9) from carrier (7). Remove thrust washer (22) and two needle roller bearings (16) with spacer (17) from carrier. Lift planetary gears (1) from carrier (7).

NOTE: Use care not to allow bearing rollers to fall free from mounting lips.

Inspect bearings for cracks, corrosion, roughness, excessive wear and any other damage. Inspect splines and gear teeth for missing teeth, chips, cracks, excessive wear and any other damage. Examine output shaft retainer and rear support plate for cracks, excessive wear and any other damage. Inspect thrust washers and bushing for cracks, excessive wear and any other damage. Inspect oil seals for hardness and any other damage. Inspect shift mechanism and coupling for excessive wear and any other damage.

NOTE: Complete reduction gear set should be renewed if several individual components must be renewed.

Use a suitable light grease to hold bearing rollers (4 and 20) in position while assembling. Eleven larger rollers (4) and 13 smaller rollers (20) are installed on each gear. Set carrier (7) on outer ring gear (13) so ring gear will act as an alignment gage for the planetary gears. Install planetary gears (1), then turn each gear so the punched tooth (master tooth) is facing exact center of carrier.

NOTE: Because the two rows of gear teeth on each planetary gear have different numbers of teeth, it is important to make sure that the marked teeth

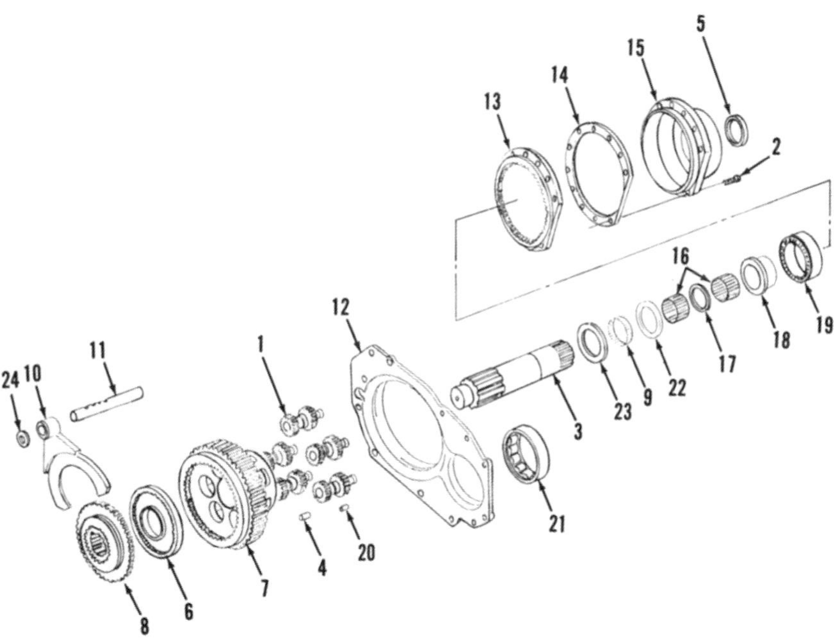

Fig. 112—Exploded view of reduction gearbox (creeper gearbox) installed on some models with constant mesh transmission. Planetary gears (1) have 16 and 12 teeth with 5.7:1 reduction ratio; 16 and 13 teeth with 10:1 reduction ratio.

1. Planetary gear
2. Socket head screw
3. Output shaft
4. Roller (9 mm)
5. Oil seal
6. Intermediate ring gear
7. Carrier
8. Coupling
9. Needle roller bearing
10. Selector fork
11. Gear shift rail
12. Support plate
13. Outer ring gear
14. Shims
15. Output shaft retainer
16. Needle roller bearing
17. Spacer
18. Bushing
19. Tapered roller bearing
20. Roller (5 mm)
21. Roller bearing
22. Thrust washer
23. Thrust washer
24. Washer

are pointing toward center as shown in Fig. 113. The marked tooth is on 12-tooth gear for 5.7:1 ratio; on the 13-tooth gear for 10:1 ratio.

Position thrust washer (22—Fig. 112) with chamfered side up in the carrier hub, then install intermediate ring gear (6) into carrier. Place coupling (8) on output shaft (3) and install thrust washer (23) on shaft with chamfered side facing coupling.

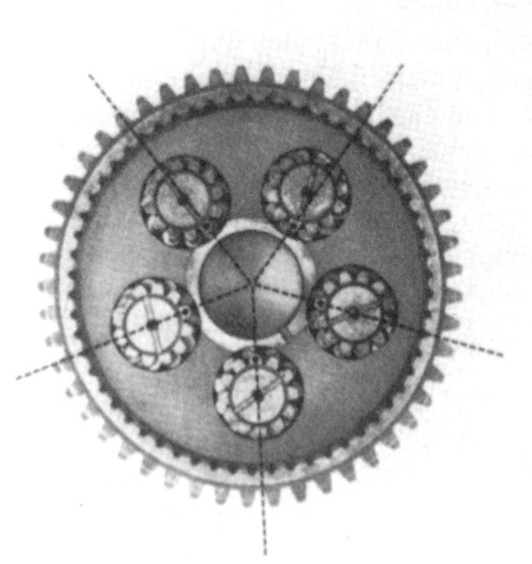

Fig. 113—Because of the different number of teeth on planetary gears, it is important that the punched master tooth on all planetary gears be aligned with center of carrier.

Install output shaft (3), coupling (8) and thrust washer (23) assembly into transmission. Be sure output shaft bearing engages the rear of countershaft and selector fork (10) engages coupling (8). Lubricate and install the narrow needle roller bearing (9) into intermediate ring gear (6). Position carrier assembly, with the outer ring gear (13) still in place, onto the output shaft. Lubricate the two wide needle roller bearings (16), support carrier and install the two wide needle roller bearings (16) and spacer (17). Spacer (17) should be located between the two bearings (16) in the carrier. Install bushing (18) on output shaft (3) against shaft shoulder, followed by bearing assembly (19). Measure clearance between flange of bushing (18) and rear of carrier (7) with a feeler gage. Clearance should be 0.3-0.8 mm (0.012-0.032 inch).

Remove outer ring gear (13) and install support plate (12) using a new gasket. Tighten screws securing support plate (12) to transmission housing to 44 N•m (32 ft.-lbs.) torque. Install outer ring gear (13), shims (14) and output shaft retainer (15), tightening retaining screws to 31 N•m (23 ft.-lbs.) torque. If output shaft binds while tightening retaining screws, add enough shims (14) to eliminate binding while maintaining some bearing preload. Check for correct output shaft bearing preload and select correct thickness of shims (14) as follows: Shift transmission to neutral and wrap a cord around output shaft (3). Attach a spring scale to the cord and measure pull required to turn the output shaft. Correct preload will require 6-10 kg (13-22 lbs. or 58-98 N) to rotate the output shaft. Bearing preload is changed by varying the thickness of shims (14).

Remainder of assembly is the reverse of disassembly procedure.

SYNCHRONIZED SHIFT TRANSMISSION (8 FORWARD - 8 REVERSE)

118. The transmission is fully synchronized and shifting may be performed while tractor is moving. Eight forward and eight reverse gear ratios may be selected by positioning the three gear selectors shown in Fig. 114.

A safety start switch is provided to prevent engine from starting unless range selection lever is in neutral (N). The tractor should be towed only with range and main shift levers both in neutral.

Refer to paragraph 128 for shuttle shift 16 × 8 equipped tractors with dual power hydraulically operated transmission.

LUBRICATION

Models With 8 × 8 Shuttle Transmission

119. Transmission lubricant capacity is 9.4 L (10 quarts). Maintain oil level to full mark on dipstick attached to filler plug (F—Fig. 115) of models without cab. For models with cab, remove the floor cover, remove the transmission access panel and plug (F—Fig. 116). The recommended lubricant is Ford M2C134-D. Transmission should be drained and refilled with new oil every 1200 hours of operation.

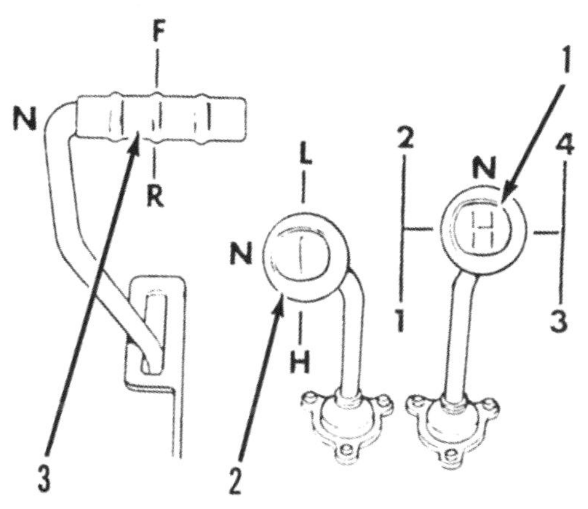

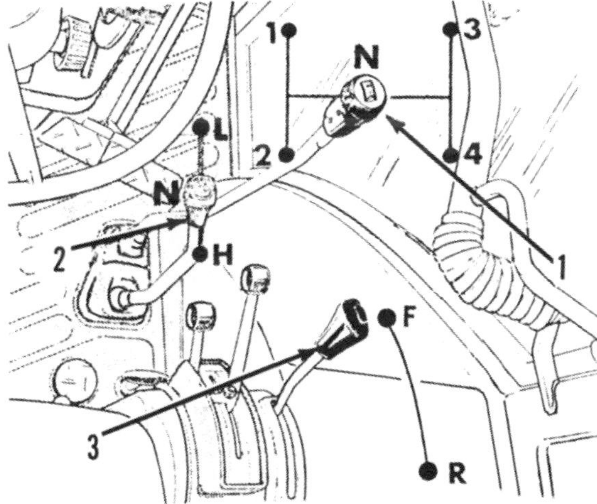

Fig. 114—View of shift levers for Shuttle Shift Synchronized transmission. Models without cab are as shown at top and models with cab are as shown in lower view.

REMOVE AND REINSTALL

Models With 8 × 8 Shuttle Shift

120. To remove shuttle transmission from models with cab, first remove cab as outlined in paragraph 212. On all models, split tractor between the engine and transmission housing as outlined in paragraph 102. Remove steering unit and fuel tank as a unit

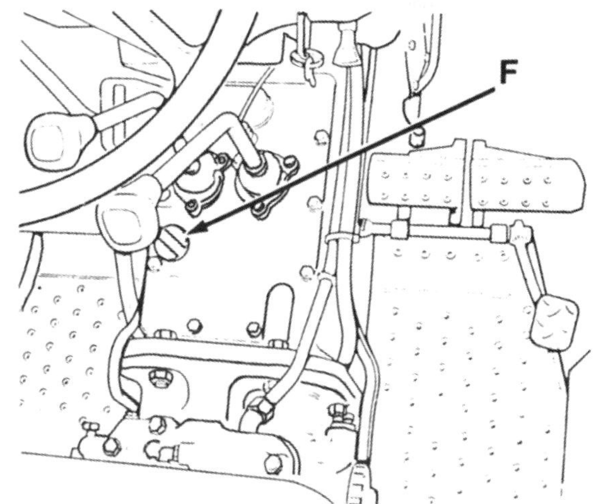

Fig. 115—View of Shuttle Shift filler plug and dipstick (F) on models without cab.

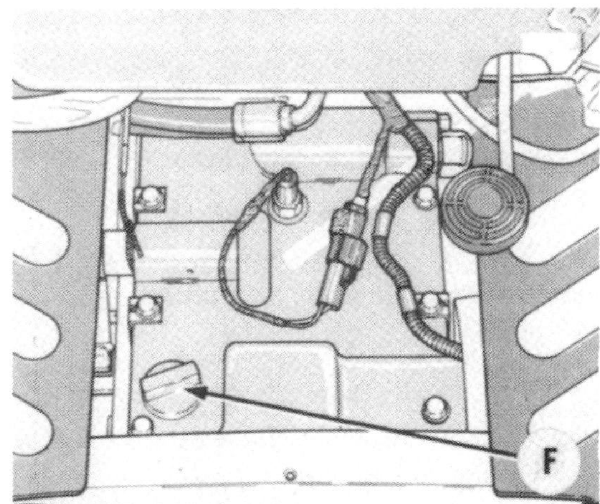

Fig. 116—View of Shuttle Shift filler plug and dipstick (F) for models with cab.

from top of transmission housing. Support transmission housing with an overhead hoist. Support rear axle center housing separately, then refer to paragraphs 139 or 140 to separate transmission from center housing.

Reverse removal procedure to install transmission. Refer to Fig. 154 for tightening torque of transmission to rear axle bolts and to Fig. 96 for tightening torques of engine to transmission bolts.

GEAR SHIFT COVER

Shuttle Shift Models Without Cab

121. REMOVE AND REINSTALL. Move all three shift levers to neutral, disconnect wires to the neutral switch (13—Fig. 117) and remove the filler plug and dipstick (F). Remove the retaining screws, then lift the cover from the transmission housing.

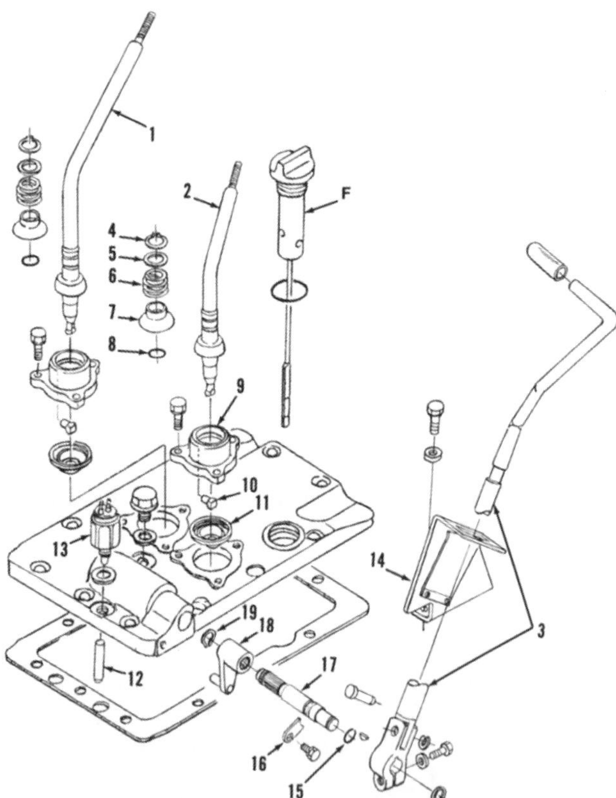

Fig. 117—Exploded view of shift cover assembly used on models with Shuttle Shift, but without cab.

F. Filler plug/dipstick
1. Main shift lever
2. Range shift lever
3. Shuttle lever
4. Snap rings
5. Washers
6. Springs
7. Cups
8. "O" rings
9. Supports
10. Pins
11. Shields
12. Plunger
13. Neutral safety switch
14. Bracket
15. "O" ring
16. Retainer
17. Forward-reverse shift shaft
18. Internal lever
19. Snap ring

Further disassembly procedure will be evident after reference to Fig. 117. When reinstalling the cover, tighten retaining screws to 36 N•m (26 ft.-lbs.) torque.

Shuttle Shift Models With Cab

122. REMOVE AND REINSTALL. Move all three shift levers to neutral, remove floor mat, then unbolt and remove transmission access plate. Disconnect shift rods from the three small levers clamped to shafts protruding from the shift cover. Disconnect both wires to the neutral safety switch and remove filler plug and dipstick. Remove retaining screws, then lift cover from transmission housing.

Procedure for further disassembly will be evident after reference to Fig. 118. When reinstalling cover, tighten retaining screws to 36 N•m (26 ft.-lbs.) torque.

SHIFT RAILS, SHIFT FORKS AND GATES

Models With 8 × 8 Shuttle Shift

123. R&R AND OVERHAUL. Remove the transmission as outlined in paragraph 120 and the shift cover as outlined in paragraph 121 or 122. Remove pin (12—Fig. 119) for neutral safety switch, plugs (15), springs (14) and detent balls (13). If springs (14)

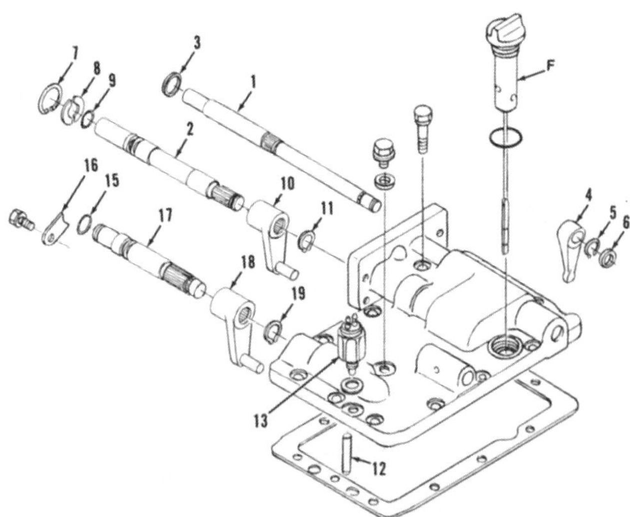

Fig. 118—Exploded view of shift cover used on tractors equipped with 8 X 8 transmission and cab.

1. Main shift shaft
2. Range shift shaft
3. Dust seal
4. Internal shift lever (main)
5. Snap ring
6. Dust seal
7. Snap ring
8. Collar
9. "O" ring
10. Internal shift lever (range)
11. Snap ring
12. Plunger
13. Neutral safety switch
15. "O" ring
16. Retainer
17. Forward-reverse shift shaft
18. Internal shift lever (Fwd.-Rev.)
19. Snap ring

and balls (13) can not be removed with a magnet, they can be withdrawn from below after rail is withdrawn. Drive pins (7) from gates and forks, then push rails (1, 2, 3 and 4) out toward rear. The output shaft must be removed to remove the 1st-2nd shift fork (17). However, the rail can be removed as follows: Remove retaining screw (24—Fig. 120), then withdraw 1st-2nd lower shift rail (5) toward rear. Be careful to catch detent spring (14—Fig. 119) and ball (13) from the fork (17) as the shaft (5) is removed. Remove plug (23) to remove the interlock plunger (6).

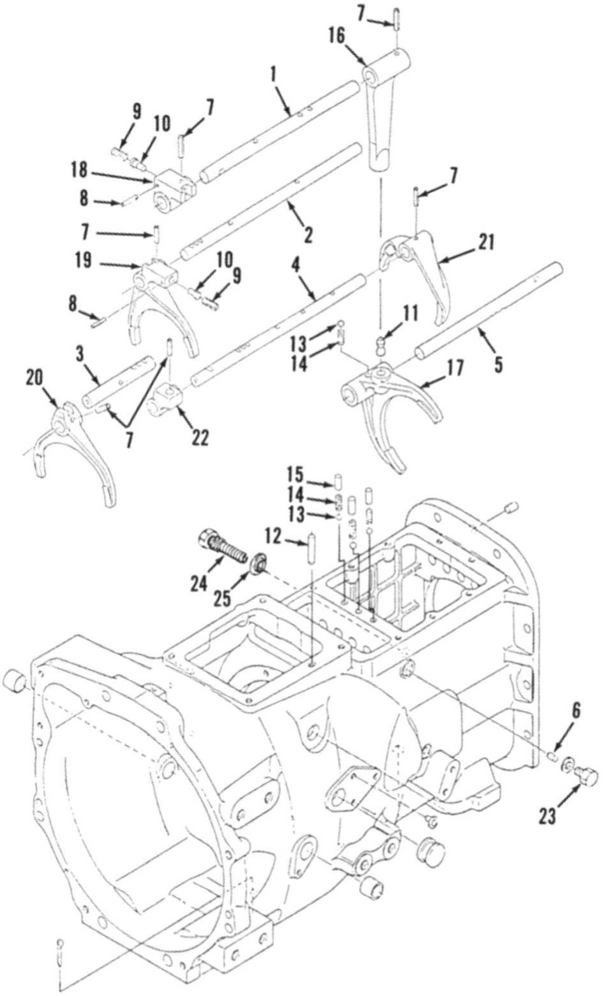

Fig. 119—View of shift rails and forks. Interlock pin (6) is located in bore between rails (1 & 2) and allows only one of the two rails to be moved at a time.

1. 1st-2nd Upper shift rail
2. 3rd-4th Shift rail
3. Forward-reverse shift rail
4. High-lo range shift rail
5. 1st-2nd Lower shift rail
6. Interlock pin
7. Retaining roll pins
8. Spring retainer pins
9. Neutral centering springs
10. Neutral centering plungers
11. Fork set pin
12. Neutral safety start pin
13. Detent balls
14. Detent springs
15. Plugs
16. 1st-2nd Selector arm
17. 1st-2nd Shift fork
18. 1st-2nd Shift gate
19. 3rd-4th Shift fork
20. Forward-reverse shift fork & gate
21. High-lo range shift fork
22. High-lo range shift gate
23. Interlock bore plug
24. 1st-2nd Lower shift rail retaining screw
25. Sealing washer

Install the lower shift rail (5) through fork (17) while compressing detent ball (13) against spring (14) in fork. A special 17 mm diameter spacer that is 50 mm long can be used to compress the detent ball and spring until the shift rail is pushed through. Install the lower rail retaining screw (24) using new sealing washer (25) and tighten screw to 25 N•m (18 ft.-lbs.) torque. Position shift forks (19, 20 and 21—Fig. 121) in their respective synchronizer assemblies. Position shift arm (16—Fig. 119) over pin (11) in shift fork (17) and align bore in shift arm with hole for shift rail (1). Slide the 1st-2nd shift rail (1) through the shift arm (16), hold shift gate (18) in position and continue

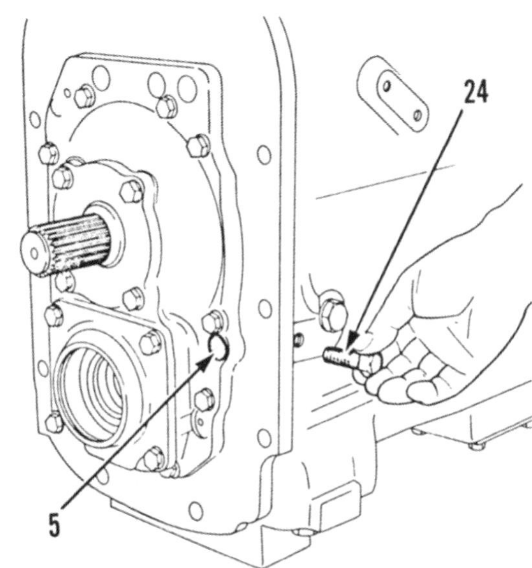

Fig. 120—The 1st-2nd lower shift rail (5) can be removed after removing retaining screw (24).

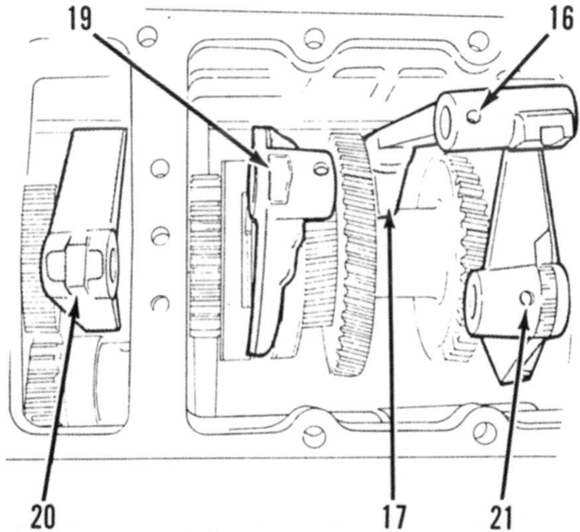

Fig. 121—View of shift arm and forks positioned so that shift rails can be installed. Refer to Fig. 119 for legend.

pushing shift rail (1) through gate. Install interlock pin (6) in bore between rail (1) and rail (2), then continue pushing rail (1) into position with detent for interlock pin (6) and holes for pins (7) correctly located. Install pins (7) in arm (16) and gate (18) and move 1st-2nd shift parts (16, 17 and 18) to the center (neutral detent) position and make sure that interlock pin (6) will permit entry of the 3rd-4th shift rail (2). Slide 3rd-4th shift rail (2) into bore of housing from the rear. Continue pushing rail through fork (19), making sure that notches for the detent ball (13) are toward top and notch for interlock pin (6) faces rail (1) and pin (6). Align hole in fork (19) for pin (7) with hole in shaft, then install pin (7). Check to be sure that interlock pin (6—Fig. 122) is correctly installed by attempting to move both shift rails (1 and 2—Fig. 119) from neutral at the same time. If assembled properly, one of the rails must remain in neutral before the other can be removed. Slide shift rail (4) into housing bore and into shift fork (21) with detent notches toward top and front. Continue sliding shift rail (4) through gate (22) and front bore in housing. Align holes in gate (22) and fork (21) with the holes in shift rail (4) and install pins (7). Slide rail (3) into housing bore with detent notches toward rear and to the top. Continue sliding rail (3) into shift fork (20), align holes for pin, then install pin (7) through fork and rail. Blanking plugs should be installed at rear of shift rail bores.

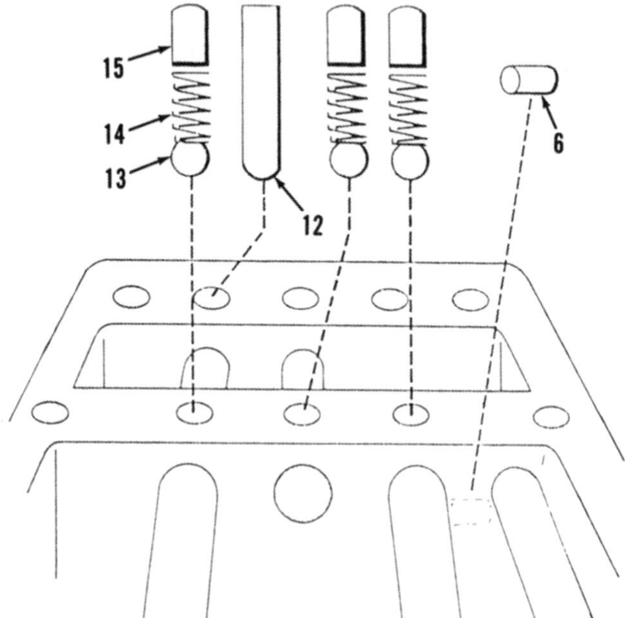

Fig. 122—Install neutral interlock pin (6), neutral safety switch plunger (12), detent balls (13), detent springs (14) and blanking plugs (15) in bores as shown.

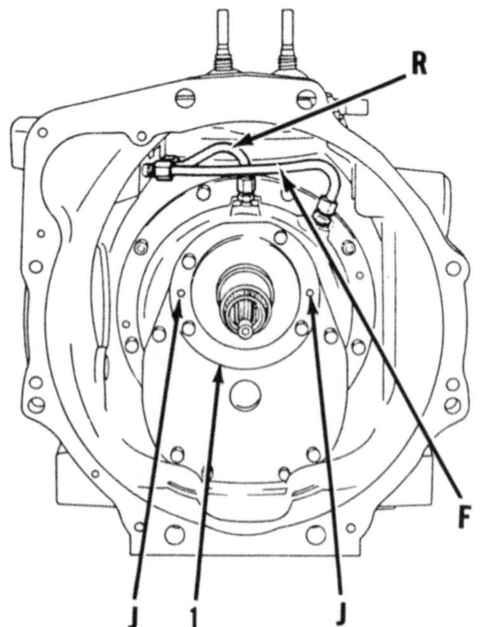

Fig. 123—View showing the oil feed pipe (F) to the oil cooler and return pipe (R) from oil cooler. Holes (J) are threaded for jack screws.

TRANSMISSION INPUT SHAFTS, PTO DRIVE, HYDRAULIC PUMP AND OIL DISTRIBUTION HOUSING

Models With 8 × 8 Shuttle Shift

124. R&R AND OVERHAUL. Remove the transmission as outlined in paragraph 120, the shift cover as outlined in paragraph 121 or 122 and the shift forks, gates and rails as outlined in paragraph 123. Remove both oil cooler pipes (F and R—Fig. 123). Remove the four screws attaching the front support (1), then thread two M8 screws approximately 30 mm (1¼ inches) long into threaded holes (J). Tighten the jack screws until front support is pushed from front case, then remove support and the jack screws.

Remove oil filter (F—Fig. 124). Remove screws attaching the front cover (7) and install two M8 screws approximately 30 mm (1¼ inches) long into the threaded holes. Tighten the two jack screws until the front cover is pushed from housing (13), then remove cover and the jack screws. Be careful not to lose the thrust collar (9).

Remove snap ring (76—Fig. 125), then pull pto shaft (72) and bearing (75) from rear of transmission housing.

Remove the screws attaching the front housing (13—Fig. 124) to the transmission housing. Two of the screws are located at (15). Install two M8 screws approximately 30 mm (1¼ inches) long in the two threaded holes. Tighten the two jack screws until the housing (13) is pushed from the transmission housing, then remove housing and the jack screws. To remove the rear input shaft (30) and tapered roller bearing (29), refer to paragraph 125 and remove the transmission output shaft and countershaft.

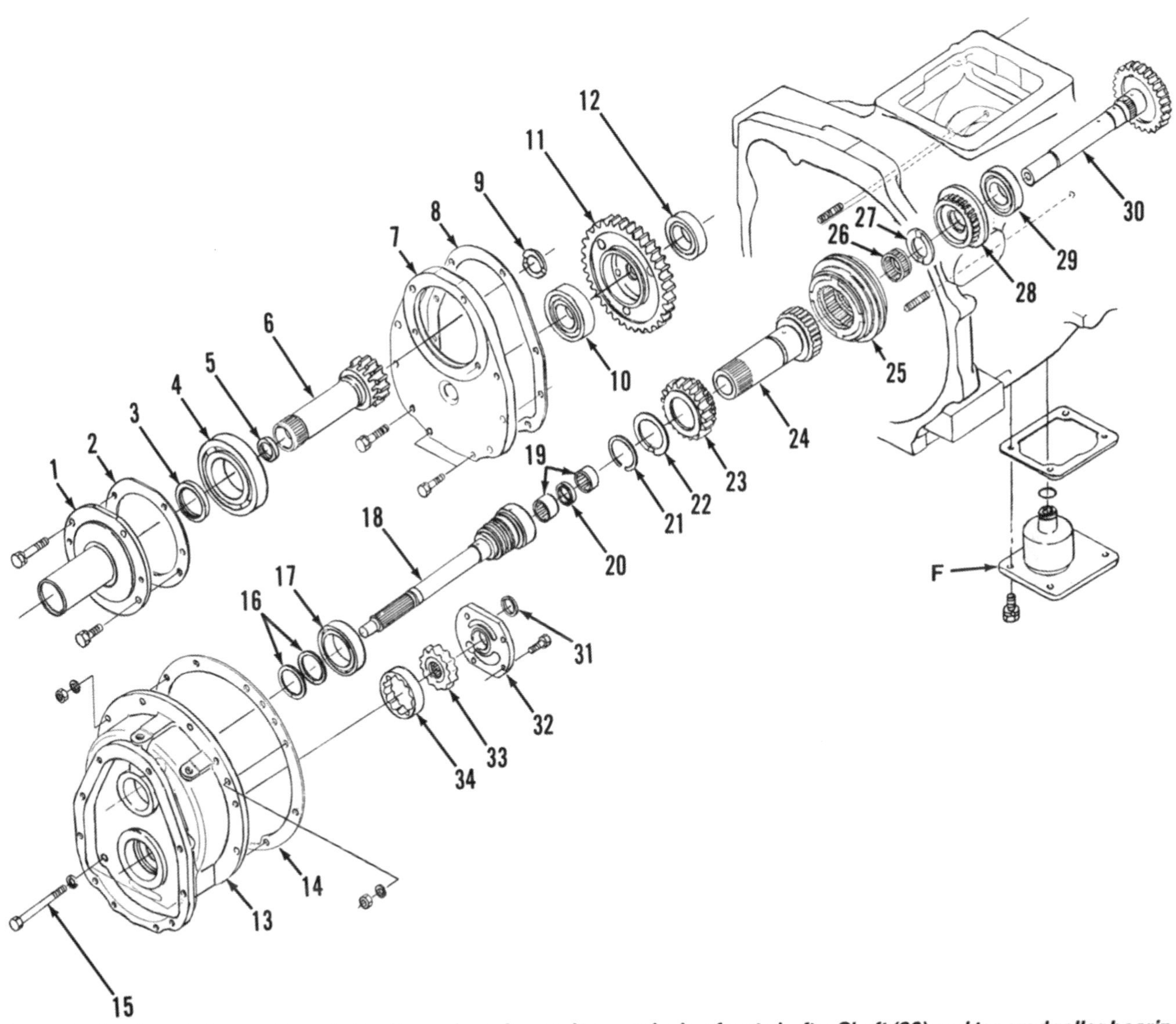

Fig. 124—Exploded views of the oil distributor housing and transmission front shafts. Shaft (30) and tapered roller bearing (29) must be installed from rear. Remaining parts can be removed from front.

F. Filter
1. Front support
2. Gasket
3. Oil seal
4. Ball bearing
5. Oil seal
6. Shaft & gear (16 teeth)
7. Front cover
8. Gasket
9. Thrust collar (various thickness)
10. Ball bearing
11. Pto gear (43 teeth)
12. Ball bearing
13. Front housing
14. Gasket
15. Longer screws (2)
16. Seal rings (2)
17. Ball bearing
18. Input shaft
19. Needle bearings (2)
20. Spacer
21. Snap ring
22. Thrust collar
23. Reverse gear
24. Input shaft & synchronizer hub
25. Synchronizer assy.
26. Needle bearing
27. Thrust washer
28. Synchronizer coupling
29. Tapered roller bearing
30. Input shaft
31. Gasket
32. Pump cover
33. Pump inner rotor
34. Pump outer rotor

Reassemble by reversing the disassembly procedure. Tighten the oil pump retaining screws to 25 N•m (18 ft.-lbs.) torque, front housing and front cover retaining screws to 52 N•m (38 ft.-lbs.) torque and the oil cooler pipes to 64 N•m (47 ft.-lbs.) torque.

REAR COVER PLATE, OUTPUT SHAFT AND COUNTERSHAFT

Models With 8 × 8 Shuttle Shift

125. R&R AND OVERHAUL. Remove transmission as outlined in paragraph 120, shift cover as outlined in paragraph 121 or 122 and, shift forks, gates and rails as outlined in paragraph 123. Remove snap ring (76—Fig. 125) and withdraw pto shaft (72) and bearing (75). Refer to paragraph 124 and remove transmission front support, front cover, front housing and forward and reverse synchronizer.

If wear is suspected, check gear backlash with a dial indicator as shown in Fig. 126. Recommended backlash is 0.1-0.2 mm (0.004-0.008 inch).

Before removing rear cover plate, tie the output shaft (Fig. 127) and countershaft (Fig. 128) together with a length of cord, so shafts will not fall and be damaged. Remove rear cover plate (51—Fig. 127 and Fig. 128). Carefully remove cord used to tie the output shaft and countershaft together, then lift top (output) shaft as an assembly from the housing.

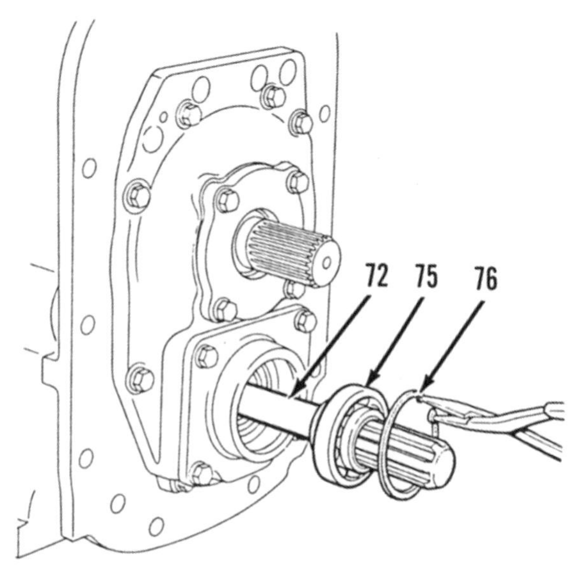

Fig. 125—Snap ring (76) must be removed before withdrawing the pto shaft (72) and bearing (75).

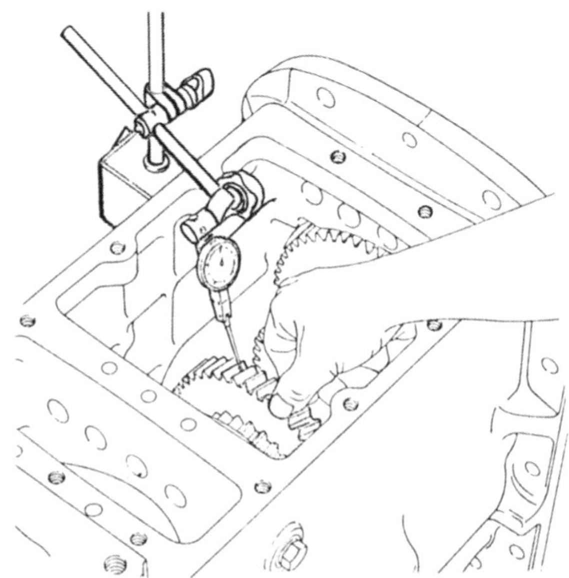

Fig. 126—Gear backlash should be checked before removing rear cover plate using a dial indicator as shown.

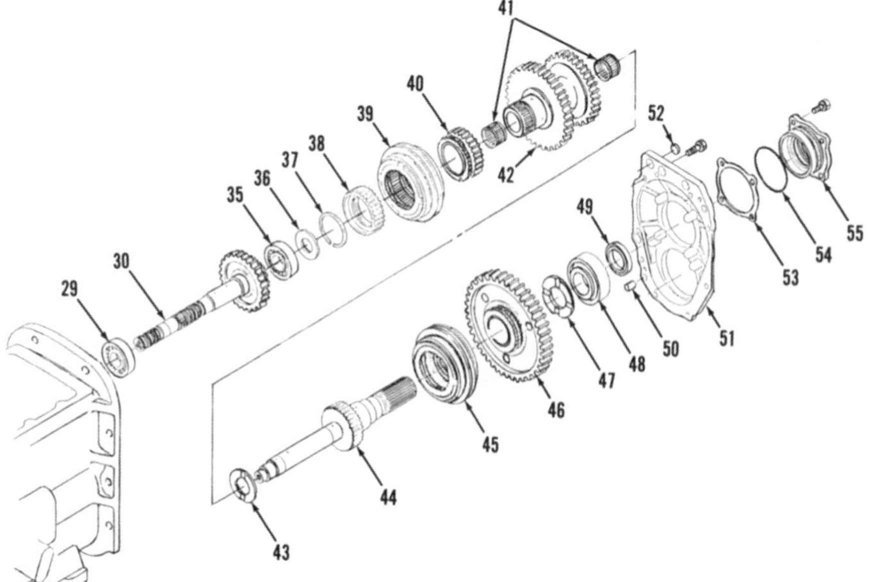

Fig. 127—Exploded view of transmission output shaft. Rear input shaft (30) is also shown in Fig. 124 and rear cover plate (51) is also shown in Fig. 128.

29. Tapered roller bearing
30. Rear input shaft
35. Tapered roller bearing
36. Thrust washer
37. Snap ring
38. Synchronizer coupling
39. Synchronizer assy. (3rd-4th)
40. Gear (22 teeth)
41. Needle bearings
42. Gear (47 & 35 teeth)
43. Thrust washer
44. Output shaft
45. Synchronizer assy. (high-lo)
46. Gear (54 teeth)
47. Thrust washer
48. Tapered roller bearing
49. Oil seal
50. Dowel pins (2)
51. Rear cover plate
52. Plugs (4)
53. Shims
54. "O" ring
55. Output shaft bearing retainer

NOTE: When removing output shaft, be sure to hold the front synchronizer assembly (39—Fig. 127) together, so it will not be bumped and fall apart.

Remove rear input shaft (30) through opening for shift cover opening in top of transmission housing. Remove countershaft (Fig. 128) from the rear of transmission housing.

Remove screw (81—Fig. 129) retaining reverse idler shaft (83—Fig. 130), remove the snap ring (87) from groove, then slide the shaft (83) out toward rear. Catch thrust washers (84), needle bearing (86) and gear (85) as shaft is withdrawn.

Synchronizers (25—Fig. 124, 45—Fig. 127 and 65—Fig. 128) are alike. Synchronizer (39—Fig. 127) is different from the others. Parts of synchronizers are not available separately and if any part is lost or damaged, the complete unit should be renewed.

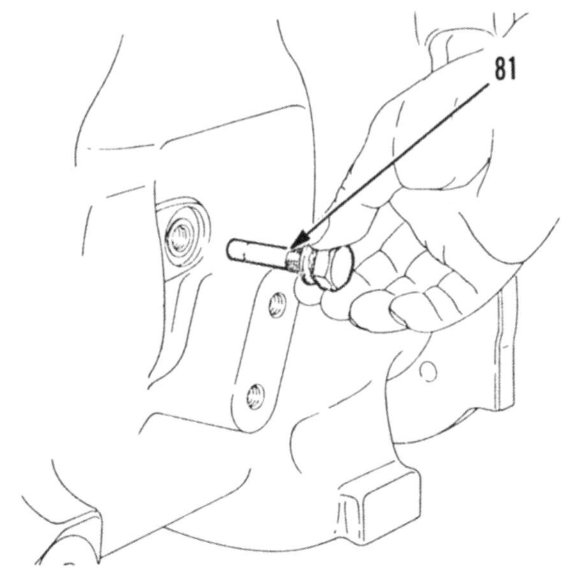

Fig. 129—The reverse idler shaft is retained by screw (81) which is also shown in Fig. 130.

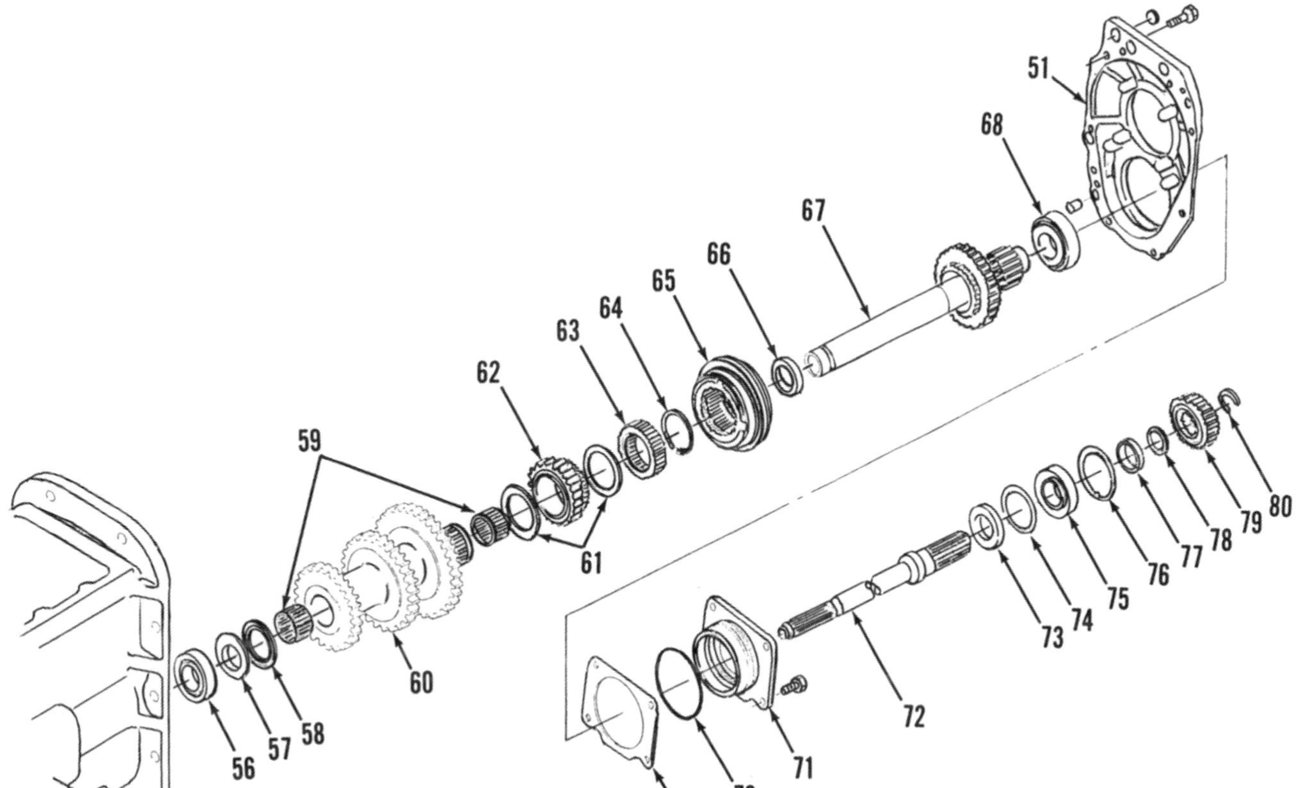

Fig. 128—Exploded view of countershaft (lower) used with 8 X 8, synchronized transmission. Rear cover plate (51) is also shown in Fig. 127.

51. Rear cover plate
56. Tapered roller bearing
57. Thrust washer
58. Thrust bearing
59. Needle bearings
60. Gear (35, 39 & 45 teeth)
61. Thrust washers
62. Gear (29 teeth)
63. Synchronizer coupling
64. Snap ring
65. Synchronizer (1st-2nd)
66. Thrust bearing
67. Shaft and gears (14 & 32 teeth)
68. Tapered roller bearing
69. Shims
70. "O" ring
71. Countershaft bearing retainer
72. Pto shaft
73. Oil seal
74. Collar
75. Ball bearing
76. Snap ring
77. Collar
78. Snap ring
79. Hydraulic pump drive gear
80. Snap ring

126. The tapered roller bearings for both output shaft and countershaft are adjusted by changing the thickness of shims (53—Fig. 127 and 69—Fig. 128). To determine the correct thickness of shims, proceed as follows:

Assemble countershaft parts (56 through 68—Fig. 128) and install in transmission housing, with no other parts installed. Unbolt and remove retainer (71), making sure that shims (69) are not damaged if they are to be reused. Install rear cover plate (51) and tighten the retaining screws to 52 N•m (38 ft.-lbs.) torque. Install the retainer (71) with cup for bearing (68) and at least the same thickness of shims (69) that were originally installed, but do not install "O" ring (70) yet. Tighten screws attaching the retainer (71) finger tight, while making sure that bearing cones correctly seat in bearing cups, then finish tightening screws to 52 N•m (38 ft.-lbs.) torque. If bearings bind when tightening screws, install additional thickness of shims (69) before continuing. Wrap a cord around shaft (67) and attach a spring scale as shown in Fig. 131. Check pull required to rotate shaft by pulling cord (C) with a spring scale. Correct bearing preload will permit shaft to rotate with 40-53 N (9-12 lbs.) of pull when measured as shown in Fig. 131. Add or remove shims (69—Fig. 128) as required, making sure that screws attaching bearing retainer (71) and rear cover plate (51) are tightened to 52 N•m (38 ft.-lbs.) torque while checking. When correct thickness of shims (69) has been selected, remove retainer (71) with correct thickness of shims, install new "O" ring (70), then reinstall and recheck bearing adjustment. Unbolt and remove the rear cover plate (51), then lift the complete countershaft assembly out of the housing. Set the countershaft aside while checking and adjusting bearings of output shaft.

Preload of output shaft bearings should be checked with only bearing (29—Fig. 127), the rear input shaft (30) and output shaft assembly (35 through 48) installed in the transmission housing. Actually, synchronizer (39) should not be installed while checking because of possible inaccurate reading and to facilitate temporary installation. Install bearing (29) and rear input shaft (30). Assemble output shaft parts (35 through 49) and install in transmission housing. Unbolt and remove retainer (55), making sure that shims (53) are not damaged if they are to be reused. Install rear cover plate (51) and tighten the retaining screws to 52 N•m (38 ft.-lbs.) torque. Install the retainer (55) with cup for bearing (48) and at least the same thickness of shims (53) that were originally installed, but do not install "O" ring (54) or seal (49) yet. Tighten screws attaching the retainer (55) finger tight, while making sure that bearing cones correctly seat in bearing cups, then finish tightening screws to 52 N•m (38 ft.-lbs.) torque. If bearings bind when tightening screws, install additional thickness of shims (53) before continuing. Wrap a cord around shaft (44) and attach a spring scale as shown in Fig. 132. Check pull required to rotate shaft by pulling cord (C) with a spring scale (S). Correct bearing preload will permit shaft to rotate with 27-40 N (6-9 lbs.) of pull when measured as shown in Fig. 132. Add or remove shims (53—Fig. 127) as required, making sure that screws attaching bearing retainer (55) and rear cover plate (51) are tightened to 52 N•m (38

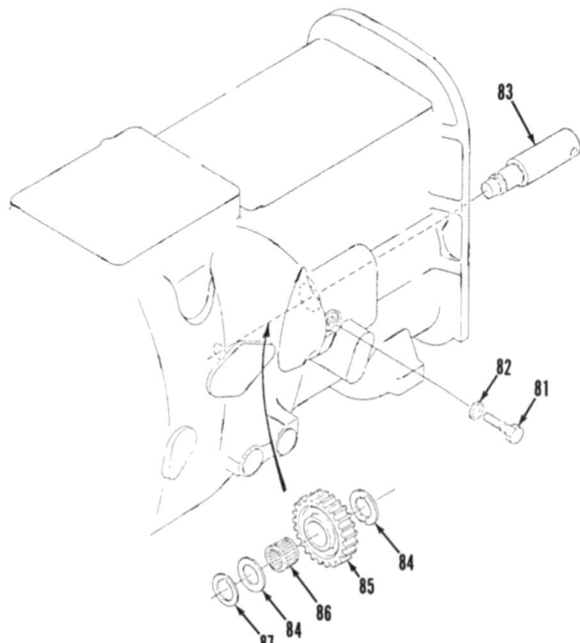

Fig. 130—Exploded view of reverse idler shaft and gear.

81. Retaining screw
82. Sealing washer
83. Reverse idler shaft
84. Thrust washers (2)
85. Reverse idler gear
86. Needle bearing
87. Snap ring

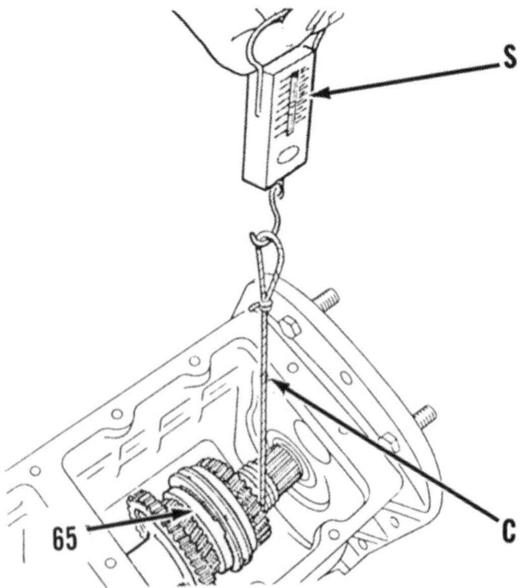

Fig. 131—Preload of the countershaft bearings should be measured using a cord (C) wrapped around the shaft and a spring scale (S) as shown. Other shafts should be removed when checking.

ft.-lbs.) torque while checking. When correct thickness of shims (53) has been selected, remove retainer (55) with correct thickness of shims, install new "O" ring (54) and new oil seal (49), then reinstall and recheck bearing adjustment. Unbolt and remove rear cover plate (51), then lift output shaft out of housing.

127. Observe the following when assembling: Install reverse idler assembly (Fig. 130) with oil grooves of thrust washers (84) toward gear (85). Install and tighten retaining screw (81) to 68 N•m (50 ft.-lbs.) torque.

Stand transmission housing on front face with rear of housing up. Position countershaft assembly (56 through 68—Fig. 128) in housing with weight resting in cup of front bearing (56). Use a 17 mm diameter shaft (D—Fig. 133) about 50 mm long to hold detent ball against spring in 1st-2nd lower shift fork (17). Position shift fork in housing and in groove of synchronizer (65), then slide lower shift rail (5—Fig. 119) through holes in housing and into shift fork. Catch dummy shaft (D—Fig. 133) as rail replaces it in fork.

Position the rear input shaft (30—Fig. 127) and bearing cone (29) in cup for bearing (29), then carefully position the mainshaft assembly (35 through 48) against the cup for bearing (35). Coat mating surface of the transmission housing with sealer and install the rear cover plate (51). Before tightening the retaining screws, make sure that parts are correctly positioned and do not bind. Tighten screws attaching the rear cover plate to 52 N•m (38 ft.-lbs.) torque and recheck to be sure that parts do not bind and that correct gear engagement is possible. Refer to paragraph 124 and install the transmission input shafts, hydraulic pump and oil distribution housing. Refer to paragraph 123 and install shift rails and forks, then check for correct shifting before proceeding further.

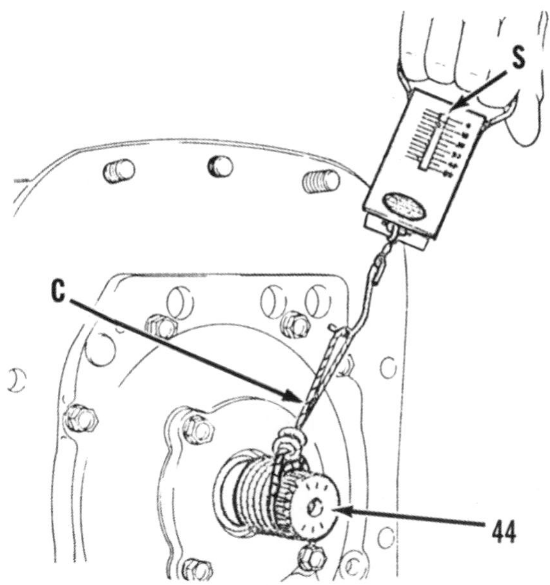

Fig. 132—Preload of the output shaft bearings should be measured using a cord (C) wrapped around the shaft and a spring scale (S) as shown. Countershaft and all input shafts except the rear input shaft should be removed when checking.

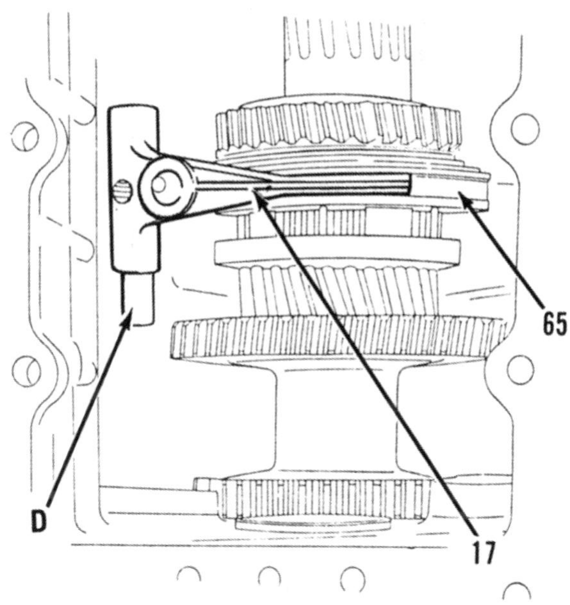

Fig. 133—Special dummy shaft (D) can be used to compress the detent in shift fork while installing.

SYNCHRONIZED SHIFT TRANSMISSION WITH DUAL POWER (16 FORWARD - 8 REVERSE)

128. The transmission is fully synchronized and shifting may be performed while tractor is moving. Eight forward and eight reverse gear ratios may be selected by positioning the three gear selectors shown in Fig. 134. While the manual shift section of the transmission is similar to the 8 × 8 speed transmission, different parts and different servicing procedures are used for models equipped with Dual Power 16 × 8 transmission.

The Dual Power hydraulically operated transmission provides an underdrive in forward ratios, doubling the number of forward speeds to sixteen. The electrical switch for the Dual Power transmission is located on the dash or on the main gear shift knob.

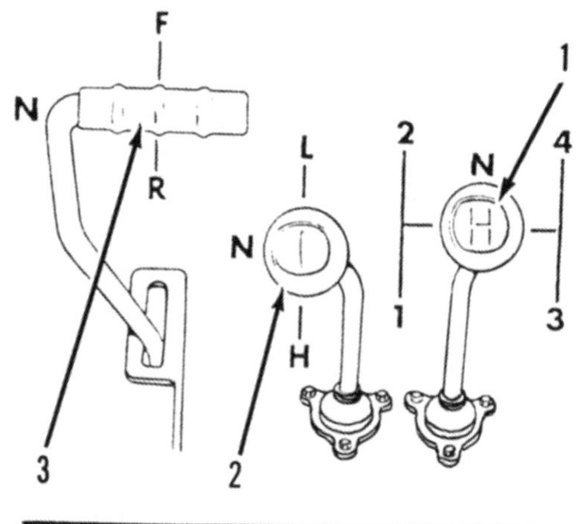

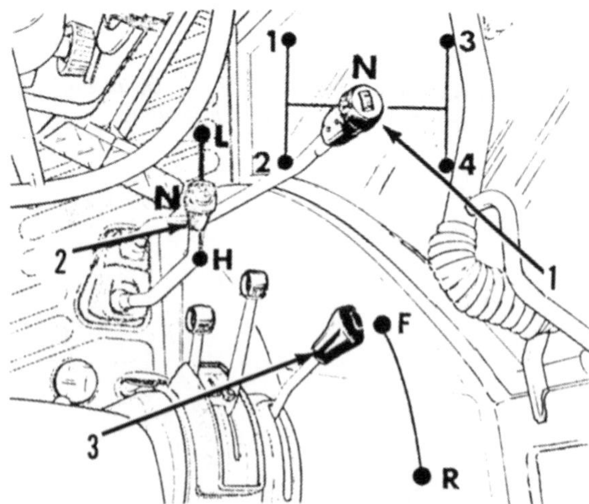

Fig. 134—View of shift levers for Shuttle Shift Synchronized transmission. Top view is for models without a cab and the lower view is for models with a cab. Dual Power electrical control switch is located on the dash or on shift lever.

Pressing the forward part of switch (indicated by a hare symbol) provides direct drive and pressing the rear part of switch (indicated by tortoise symbol) engages the underdrive.

A safety start switch is provided to prevent engine from starting unless range selection lever is in neutral (N). The tractor should be towed only with both range and main shift levers in neutral.

Refer to paragraph 118 for tractors equipped with 8 × 8 Shuttle Shift transmission without Dual Power.

LUBRICATION

Dual Power Shuttle Shift Models

129. Transmission lubricant capacity for models with Dual Power Shuttle transmission is 8.4 L (8.9 quarts). Maintain oil level to full mark on dipstick attached to filler plug (F—Fig. 135) of models without cab. For models with cab, remove the floor cover, remove the transmission access panel and plug (F—Fig. 136). The recommended lubricant is Ford M2C134-D. Transmission should be drained and refilled with new oil every 1200 hours of operation.

REMOVE AND REINSTALL

Dual Power Shuttle Shift Models

130. To remove the shuttle transmission from models with cab, first remove the cab as outlined in paragraph 212. On all models, split tractor between the engine and transmission housing as outlined in paragraph 102. Remove the steering unit and fuel

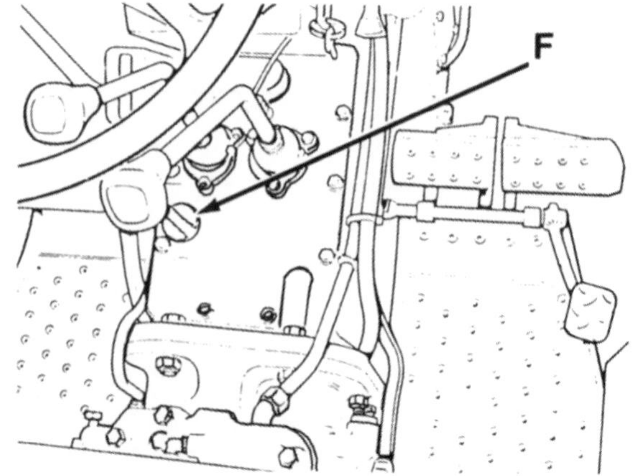

Fig. 135—View of Shuttle Shift filler plug and dipstick (F) on models without cab.

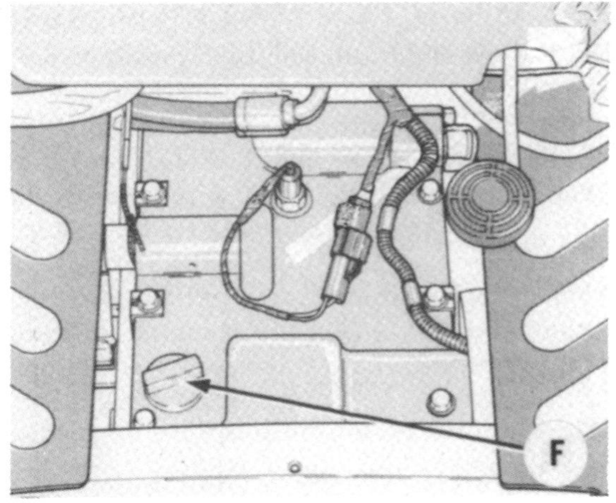

Fig. 136—View of Shuttle Shift filler plug and dipstick (F) for models with cab.

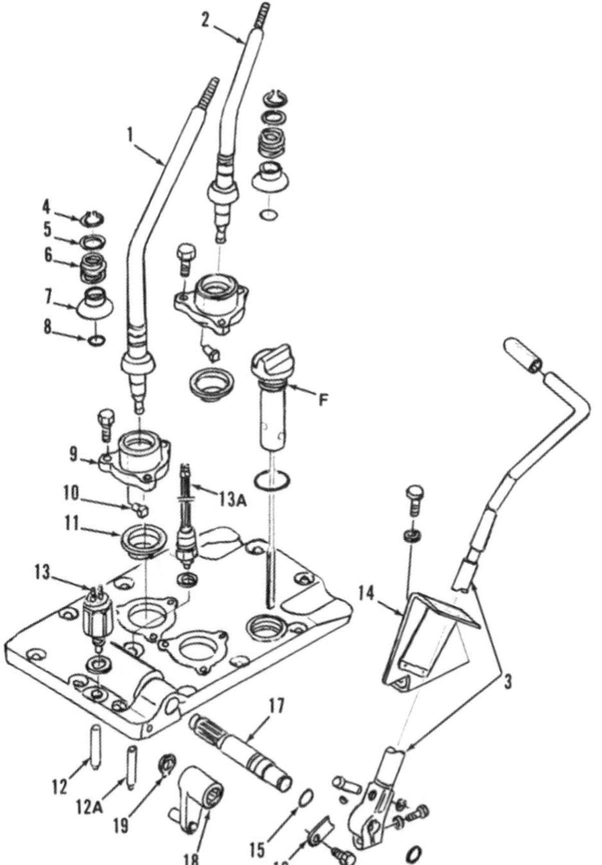

Fig. 137—Exploded view of shift cover assembly used on models with Dual Power Shuttle Shift, but without cab.

F. Filler plug/dipstick
1. Main shift lever
2. Range shift lever
3. Shuttle lever
4. Snap rings
5. Washers
6. Springs
7. Cups
8. "O" rings
9. Supports
10. Pins
11. Shields
12. Plunger for (13)
12A. Plunger for (13A)
13. Neutral safety switch
13A. High-low switch
14. Bracket
15. "O" ring
16. Retainer
17. Forward-reverse shift shaft
18. Internal lever
19. Snap ring

tank as a unit from top of transmission housing. Support transmission housing from an overhead hoist. Support the rear axle center housing separately, then refer to paragraph 139 or 140 to separate transmission from the center housing.

Reverse removal procedure to install transmission. Refer to Fig. 154 for tightening torque of transmission to rear axle bolts and to Fig. 96 for tightening torques of engine to transmission bolts.

GEAR SHIFT COVER

Dual Power Shuttle Shift Models Without Cab

131. REMOVE AND REINSTALL. Move all three shift levers to neutral, disconnect wires to the neutral switch (13—Fig. 137) and remove the filler plug and dipstick (F). Remove the retaining screws, then lift the cover from the transmission housing.

Procedure for further disassembly will be evident after reference to Fig. 137. When reinstalling cover, tighten retaining screws to 36 N•m (26 ft.-lbs.) torque.

Dual Power Shuttle Shift Models With Cab

132. REMOVE AND REINSTALL. Move all three shift levers to neutral, remove the floor mat, then unbolt and remove the transmission access plate. Disconnect shift rods from the three small levers clamped to shafts protruding from the shift cover. Disconnect both wires to the neutral safety switch and remove the filler plug and dipstick. Remove the retaining screws, then lift the cover from the transmission housing.

Procedure for further disassembly will be evident after reference to Fig. 138. When reinstalling cover, tighten retaining screws to 36 N•m (26 ft.-lbs.) torque.

SHIFT RAILS, SHIFT FORKS AND GATES

Dual Power Shuttle Shift Models

133. R&R AND OVERHAUL. Remove the transmission as outlined in paragraph 130 and the shift cover as outlined in paragraph 131 or 132. Remove plungers (12 and 12A—Fig. 139), pins (15), springs (14) and detent balls (13). If springs (14) and balls (13) cannot be removed with a magnet, they can be withdrawn from below after rail is withdrawn. Drive pins (7) from gates and forks, then push rails (1, 2, 3 and 4) out toward rear. Remove retaining screw (24—Fig. 140), then withdraw lower shift rail (5) toward rear. Be careful to catch detent spring (14—Fig. 139) and ball (13) from the fork (17) as the shaft (5) is removed. Remove plug (23) to remove the interlock plunger (6).

Position shift forks (17 and 20) in respective synchronizer and install lower shift rail (5) through forks while compressing detent ball (13) against spring (14) in fork. A special 17 mm diameter spacer that is 50 mm long can be used to compress the detent ball and spring until the shift rail is pushed through. Install the lower rail retaining screw (24) using new sealing washer (25) and tighten screw to 25 N.m (18 ft.-lbs.) torque. Position the shift forks (19 and 21) in their respective synchronizer assemblies. Position shift arm (20A) over pin (11) in shift fork (20), align bore in shift arm with hole for shift rail (3), then slide shift rail (3) through the shift arm (20A) with slot for plunger (12A) toward rear and to the top. Continue sliding rail (3) into shift arm (20A), align holes for pin, then install pin (7) through arm and rail. Position shift arm (16) over pin (11) in shift fork (17) and align bore in shift arm with hole for shift rail (1). Slide the 1st-2nd shift rail (1) through the shift arm and gate (16). Install interlock pin (6) in bore between rail (1) and rail (2), then continue pushing rail (1) into position with detent for interlock pin (6) and holes for pin (7) correctly located. Move 1st-2nd shift arm and fork (16 and 17) to the center (neutral detent) position and make sure that interlock pin (6) will permit entry of the 3rd-4th shift rail (2). Slide 3rd-4th shift rail (2) into bore of housing from the rear and install shift gate (19G) and fork (19). Continue pushing rail forward, making sure that notches for the plunger/detent ball (12A) is toward top and notch for interlock pin (6) faces rail (1) and pin (6). Align holes in fork (19) and gate (19G) for pins (7) with holes in shaft, then install both pins (7). Check to see that interlock pin is correctly installed by attempting to move both shift rails (1 and 2) from neutral at the same time. If assembled properly, one of the rails must remain in neutral before the other can be removed. Slide shift rail (4) into housing bore and into shift fork (21) with

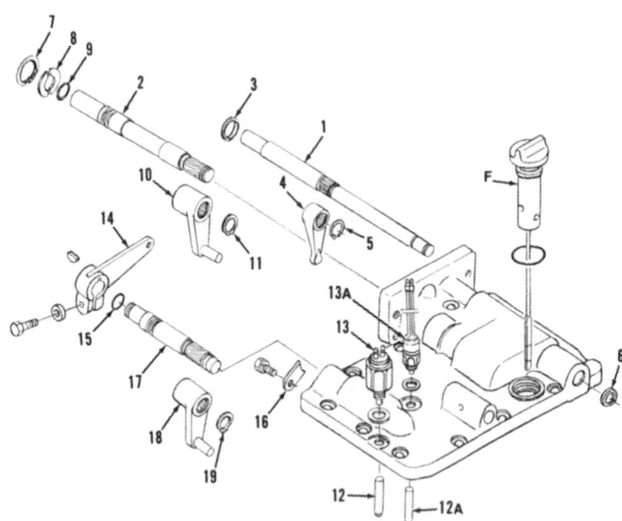

Fig. 138—Exploded view of shift cover used on tractors equipped with Dual Power transmission and cab.

1. Main shift shaft
2. Range shift shaft
3. Dust seal
4. Internal shift lever (main)
5. Snap ring
6. Dust seal
7. Snap ring
8. Collar
9. "O" ring
10. Internal shift lever (range)
11. Snap ring
12. Plunger for (13)

12A. Plunger for (13A)

13. Neutral safety switch

13A. High-low switch

14. Lever
15. "O" ring
16. Retainer
17. Forward-reverse shift shaft
18. Internal shift lever (Fwd.-Rev.)
19. Snap ring

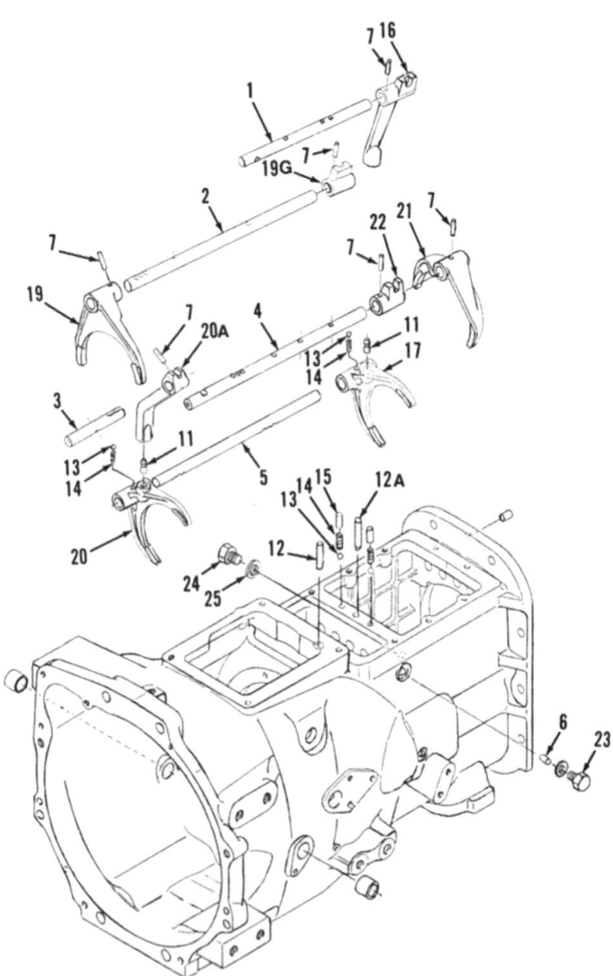

Fig. 139—View of shift rails and forks. Interlock pin (6) is located in bore between rails (1 & 2) and allows only one of the two rails to be moved at a time. Pins (12 & 12A) are also shown in Fig. 137, Fig. 138 and Fig. 142.

1. 1st-2nd Upper shift rail
2. 3rd-4th Shift rail
3. Forward-reverse shift rail
4. High-lo range shift rail
5. Lower shift rail
6. Interlock pin
7. Retaining roll pins
11. Fork set pin (2)
12. Neutral safety start pin

12A. Plunger

13. Detent balls
14. Detent springs
15. Plugs
16. 1st-2nd Selector arm
17. 1st-2nd Shift fork
19. 3rd-4th Shift fork

19G. 3rd-4th Shift gate

20. Forward-reverse shift fork

20A. Forward-reverse shift gate

21. High-lo range shift fork
22. High-lo range shift gate
23. Interlock bore plug
24. 1st-2nd Lower shift rail retaining screw
25. Sealing washer

detent notches toward top and front. Continue sliding shift rail (4) through gate (22) and front bore in housing. Align holes in gate (22) and fork (21) with the holes in shift rail (4) and install pins (7). Refer to Fig. 141. Blanking plugs should be installed at rear of shift rail bores. Refer to Fig. 142 for installation of detent assemblies (13, 14 and 15), for correct positioning of plunger (12), which is rounded, and for positioning and identification of plunger (12A).

OIL DISTRIBUTION HOUSING, PTO DRIVE AND HYDRAULIC PUMP

Dual Power Shuttle Shift Models

134. R&R AND OVERHAUL. Remove transmission as outlined in paragraph 120, then remove both oil cooler pipes (F and R—Fig. 143). Disconnect wiring from solenoid valve (S), then unbolt and remove valve. Remove the four screws attaching front support (1), then thread two M8 screws approximately 30 mm (1¼ inches) long into threaded holes (J). Tighten jack screws until front support is pushed from front case, then remove support and jack screws.

Remove oil filter (F—Fig. 144). Remove screws attaching the front cover (7) and install two M8 screws approximately 30 mm (1¼ inches) long into the threaded holes. Tighten the two jack screws until the front cover is pushed from housing (13), then remove cover and jack screws. Be careful not to lose the thrust collar (9).

Remove snap ring (76—Fig. 145), then pull pto shaft (72) and bearing (75) from rear of transmission housing.

Remove the screws attaching the front housing (13—Fig. 144) to the transmission housing. Two of the screws are located under the cover (7) at (15).

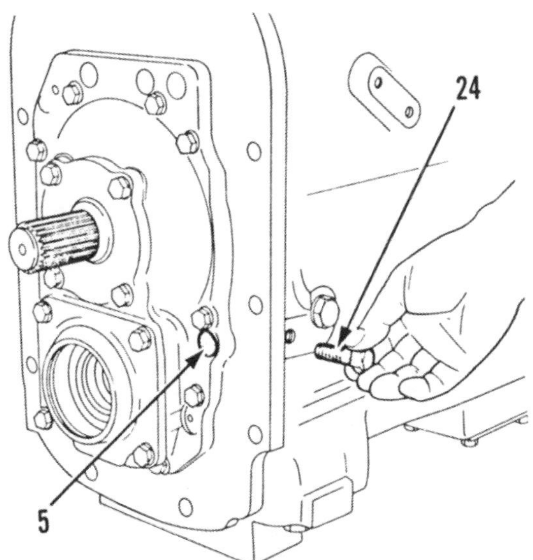

Fig. 140—The lower shift rail (5) can be removed after removing retaining screw (24).

Install two M8 screws approximately 30 mm (1¼ inches) long in the two threaded holes. Tighten the two jack screws until the housing (13) is pushed from the transmission housing, then remove housing and the jack screws.

Refer to paragraph 135 for removal of input shaft (18) and service to the Dual Power clutch (22 through 40). To remove rear input shaft (30) and tapered roller bearing (29), refer to paragraph 136 and remove the transmission output shaft and countershaft.

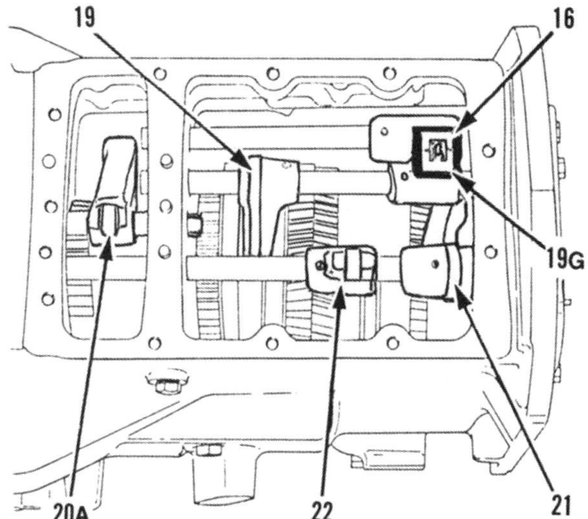

Fig. 141—View of the shift forks, shift arms and shift gates properly installed and in neutral. Refer to Fig. 139 for legend.

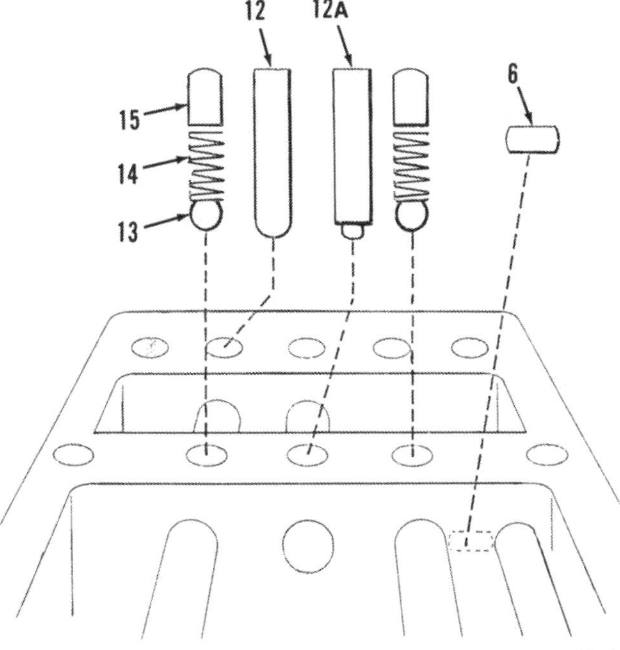

Fig. 142—Install neutral interlock pin (6), plungers (12 & 12A), detent balls (13), detent springs (14) and blanking plugs (15) in bores as shown. Notice the difference between plunger (12) and plunger (12A).

Individual parts of the control valve (60) and hydraulic pump (56) are not available. Renew the complete assembly if damaged. Tighten screws retaining oil pump (56) to 25 N•m (18 ft.-lbs.) torque. Tighten screws retaining solenoid (62) to 8.8 N•m (6.5 ft.-lbs.)and screws retaining valve (60) to 52 N•m (38 ft.-lbs.) torque.

Free length of regulating valve spring (53) should be 69 mm (2.72 inches) and free length of by-pass spring (47) should be 66.3 mm (2.61 inches). By-pass valve setting pressure is changed by adding or removing shims (48). By-pass pressure should be 1027 kPa (149 psi). Regulating valve pressure is changed by changing thickness of shims (52). Regulating valve pressure should be 1524 kPa (221 psi). Tighten caps (50 and 55) to 83 N•m (61 ft.-lbs.) torque. Tighten upper plug of regulator valve (51 through 55) to 52 N•m (38 ft.-lbs.) torque.

Install thrust washer (17A) and seal rings (16) on input shaft. Make sure that seal rings are latched and use light grease to lubricate and center the seals on shaft. Carefully install the front housing (13) over shaft (18), seal rings (16) and bearing (17).

NOTE: DO NOT force. If difficulty is encountered, make sure that seal ring has not fallen from groove. Attempting to force will only damage parts and will not correct problem.

Front housing (13) and front cover (7) retaining screws should be tightened to 52 N•m (38 ft.-lbs.) torque. Tighten fittings for oil cooler pipes to 64 N•m (47 ft.-lbs.) torque.

TRANSMISSION FRONT INPUT SHAFT AND DUAL POWER CLUTCH

Dual Power Shuttle Shift Models

135. R&R AND OVERHAUL. Remove the transmission as outlined in paragraph 120, then refer to paragraph 134 and remove the oil distribution housing (13—Fig. 144). Remove snap ring (36F) and lift front input shaft (18) from clutch. Remove direct drive clutch hub (22), plates (23F) and friction discs (24F). Remove snap ring (21) and seal rings (37 and 38) from rear input shaft (30), then pull clutch drum (33) and associated parts from shaft.

Many parts for the rear clutch are the same as for the front, but if reused, parts should not be interchanged. Remove snap ring (36R), then lift Dual Power clutch (23R, 24R and 35) from drum (33). Compress clutch release springs (27F and 27R) and remove retaining snap rings (25F and 25R). Remove spring seats (26F and 26R) and springs (27F and 27R). Clutch engaging pistons (28F and 28R) can be pushed from drum using compressed air directed into ports in center of clutch drum (33). Hub (40), gear (43) and associated parts can be withdrawn from rear input shaft (30). To remove rear input shaft (30) and tapered roller bearing (29), refer to paragraph 136 and remove transmission output shaft and countershaft.

Inspect all drive plates (23F and 23R) and friction discs (24F and 24R) for wear, scoring, overheating and warping. Clutch drive plates (23F and 23R) are normally 1.6 mm (0.063 inch) thick, but 1.75 mm (0.069 inch) thick plates are available to adjust clutch pack clearance. Friction discs (24F and 24R) are 2.44 mm (0.096 inch) thick when new.

When assembling rear clutch, install "O" ring (31R) in groove of piston (28R) and "O" ring (32R) in groove of clutch drum (33). Lubricate the "O" rings liberally and install piston in clutch drum. Notice that spacer (34) is installed under rear (Dual Power) piston (28R). Position a release spring (27R) and retainer (26R) over center of clutch drum, compress the spring and install snap ring (25R). Install five drive plates (23R) and five friction discs (24R) beginning with one drive plate (23R). Install the thick drive plate (35) last, then install snap ring (36R). Measure clearance between drive plate (35) and the snap ring (36R) with a feeler gage. Measure at three locations around clutch to be sure of clearance. Correct clutch pack clearance for the rear (Dual Power) clutch pack should be 0.8-1.0 mm (0.031-0.040 inch). Clutch drive plates (23R) are available in thicknesses of 1.6 mm (0.063 inch) and 1.75 mm (0.069 inch). Install drive plates of combined thickness to adjust clutch pack clearance within the desired clearance. Desired clearance is different for front clutch.

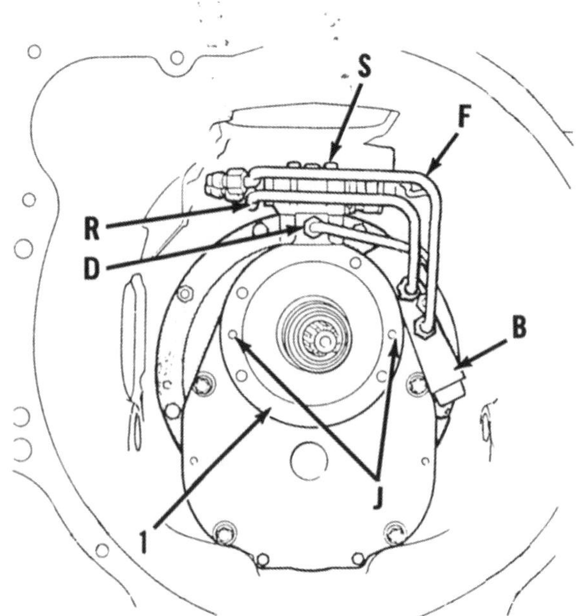

Fig. 143—View of front of transmission with Dual Power showing location of oil delivery tubes.

B. Bypass valve
D. Delivery tube
F. Oil cooler feed tube
R. Oil cooler return tube
S. Solenoid valve

When assembling front clutch, install "O" ring (31F) in groove of piston (28F) and "O" ring (32F) in groove of clutch drum (33). Lubricate the "O" rings liberally and install piston in clutch drum. Position a release spring (27F) and retainer (26F) over center of clutch drum, compress the spring and install snap ring (25F). Install six drive plates (23F) and four friction discs (24F) beginning with one drive plate (23F). Install two drive plates (23F) last, install input shaft (18), then install snap ring (36F). Measure clearance between input shaft (18) and the snap ring (36F) with a feeler gage. Measure at three locations around clutch to be sure of clearance. Correct clutch pack clearance for the front (Direct Drive) clutch pack should be 2.3-2.5 mm (0.091-0.098 inch). Clutch drive plates (23F) are available in thicknesses of 1.6 mm (0.063 inch) and 1.75 mm (0.069 inch). Install drive plates of combined thickness to adjust clutch pack clearance within the desired clearance. Desired clearance is different for rear clutch. After selecting parts

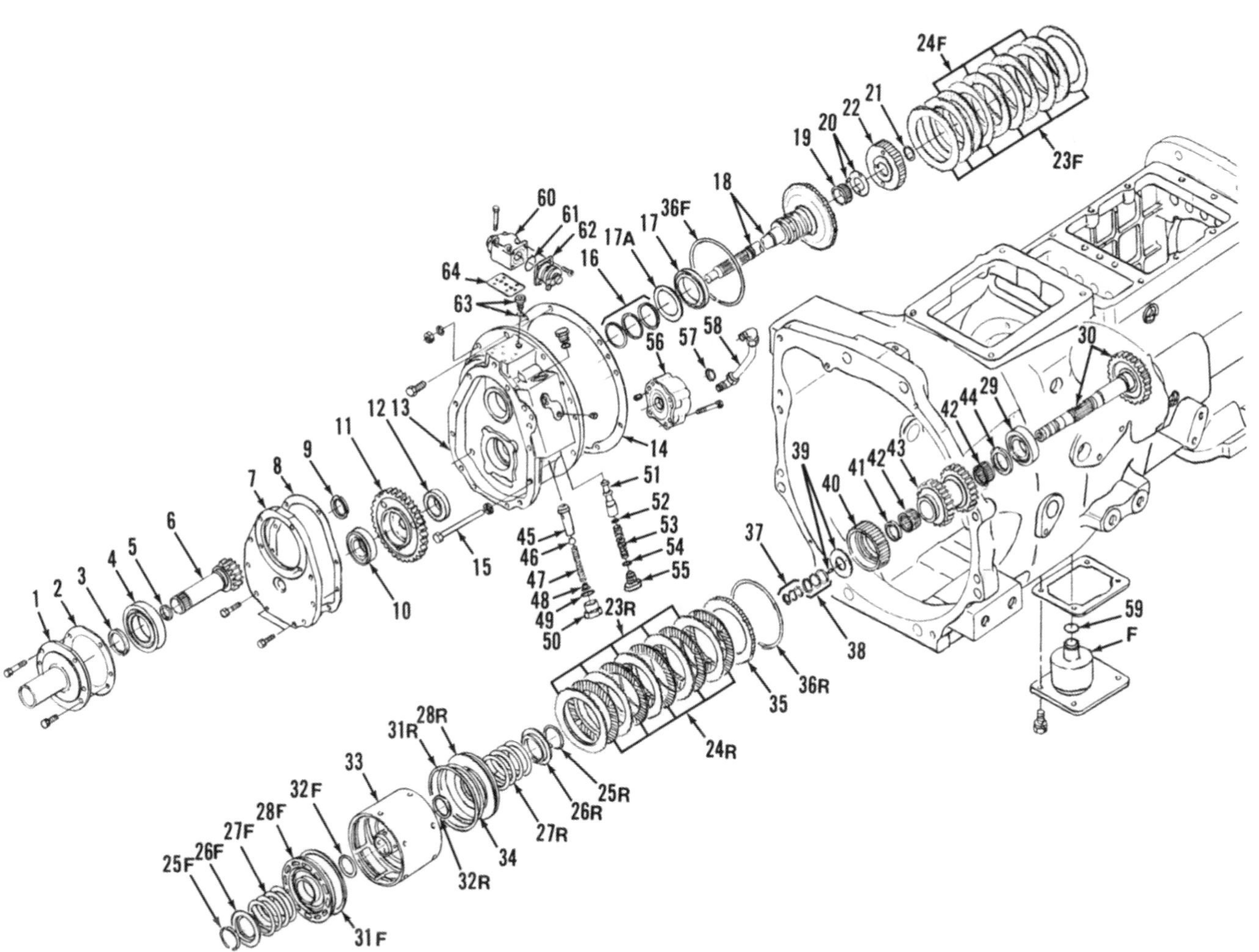

Fig. 144—Exploded views of the oil distributor housing, transmission front shafts and Dual Power clutch. Shaft (30) and tapered roller bearing (29) must be installed from rear.

F. Filter
1. Front support
2. Gasket
3. Oil seal
4. Ball bearing
5. Oil seal
6. Shaft & gear (16 teeth)
7. Front cover
8. Gasket
9. Thrust collar (various thickness)
10. Ball bearing
11. Pto gear (43 teeth)
12. Ball bearing
13. Front housing
14. Gasket
15. Longer screws (2)
16. Seal rings (3)
17. Ball bearing
17A. Thrust washer
18. Input shaft
19. Needle bearing
20. Thrust washer (w/2 pins)
21. Snap ring
22. Direct drive clutch hub
23F. Drive plates (6-F)
23R. Drive plates (5-R)
24F. Friction discs (4-F)
24R. Friction discs (5-R)
25F. Snap ring
25R. Snap ring
26F. Spring retainer
26R. Spring retainer
27F. Release spring
27R. Release spring
28F. Piston
28R. Piston
29. Tapered roller bearing
30. Rear input shaft
31F. Seal ring
31R. Seal ring
32F. Seal ring
32R. Seal ring
33. Clutch drum (body)
34. Spacer plate
35. Clutch plate
36F. Snap rings
36R. Snap rings
37. Seal rings (3)
38. Seal rings (3)
39. Thrust washer (w/2 pins)
40. Clutch hub
41. Spacer collar
42. Needle bearings
43. Gear (25-25 teeth)
44. Thrust washer
45. By-pass valve seat
46. By-pass valve ball
47. By-pass valve spring
48. Shims
49. "O" ring
50. Cap
51. Regulating valve spool
52. Shims
53. Regulating valve spring
54. "O" ring
55. Cap
56. Hydraulic pump
57. Seal
58. Delivery tube
59. "O" ring
60. Valve
61. "O" ring
62. Solenoid
63. Plug & "O" ring
64. Gasket

that provide correct clutch clearance, remove snap ring (36F), input shaft (18) and clutch plates and discs (23F and 24F) so that clutch can be installed.

Install parts (41 through 44) on rear input shaft (30). Position thrust washer (39) on pins in center of clutch drum (33) and install hub (40), engaging hub with all of the friction discs (24R). Install new seal rings (38) in the three grooves. Make sure that seal rings are latched and use light grease to lubricate and center the seals on shaft. Carefully slide the partially assembled clutch onto shaft, over seal rings.

NOTE: DO NOT force. If difficulty is encountered, make sure that seal ring has not fallen from groove or thrust washer is not engaging pins correctly. Attempting to force will only damage parts and will not correct problem.

Install snap ring (21) and hub (22). Install bearing (17) on front of input shaft (18), then install bearing (19) and thrust washer (20) in rear of input shaft. Install seal rings (37) in grooves of shaft (30). Make sure that seal rings are latched and use light grease to lubricate and center the seals on shaft. Carefully slide the input shaft (18) over seal rings (37).

NOTE: DO NOT force. If difficulty is encountered, make sure that seal ring has not fallen from groove or thrust washer is not engaging pins correctly. Attempting to force will only damage parts and will not correct problem.

Install snap ring (36F). Install thrust washer (17A) and seal rings (16) on input shaft. Make sure that seal rings are latched and use light grease to lubricate and center the seals on shaft. Carefully install the front housing (13) over shaft (18), seal rings (16) and bearing (17). Refer to paragraph 134 for remainder of reassembly.

REAR COVER PLATE, OUTPUT SHAFT AND COUNTERSHAFT

Dual Power Shuttle Shift Models

136. R&R AND OVERHAUL. Remove the transmission as outlined in paragraph 120, the shift cover as outlined in paragraph 131 or 132 and the shift forks, gates and rails as outlined in paragraph 133. Remove the snap ring (76—Fig. 145) and withdraw pto shaft (72) and bearing (75). Refer to paragraph 134 and remove the transmission front support, front cover and front housing. Refer to paragraph 135 and remove the Dual Power clutch and idler gear (43—Fig. 144).

If wear is suspected, check gear backlash with a dial indicator as shown in Fig. 146. Recommended backlash is 0.1-0.2 mm (0.004-0.008 inch).

Before removing the rear cover plate, tie the output shaft (Fig. 147) and countershaft (Fig. 148) together with a length of cord so shafts will not fall and be damaged. Remove the rear cover plate (51—Fig. 147 and Fig. 148). Carefully remove the cord used to tie the output shaft and countershaft together, then lift top (output) shaft as an assembly from the housing.

NOTE: Be careful when removing the output shaft to hold the front synchronizer assembly (39—Fig. 147) together, so that it will not be bumped and fall apart.

Remove the rear input shaft (30) through opening for shift cover opening in top of transmission housing.

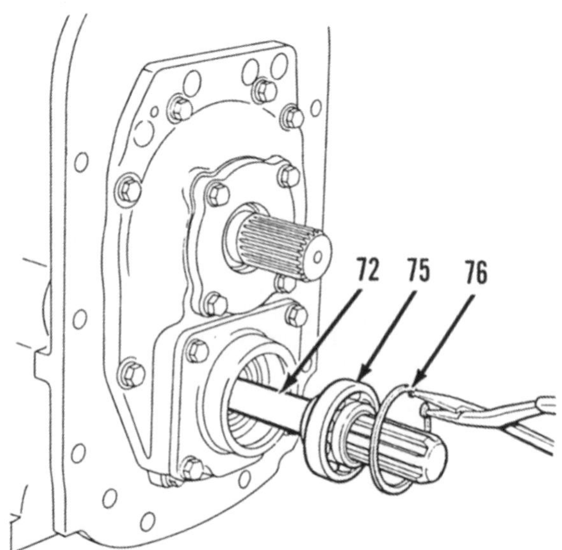

Fig. 145—Snap ring (76) must be removed before withdrawing the pto shaft (72) and bearing (75).

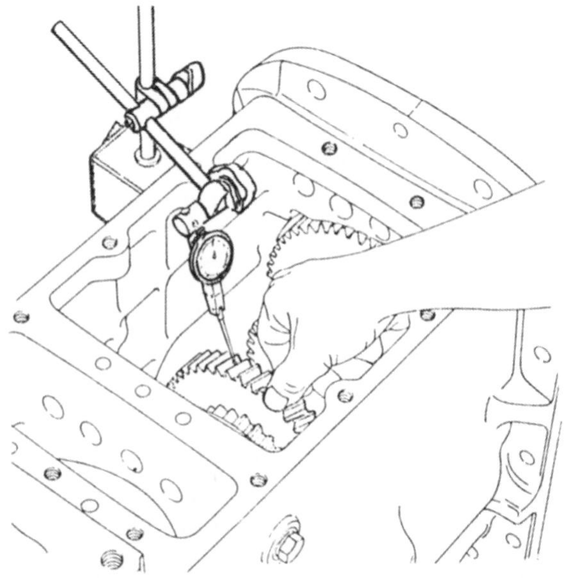

Fig. 146—Gear backlash should be checked before removing rear cover plate using a dial indicator as shown.

Remove the countershaft (Fig. 148) from the rear of transmission housing.

Remove screw (81—Fig. 149) retaining reverse idler shaft (95—Fig. 150), remove the snap rings (91 and 92F), then slide the shaft (95) out toward rear. Catch the gear (94), bearings (93) and snap ring (92R) as shaft is withdrawn.

Synchronizers (45—Fig. 147, 65—Fig. 148 and 87—Fig. 148) are alike. Synchronizer (39—Fig. 147) is different from the others. Parts of synchronizers are not available separately and if any part is lost or damaged, the complete unit should be renewed.

137. The tapered roller bearings for both output shaft and countershaft are adjusted by changing the thickness of shims (53—Fig. 147 and 69—Fig. 148). To determine the correct thickness of shims, proceed as follows:

Assemble countershaft parts (56 through 68 and 82 through 89—Fig. 148) and install in transmission housing, with no other parts installed. Unbolt and remove retainer (71), making sure that shims (69) are not damaged if they are to be reused. Install rear cover plate (51) and tighten the retaining screws to 52 N•m (38 ft.-lbs.) torque. Install the retainer (71) with cup for bearing (68) and at least the same thickness of shims (69) that were originally installed, but do not install "O" ring (70) yet. Tighten screws attaching the retainer (71) finger tight, while making sure that bearing cones correctly seat in bearing cups, then finish tightening screws to 52 N•m (38 ft.-lbs.) torque. If bearings bind when tightening screws, install additional thickness of shims (69) before continuing. Wrap a cord around shaft (67) and attach a spring scale as shown in Fig. 151. Check pull required to rotate shaft by pulling cord (C) with a spring scale. Correct bearing preload will permit shaft to rotate with 40-53 N (9-12 lbs.) of pull when measured as shown in Fig. 151. Add or remove shims (69—Fig. 148) as required, making sure that screws attaching bearing retainer (71) and rear cover plate (51) are tightened to 52 N•m (38 ft.-lbs.) torque while checking. When correct thickness of shims (69) has been selected, remove retainer (71) with correct thickness of shims, install new "O" ring (70), then reinstall and recheck bearing adjustment. Unbolt and remove the rear cover plate (51), then lift the complete countershaft assembly out of the housing. Set the countershaft aside while checking and adjusting bearings of output shaft.

Preload of output shaft bearings should be checked with only bearing (29—Fig. 147), the rear input shaft

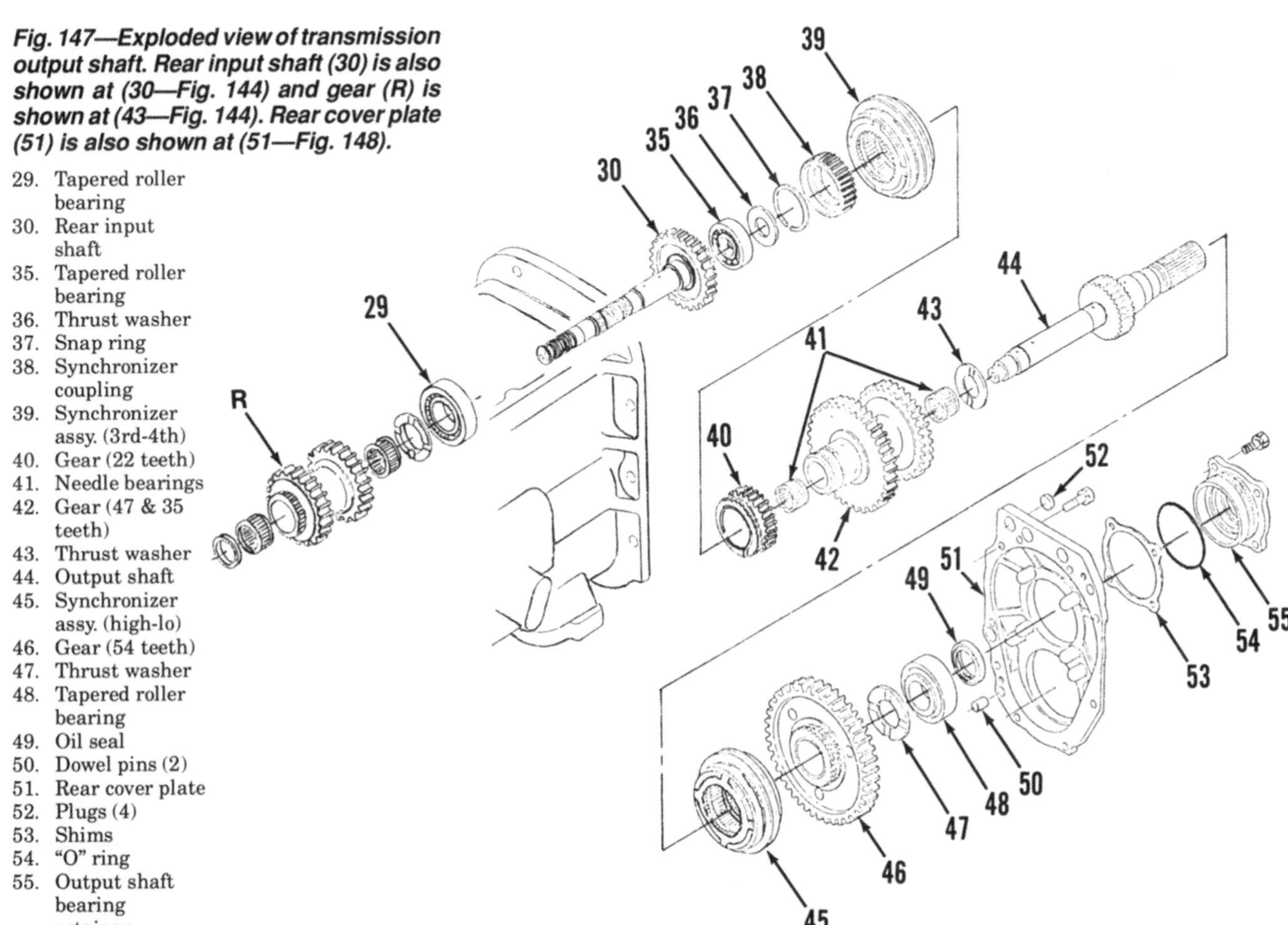

Fig. 147—Exploded view of transmission output shaft. Rear input shaft (30) is also shown at (30—Fig. 144) and gear (R) is shown at (43—Fig. 144). Rear cover plate (51) is also shown at (51—Fig. 148).

29. Tapered roller bearing
30. Rear input shaft
35. Tapered roller bearing
36. Thrust washer
37. Snap ring
38. Synchronizer coupling
39. Synchronizer assy. (3rd-4th)
40. Gear (22 teeth)
41. Needle bearings
42. Gear (47 & 35 teeth)
43. Thrust washer
44. Output shaft
45. Synchronizer assy. (high-lo)
46. Gear (54 teeth)
47. Thrust washer
48. Tapered roller bearing
49. Oil seal
50. Dowel pins (2)
51. Rear cover plate
52. Plugs (4)
53. Shims
54. "O" ring
55. Output shaft bearing retainer

(30) and output shaft assembly (35 through 48) installed in the transmission housing. Actually, synchronizer (39) should not be installed while checking, because of possible inaccurate reading and to facilitate temporary installation. Install bearing (29) and rear input shaft (30). Assemble output shaft parts (35 through 49) and install in transmission housing. Unbolt and remove retainer (55), making sure that shims (53) are not damaged if they are to be reused. Install rear cover plate (51) and tighten the retaining screws to 52 N•m (38 ft.-lbs.) torque. Install the retainer (55) with cup for bearing (48) and at least the same thickness of shims (53) that were originally installed, but do not install "O" ring (54) or seal (49) yet. Tighten screws attaching the retainer (55) finger tight, while making sure that bearing cones correctly seat in bearing cups, then finish tightening screws to 52 N•m (38 ft.-lbs.) torque. If bearings bind when tightening screws, install additional thickness of shims (53) before continuing. Wrap a cord around shaft (44) and attach a spring scale as shown in Fig. 152. Check pull required to rotate shaft by pulling cord (C) with a spring scale (S). Correct bearing preload will permit shaft to rotate with 27-40 N (6-9 lbs.) of pull when measured as shown in Fig. 152. Add or remove shims (53—Fig. 147) as required, making sure that screws attaching bearing retainer (55) and rear cover plate (51) are tightened to 52 N•m (38 ft.-lbs.) torque while checking. When correct thickness of shims (53) has been selected, remove retainer (55) with correct thickness of shims, install new "O" ring (54) and new oil seal (49), then reinstall and

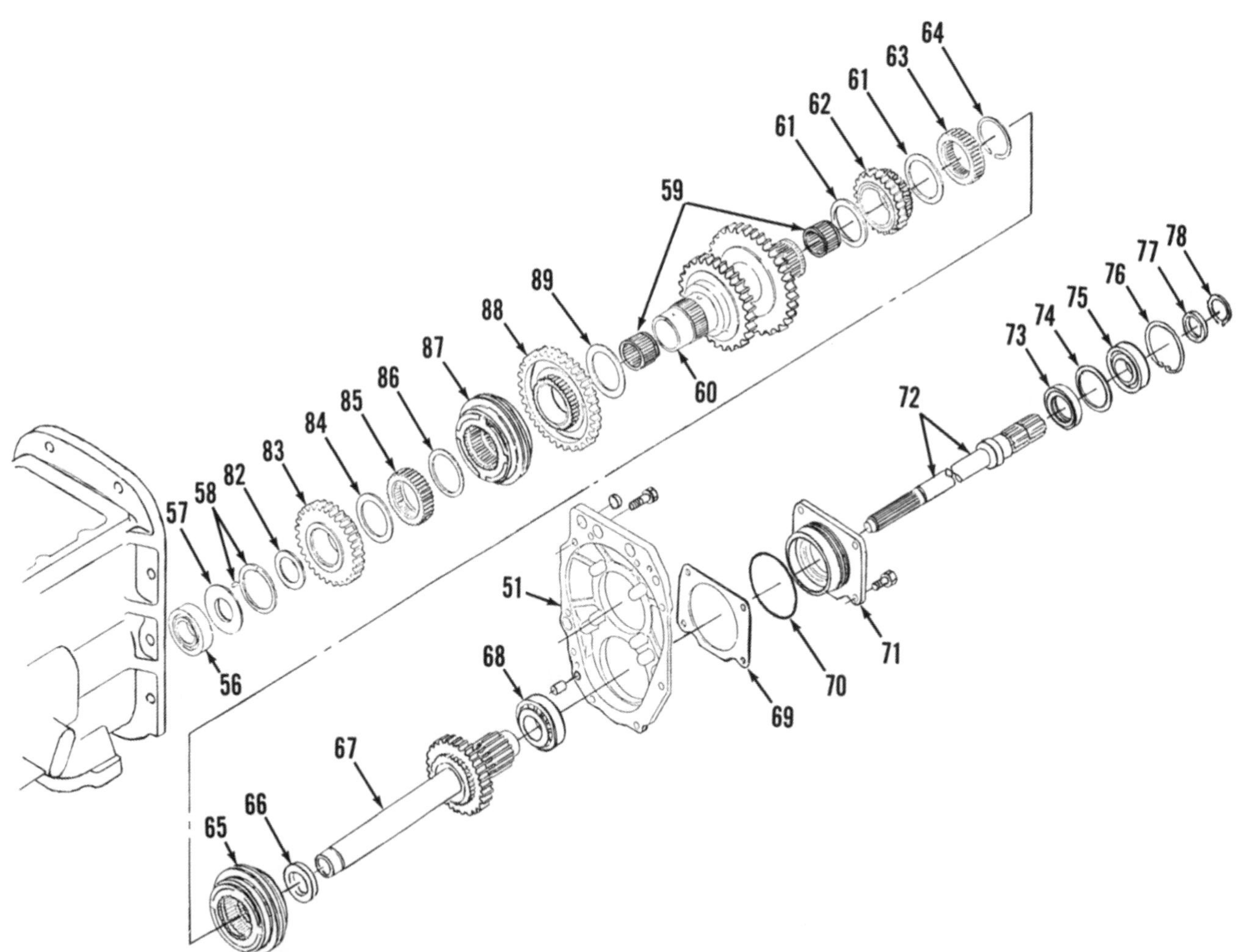

Fig. 148—Exploded view of countershaft (lower) used with Dual Power (16 X 8), synchronized transmission. Rear cover plate (51) is also shown at (51—Fig. 147).

51. Rear cover plate
56. Tapered roller bearing
57. Thrust washer
58. Thrust bearing (w/2 pins)
59. Needle bearings
60. Gear (35 & 45 teeth)
61. Thrust washers
62. Gear (29 teeth)
63. Synchronizer coupling
64. Snap ring
65. Synchronizer (1st-2nd)
66. Thrust bearing
67. Shaft and gears (14 & 32 teeth)
68. Tapered roller bearing
69. Shims
70. "O" ring
71. Countershaft bearing retainer
72. Pto shaft
73. Oil seal
74. Collar
75. Ball bearing
76. Snap ring
77. Collar
78. Snap ring
82. Thrust washer
83. Gear (35 teeth)
84. Thrust washer
85. Synchronizer coupling
86. Thrust washer
87. Forward-reverse synchronizer
88. Gear (43 teeth)
89. Thrust washer

recheck bearing adjustment. Unbolt and remove the rear cover plate (51), then lift the output shaft out of the housing.

138. Observe the following when assembling. Install the reverse idler assembly (Fig. 150), with hole in shaft (95) for retaining screw aligned with hole in housing for screw (81). Install and tighten retaining screw (81) to 68 N.m (50 ft.-lbs.) torque.

Stand the transmission housing on front face with rear of housing up. Position the countershaft assembly (56 through 68 and 82 through 89—Fig. 148) in housing with weight resting in cup of front bearing (56). Use two 17 mm diameter shafts (D—Fig. 153) about 50 mm long to hold detent ball against spring in the 1st-2nd lower shift fork (17—Fig. 139) and the forward-reverse shift fork (20). Position the shift

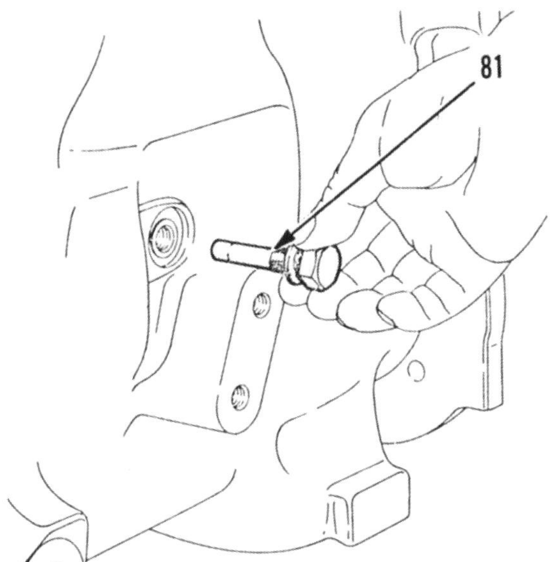

Fig. 149—The reverse idler shaft is retained by screw (81) which is also shown in Fig. 150.

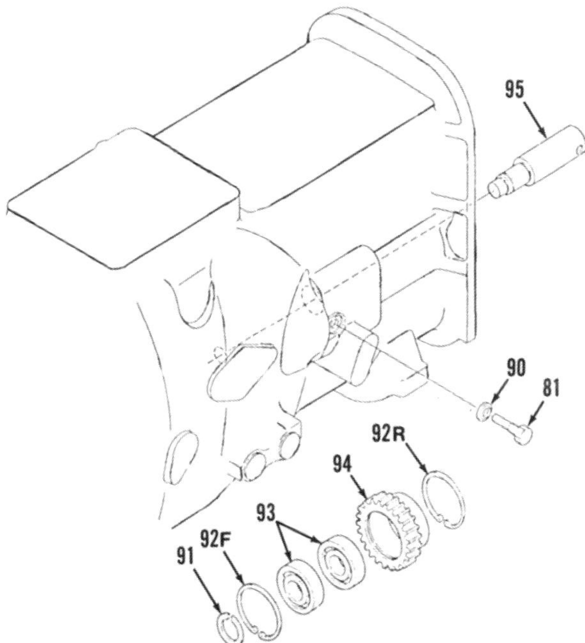

Fig. 150—Exploded view of reverse idler assembly. Gear (94) operates against front gear (R—Fig. 147) and gear (83—Fig. 148).

81. Retaining screw
90. Sealing washer
91. Snap ring
92F. Snap ring
92R. Snap ring
93. Ball bearings
94. Reverse idler gear
95. Reverse idler shaft

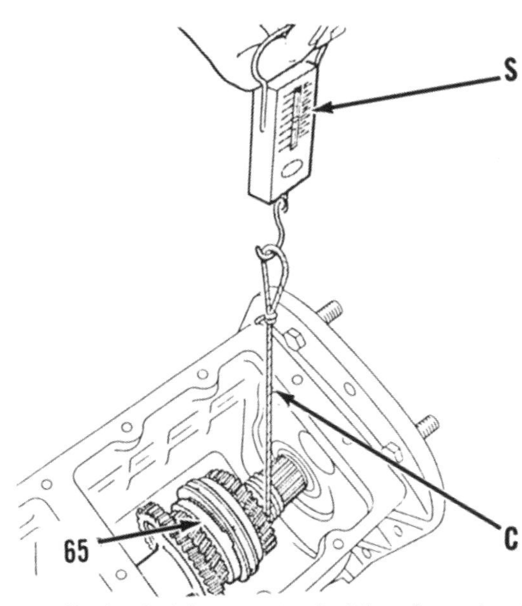

Fig. 151—Preload of the countershaft bearings should be measured using a cord (C) wrapped around the shaft and a spring scale (S) as shown. Other shafts should be removed when checking.

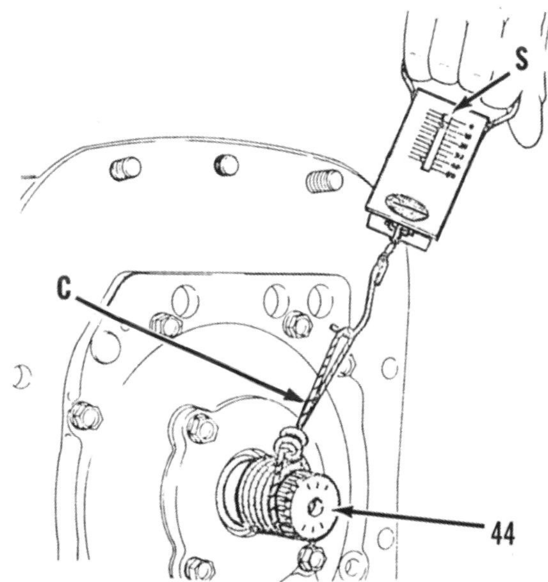

Fig. 152—Preload of the output shaft bearings should be measured using a cord (C) wrapped around the shaft and a spring scale (S) as shown. Countershaft should be removed when checking.

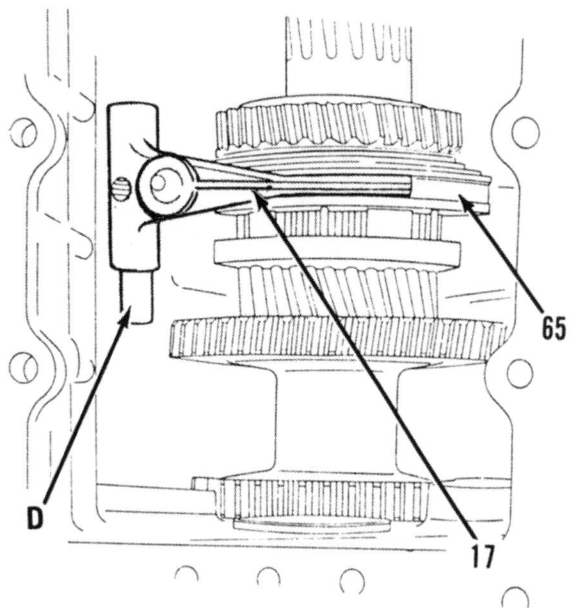

Fig. 153—Special dummy shaft (D) can be used to compress the detent in shift fork while installing.

forks (17 and 20) in housing and in groove of synchronizers (65 and 87—Fig. 148), then slide the lower shift rail (5—Fig. 139) through holes in housing and into both shift forks. Catch the dummy shaft as it is replaced in fork by rail.

Position the rear input shaft (30—Fig. 147) and bearing cone (29) in cup for bearing (29), then carefully position the mainshaft assembly (35 through 48) against the cup for bearing (35). Coat mating surface of the transmission housing with sealer and install the rear cover plate (51). Before tightening the retaining screws, make sure that parts are correctly positioned and do not bind. Tighten screws attaching the rear cover plate to 52 N•m (38 ft.-lbs.) torque and recheck to be sure that parts do not bind and that correct gear engagement is possible. Refer to paragraph 135 and install the Dual Power clutch and idler gear (43—Fig. 144). Refer to paragraph 134 and install the transmission front support, front cover and front housing. Refer to paragraph 133 and install the shift rails and forks, then check for correct shifting before proceeding further.

DIFFERENTIAL, MAIN DRIVE BEVEL GEARS, FINAL DRIVE AND REAR AXLE

SPLIT BETWEEN TRANSMISSION AND REAR AXLE CENTER HOUSING

Models Without Cab

139. To separate the tractor between the rear of the transmission and the rear axle center housing, disconnect the battery and drain oil from the rear axle center housing. Disconnect rear wiring harness at coupling located on right side at base of steering wheel shaft. Remove clips securing rear wiring harness to transmission housing and lay harness out of the way. Unbolt right and left step plates from the transmission housing. Disconnect clutch operating linkage from release arm and low exhaust pipe (where fitted) from the exhaust manifold.

Attach splitting fixture or support rear assembly and front assembly separately. Block front assembly securely to prevent tipping or rolling. Rear assembly should be supported using a hoist, splitting fixture or other suitable device to permit the rear assembly to be safely moved away from the front assembly. Remove bolts attaching the rear axle center housing to the transmission housing and carefully roll the rear assembly rearward.

To reconnect the rear assembly to the transmission housing, reverse the splitting procedure. Tighten bolts (4—Fig. 154) securing the rear axle center housing to the transmission housing to 104-142 N•m (77-105 ft.-lbs.) torque and bolts (6) to 65-87 N•m (48-64 ft.-lbs.) torque.

Models With Cab

140. It is necessary to remove the tractor cab as outlined in paragraph 212 or tip the cab forward as described in paragraph 211, before the rear axle center housing can be separated from the transmission.

Deciding whether to tilt the cab or to remove the cab will depend upon: What equipment and space is available to remove and store the cab while it is off. Is sufficient equipment available to safely support the cab while it is still attached to the front assembly. Be extremely careful, because either tipped or removed, the cab is extremely susceptible to extensive damage. Another consideration is the specific work to be performed following the split.

Refer to paragraph 139 to complete separation between rear axle center housing and transmission.

DIFFERENTIAL AND BEVEL GEARS

All Models

141. R&R DIFFERENTIAL ASSEMBLY. To remove the differential assembly, the left axle and housing must be removed as outlined in paragraph 149.

With left axle removed, remove the differential assembly from the rear axle center housing. Refer to paragraph 150 for overhaul procedure. Refer to paragraph 142 if carrier bearing adjustment is necessary. Refer to paragraph 144 to remove the bevel pinion.

142. DIFFERENTIAL CARRIER BEARING ADJUSTMENT. The differential carrier bearing preload of 0.53-0.71 mm (0.021-0.028 inch) is adjusted by changing the thickness of shims (22—Fig. 155). Correct measurement requires the use of special tools **Churchill No. FT.4501 (SW.7) or Nuday No. 2141 (SW.505).** Correct selection of shims will be difficult if the special tools are not available.

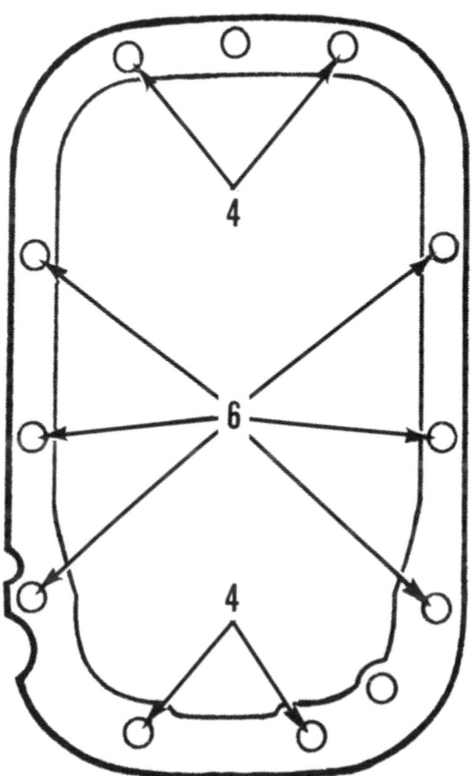

Fig. 154—Bolts (4) securing the rear axle center housing to the transmission housing should be tightened to 104-142 N•m (77-105 ft.-lbs.) torque and bolts (6) should be tightened to 65-87 N•m (48-64 ft.-lbs.) torque.

Carrier position is established by the position of the left bearing, which is installed without shims. To determine correct thickness of shims (22), assemble the right axle and inner brake housing (21) without bearing cup (38). Position the gage block in bearing bore and attach the special gage as shown in Fig. 156. Measure clearance between the gage block and special tool with a feeler gage as shown, then remove the special tool and gage block. Install shims (22—Fig. 155) equal to the measured clearance plus an additional 0.53-0.71 mm (0.021-0.028 inch) and bearing race (38).

143. OVERHAUL DIFFERENTIAL. Differential can be removed as outlined in paragraph 141. Remove snap ring (43—Fig. 155), washer (42), coupling (41), spring (40) and adapter (39). Mark the two halves of the differential case (25 and 36) so they can be reassembled in same relative position, then unbolt and remove the right case half (36). Remove right thrust washer (30), right side gear (35), spider (34), pinions (33), washers (32), left side gear (31) and left thrust washer (30). Check differential carrier bearings (23, 24, 37 and 38) and renew if damaged.

The bevel ring gear (28) may be riveted to the differential left half (25) during factory assembly, of some models. If necessary to renew either case or gear, remove rivet heads with a ½-inch (12.7 mm) drill bit, then remove rivets with a punch. Assemble new ring gear and/or differential case with special screws (27) and nuts (26). Be sure that ring gear is not cocked and is fully seated on differential case, then tighten nuts to 54-61 N•m (40-45 ft.-lbs.) torque.

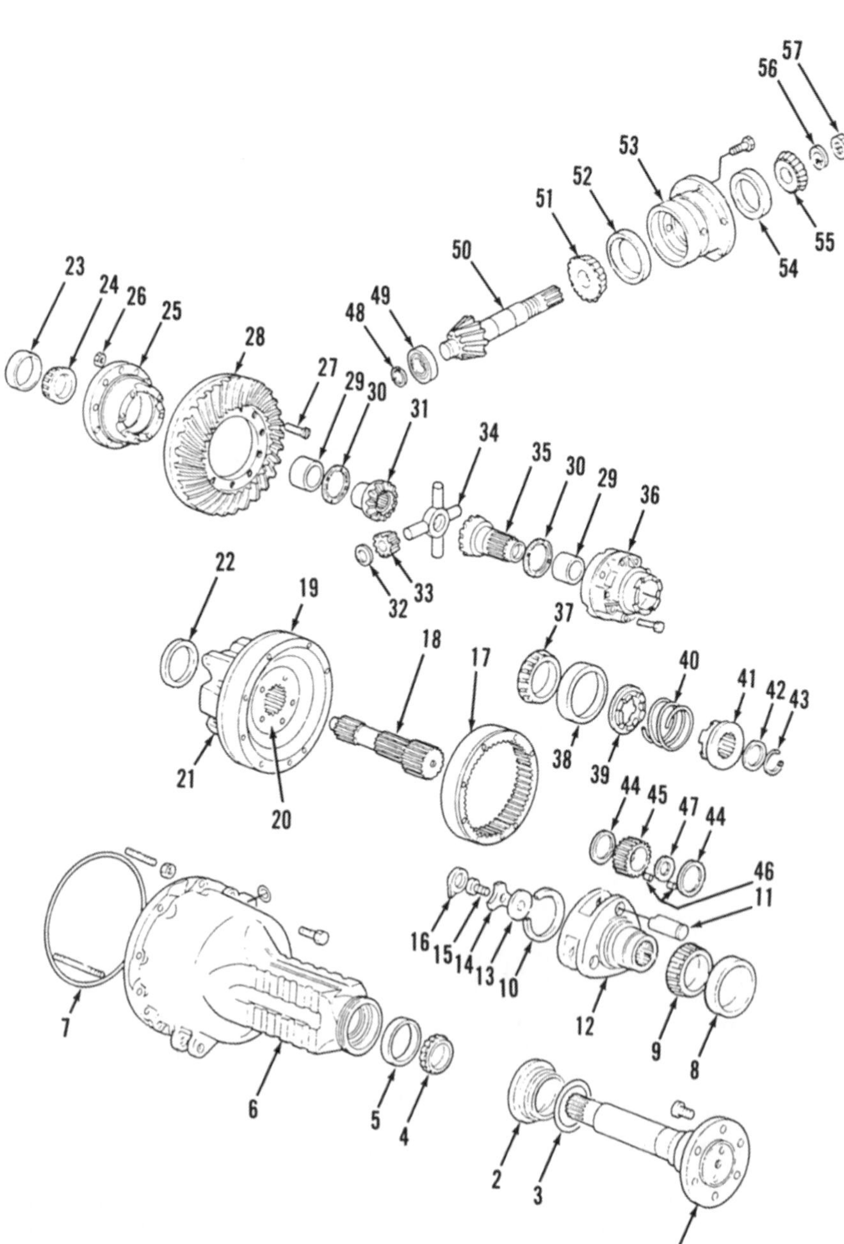

Fig. 155—Exploded view of differential, final drive, brake and rear axle assemblies typical of all models. Parts (19, 20 & 21) are shown separated in Fig. 168.

1. Axle shaft
2. Oil seal
3. Seal
4. Bearing cone
5. Bearing cup
6. Axle housing
7. "O" ring
8. Bearing cup
9. Bearing cone
10. Snap ring
11. Planetary pinion shaft
12. Planetary carrier
13. Spacer shim
14. Retaining washer
15. Cap screw
16. Lock plate
17. Planetary ring gear
18. Planetary sun gear
19. Outer brake housing
20. Brake assembly
21. Inner brake housing
22. Shim
23. Bearing cup
24. Bearing cone
25. L.H. differential case half
26. Ring gear retaining nuts
27. Ring gear retaining screws
28. Bevel ring gear
29. Bushings (2)
30. Thrust washers (2)
31. Differential side gear
32. Thrust washers (4)
33. Differential pinions (4)
34. Differential spider
35. Differential side gear with lock
36. R.H. differential case with lock
37. Bearing cone
38. Bearing cup
39. Adapter
40. Spring
41. Coupling
42. Washer
43. Snap ring
44. Thrust washers
45. Planetary pinions
46. Needle rollers
47. Spacer washers
48. Snap ring
49. Pilot bearing
50. Bevel pinion
51. Bearing cone
52. Bearing cup
53. Pinion bearing retainer
54. Bearing cup
55. Bearing cone
56. Locking washer
57. Hex nut

Bevel ring gear (28) and bevel pinion (50) are available as a matched set only. When installing new ring gear, refer to paragraph 144 and install the bevel pinion.

Install new thrust washers (30 and 32) if worn or scored. Differential case bushings (29) may be renewed if case halves are otherwise serviceable. Bushings are presized and should not require reaming if carefully installed.

Reassemble differential by reversing disassembly procedure and tighten differential case retaining screws to 88-101 N•m (65-75 ft.-lbs.) torque.

144. BEVEL PINION. To remove the main drive bevel pinion, first remove the differential assembly as outlined in paragraph 141, hydraulic lift cover as outlined in paragraph 178 and hydraulic pump as outlined in paragraph 197. Split the tractor between transmission and rear axle center housing as outlined in paragraph 139 or 140.

Remove cap screws and lockwashers retaining bevel pinion bearing carrier (53—Fig. 155 or Fig. 157) to rear axle center housing, then remove the carrier and pinion assembly using jackscrews threaded into holes in the flange of bevel pinion bearing carrier.

Straighten locking washer (56) and remove nut (57). Press shaft (50) and bearing (51) out of bearing (55), carrier (53) and associated parts. The bevel pinion is available only in a matched set with ring gear (28—Fig. 155). If pinion is renewed, refer to paragraph 143 and install matching ring gear.

Install new bearings (51, 52, 54 and 55—Fig. 155 or Fig. 157) if rough or otherwise damaged. Adjust preload of pinion bearings by tightening nut (57). Measure preload by wrapping a cord around the pinion shaft, attach a spring scale to the cord, then measure the amount of pull to rotate the pinion shaft. Correct bearing preload will require 7.2-9.5 kg. (16-21 lbs.) pull to rotate pinion shaft. When bearing preload is correct, lock position of nut (57) by bending washer (56) against one of the flats. Recheck bearing adjustment after locking with washer (56). Install pinion and carrier assembly, tightening retaining screws to 136-167 N•m (100-125 ft.-lbs.) torque.

145. ADJUST DIFFERENTIAL LOCK. To adjust the differential lock linkage on models not equipped with cab, refer to paragraph 146. Refer to paragraph 147 for models equipped with a cab.

146. To adjust the differential lock linkage on models without cab, refer to Fig. 158. Disconnect spring-loaded link (77) from operating lever (76) and loosen locknut (78). Jack up the right rear wheel, push down on operating lever (76), turn rear wheel until clutch dog is aligned and differential lock is fully engaged. Let foot pedal (80) rest on platform (foot rest) and adjust length of link (77) so that link can just be reconnected. Shorten link (77) one turn and reinstall pin, install cotter pin and tighten locknut (78).

147. To adjust differential lock on models with cab, refer to Fig. 159. Disconnect spring-loaded link (77) and loosen locknut (78). Turn clevis (79) until distance between attaching holes in spring-loaded link (77) and clevis (79) is 263.5 mm (10⅜ inches). Tighten locknut (78) and attach link (77). Loosen locknut (84) and turn pedal (87) until clearance between bottom of pedal and top of sleeve (85) is 32 mm (1¼ inches).

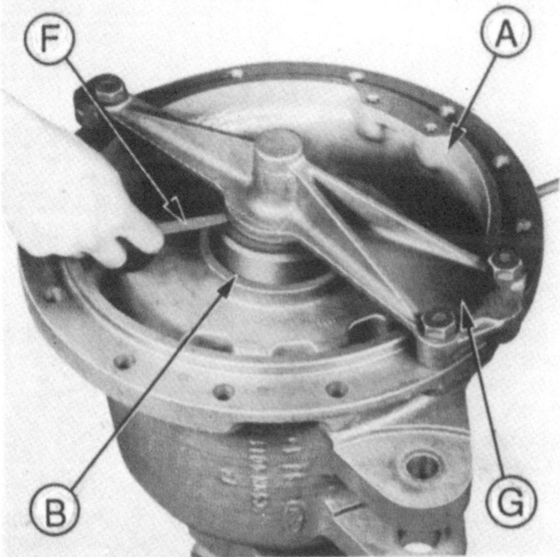

Fig. 156—Special tools are required to measure clearance shown for selecting proper thickness of shims when setting carrier bearing preload. Refer to text.

A. Right axle housing
B. Gage block
F. Feeler gage
G. Special tool SW.7

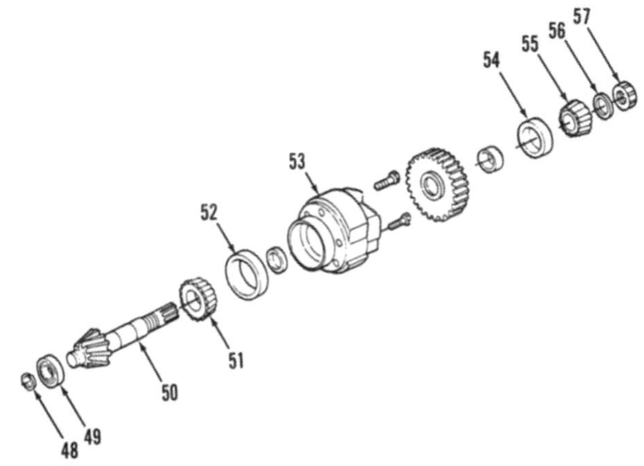

Fig. 157—Exploded view of bevel pinion used on some models with four-wheel drive. Refer to Fig. 155 for models without four-wheel drive.

48. Snap ring
49. Pilot bearing
50. Bevel pinion
51. Bearing cone
52. Bearing cup
53. Pinion bearing retainer
54. Bearing cup
55. Bearing cone
56. Locking washer

Tighten locknut (84) and recheck clearance between bottom of pedal and sleeve (85).

148. OVERHAUL DIFFERENTIAL LOCK. The differential lock can be removed from differential assembly after removing the right axle and housing as outlined in paragraph 149. Parts (39 through 42—Fig. 155) can be removed after removing snap ring (43). Refer to paragraph 143 if differential overhaul is required. Refer to Fig. 158 or Fig. 159 for service to operating linkage.

Tighten screw (70) to 33-40 N•m (24-30 ft.-lbs.) torque, then lock by tightening locknut (71). Refer to paragraph 146 (models without cab) or paragraph 147 (models with cab) for adjusting operating linkage.

REAR AXLE AND FINAL DRIVE

All Models

149. REMOVE AND REINSTALL. Remove fender from models without cab. On models with cab, support rear of cab and remove cab mounting bracket from rear axle being removed. If both axle housings are to be removed from tractor equipped with a cab, it is suggested that cab be tilted as outlined in paragraph 211 or removed as outlined in paragraph 212.

On all models, drain lubricant and disconnect brake linkage. Support rear axle center housing and remove rear wheel. Support axle and housing, remove retaining screws and stud nuts, then carefully remove axle and housing assembly. If the left housing is removed, the differential should be carefully removed and stored to prevent assembly from accidentally falling.

Position new "O"ring seal (7—Fig. 155) on axle housing before installing. Tighten axle housing retaining screws to 176-230 N•m (130-170 ft.-lbs.) torque and stud nuts to 210-271 N•m (155-200 ft.-lbs.) torque.

150. OVERHAUL. Remove the rear axle as outlined in paragraph 149 and proceed as follows. If the left housing is removed, the differential should be carefully removed and stored to prevent assembly from accidentally falling. If right axle and housing are to be overhauled, first loosen jam nut (71—Fig. 158 or Fig. 159) and remove screw (70), then remove shaft (73) and fork (72).

Proceed with service as follows for both sides. Remove stud nuts attaching the brake inner housing (21—Fig. 155) to the axle housing (6) and remove inner brake housing, brake assembly (20) and brake outer housing (19). Refer to paragraph 154 if brake assembly is to be serviced. Lift out final drive sun

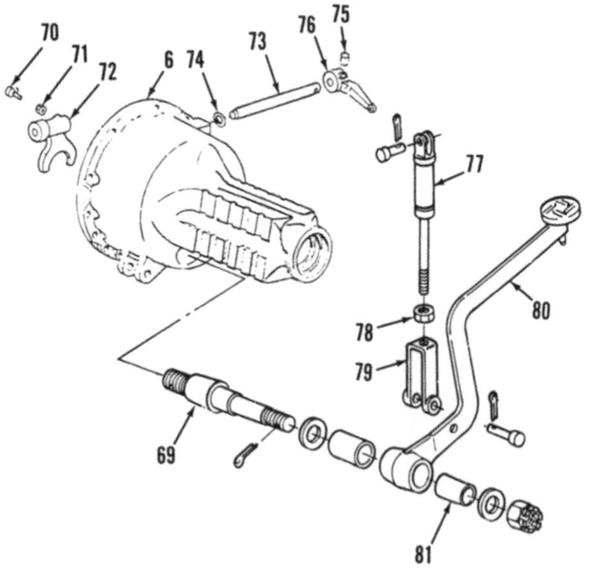

Fig. 158—View of differential lock linkage for models without cab.

6. Axle housing
69. Pivot shaft
70. Set screw
71. Jam nut
72. Fork
73. Shaft
74. Oil seal
75. Pin
76. Arm
77. Spring-loaded link
78. Jam nut
79. Clevis
80. Pedal
81. Bushing

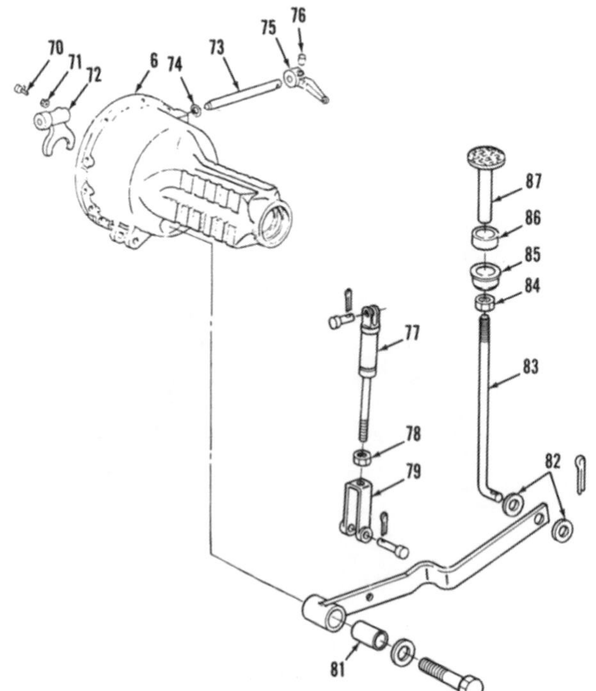

Fig. 159—View of differential lock linkage used on models with cab. Refer to Fig. 158 for legend except the following.

82. Washers
83. Rod
84. Locknut
85. Sleeve
86. Bushing
87. Pedal

gear (18) and locking plate (16), then remove screw (15). Remove the planetary assembly (9, 10, 11, 12, 44, 45, 46 and 47), retainer (14) and spacer shim (13) from axle and housing. Drive seal (2) from housing, then withdraw axle.

Remove snap ring (10) from planetary, then bearing (9) and shafts (11) can be pressed from planetary housing as shown in Fig. 160. Be careful not to lose, damage or mix bearing rollers (46—Fig. 155) as planetary gears (45), washers (44 and 47) and rollers (46) are removed. There are two rows of 16 rollers in each planet gear (total of 32 in each gear).

Assemble planetary unit by reversing the disassembly procedure. Use grease to hold 16 bearing rollers (46) on each side of spacer washer (47). Make sure that rollers are against gear (45) so that shaft (11) can be pressed through. Locate gear (45), rollers (46), washer (47) and washers (44) in planet carrier and press shaft (11) into position with groove for snap ring (10) toward center. Install snap ring (10) in groove of the planet carrier and each of the three shafts (11), then bend the ends down into the depression provided in the planet carrier. Lubricate bearing (9) liberally when assembling.

If the final drive ring gear (17) must be removed, use an expanding removal tool such as the Nuday SW.6 shown in Fig. 161. Install tool under the ring gear and press the tool and ring gear from the outer end of the axle. The tool can be used for pressing new gear into position. Be sure gear is fully seated against housing and not cocked. Use a thin feeler gage to check for complete seating.

Use suitable bearing pullers to remove bearing cone (4—Fig. 155) from axle shaft and cups (5 and 8) from bores in housing. Remove all rust, paint, dirt or burrs from seal surfaces and inspect axle sealing surfaces. Coat seal contacting surfaces of axle and lip of new seal with grease, then install seal ring (3) and lip seal (2) over shaft. Lubricate bearing cone (4) liberally and press tightly on shaft against shoulder of axle. Seat bearing cups (5 and 8) firmly in bores of axle housing.

NOTE: Assembly may be easier if axle is placed on floor with splined inner end up. Place seals (2 and 3) over axle, then lower housing (6) with cups (5 and 8) in place over the axle, outer bearing and seals. Assembly in this manner permits easier and more accurate measurement of axle shaft end play necessary for correct adjustment.

Make sure bearing cone (9) is installed on planet carrier and lubricated, then install the planetary assembly into housing and onto axle shaft splines. Axle shaft end play (bearing preload) must be adjusted at this time as follows. Obtain the thickest shim (13) available and install this shim with retainer (14) and screw (15) on inner end of axle shaft. Tighten screw (15) to 217-745 N•m (160-550 ft.-lbs.) torque. Mount a dial indicator on axle housing with an extended plunger contacting the head of screw (15) as shown in Fig. 162 and zero the indicator. Lift the axle housing and note the dial indicator reading, which is the current end play. Lower axle housing and remove cap screw (15—Fig. 155), retainer (14) and

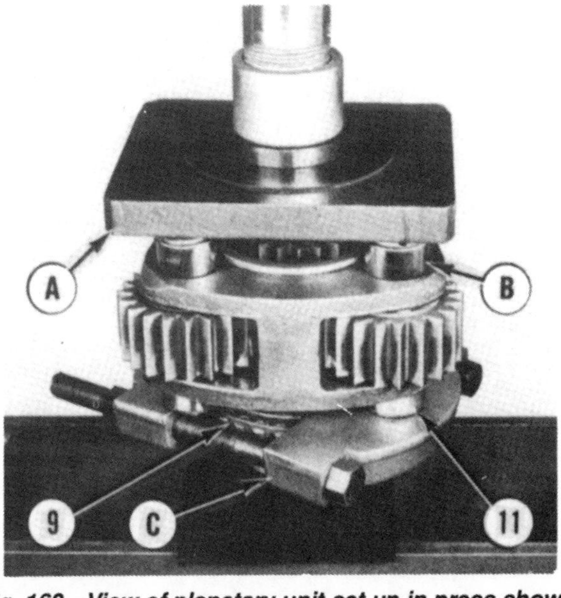

Fig. 160—View of planetary unit set up in press showing suggested method of removing the bearing cone (9) and shafts (11). Press against plate (A) which contacts the three spacers (B). The spacers (B) press against shafts (11), which press against bearing puller (C).

A. Plate
B. Spacers (3)
C. Bearing puller attachment

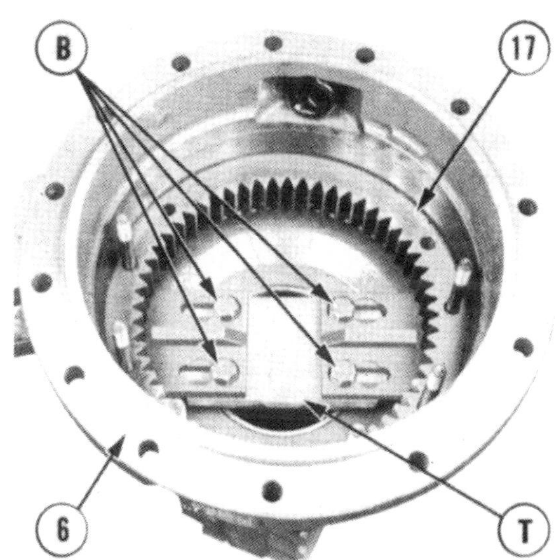

Fig. 161—View of special planetary ring gear tool (Nuday SW.6) positioned behind ring gear so that ring gear can be pressed from housing.

B. Bolts
T. Special tool (SW.6)
6. Housing
17. Ring gear

shim (13). Measure thickness of shim (13) used during test, then select a thinner shim whose thickness will reduce axle end play to a maximum of 0.025 mm (0.001 inch). Slight bearing preload is desirable, but should be less than 0.076 mm (0.003 inch). Shims of various thickness are available, from 1.245 mm (0.049 inch) to 2.337 mm (0.092 inch).

Install the selected shim, retainer and screw, tighten screw to 217-745 N•m (160-550 ft.-lbs.) torque and install locking plate (16). The wide range of torque is to permit tightening or loosening the screw as required to align locking plate.

Be sure that seal (2) is fully over the end of axle housing (6), then stake rim of seal into groove in at least four places equally spaced around housing. Refer to paragraph 154 to service brakes if required. Reinstall final drive sun gear, outer brake housing, brake assembly and inner brake housing. Tighten inner brake housing retaining screws to 97-122 N•m (72-90 ft.-lbs.) torque. Refer to paragraph 142 and adjust differential carrier bearings if ring gear (17), outer brake housing (19) or inner brake housing (21) on right side is renewed. Tighten screw (70—Fig. 158 or Fig. 159) to 33-40 N•m (24-30 ft.-lbs.) torque, then lock by tightening lock nut (71). Refer to paragraph 149 when reinstalling axle housing. Tighten axle housing retaining screws to 176-230 N•m (130-170 ft.-lbs.) torque and stud nuts to 210-271 N•m (155-200 ft.-lbs.) torque.

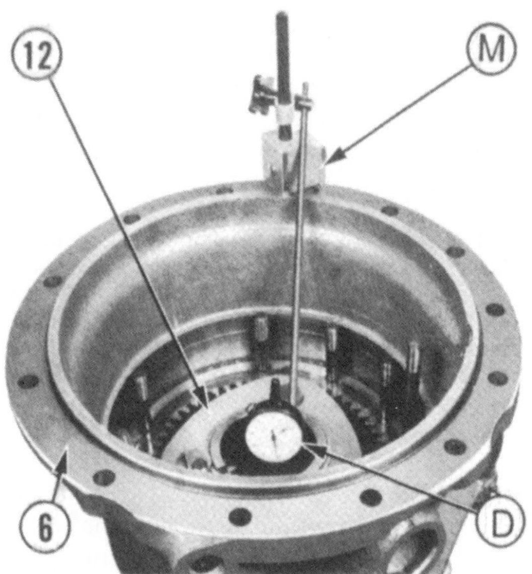

Fig. 162—View of dial indicator (D) mounted on the housing (M) for checking axle shaft end play.

6. Housing
12. Planet carrier

BRAKES

151. Rear wheel brakes are multiple disc wet-type located in the rear axle housings. The hand brake operates the same rear wheel brakes either through cables or connecting linkage. A disc brake is located in the front-wheel-drive transfer gearbox as shown in Fig. 22 of some models. The brake located in the front-wheel-drive gearbox is operated by a hand-operated lever.

ADJUSTMENT

All Models

152. PEDAL-OPERATED REAR WHEEL BRAKES. Various linkages may be used depending upon the model and whether a cab is installed, but all are similarly adjusted. Depress the right side pedal and measure the distance that pedal moves from released to fully depressed. Pedal travel should be approximately 32 mm (1¼ inches). If incorrect, loosen locknut (16—Fig. 163 or Fig. 164) and turn adjusting nut (17) until right side brake pedal travel is correct, then tighten locknut (16). Adjust travel for left brake pedal in a similar way, but lock and adjusting nuts are located on the left side. After left side has been adjusted, lock the pedals together and test for straight stopping. Change adjustment for left side brake if necessary.

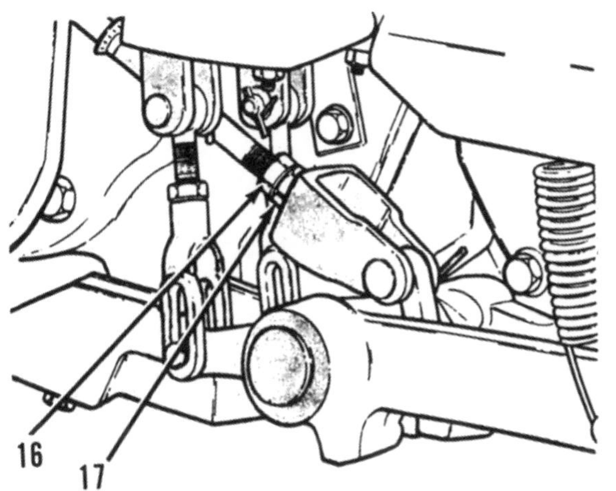

Fig. 163—View of right side brake adjusting locknut (16) and adjusting nut (17) typical of some models. Clevises with slots, also shown, are for mechanical (not cable) hand brake.

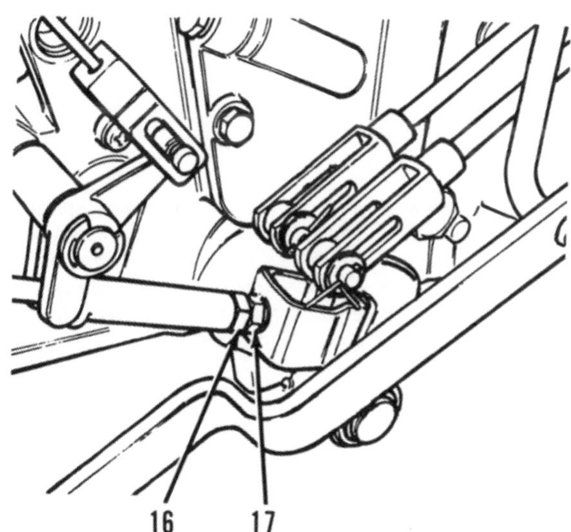

Fig. 164—View of right side brake adjusting locknut (16) and adjusting nut (17) typical of some models with cab.

153. HAND BRAKE. The hand brake operates the tractor rear brakes, but operation may be via mechanical link, one cable or two cables, depending upon model. Because the foot pedal and hand brakes interact, it is important to adjust the pedal-operated rear brakes as outlined in paragraph 152 before changing adjustment of the hand brake linkage.

To adjust mechanical linkage shown in Fig. 165, block the wheels so that tractor will not roll, release hand brake and proceed as follows. Loosen nuts (N), then remove cotter pins and clevis pins (P). Be sure that bellcranks (B) are pulled down and contact bosses on pedals, then turn clevis until the bottom of slot for pin (P) is aligned with the hole in bellcrank and install pin. Adjust the other clevis in the same way and install cotter pins in pins (P). Check to be sure that brakes for both sides are adjusted evenly.

To adjust hand brake operated by a single cable as shown in Fig. 166, block the wheels so that tractor will not roll, release hand brake and proceed as follows. Loosen nuts (N), then turn nuts as required to

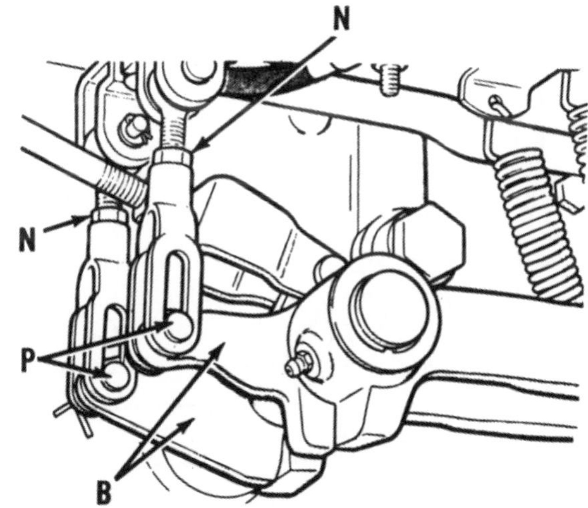

Fig. 165—View of mechanical hand brake linkage used on some models. Refer to text to adjust.

provide correct operation of the hand brake. Both nuts should be tightened against the bracket to maintain adjustment.

To adjust the hand brake operated by two cables as shown in Fig. 167, block the wheels so that tractor will not roll, release hand brake and proceed as follows: Loosen lower nuts (L), then turn upper nuts (N) as required to provide correct operation of the hand brake. Adjust both cables and check to be sure that brake operation is equal for both sides. Tighten lower nut (L) against the bracket to lock adjustment.

REAR BRAKE DISCS AND ACTUATING ASSEMBLY

All Models

154. R&R AND OVERHAUL. The multiple disc wet-type rear brakes are located in the rear axle housings. To gain access to brakes, remove axle assemblies as outlined in paragraph 149. When the left axle housing is removed, carefully remove and store the differential to prevent assembly from accidentally falling. On right axle housing, loosen jam nut (71—Fig. 158 or Fig. 159) and remove screw (70), then remove shaft (73) and differential lock fork (72).

Proceed with service for both sides as follows: Remove adjuster nut (17—Fig. 168), clevis (15) and nut (16). Remove stud nuts attaching brake inner housing (21) to axle housing and remove inner brake housing, brake assembly and brake outer housing (19). Remove seal (13) from axle housing if leaking or if new seal is to be installed. Some models are equipped with only one stationary disc (4) and three brake discs (3) instead of number shown in Fig. 168.

NOTE: Shims (22—Fig. 155) located in the right side inner brake housing (21) are used to adjust the differential carrier bearings. Do not lose shims if

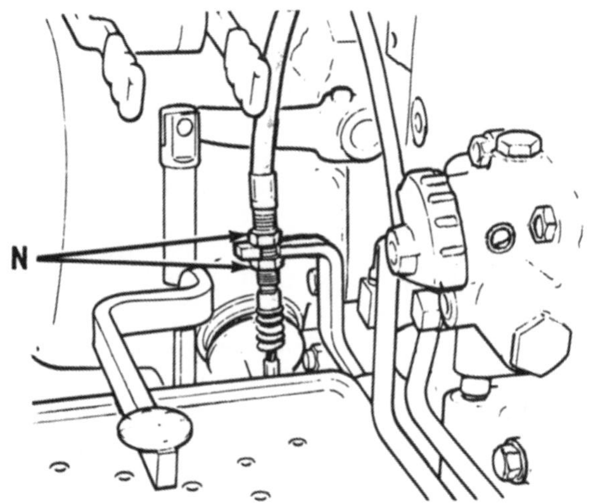

Fig. 166—View of hand brake adjustment nuts (N) for models which use only one cable. Refer to text.

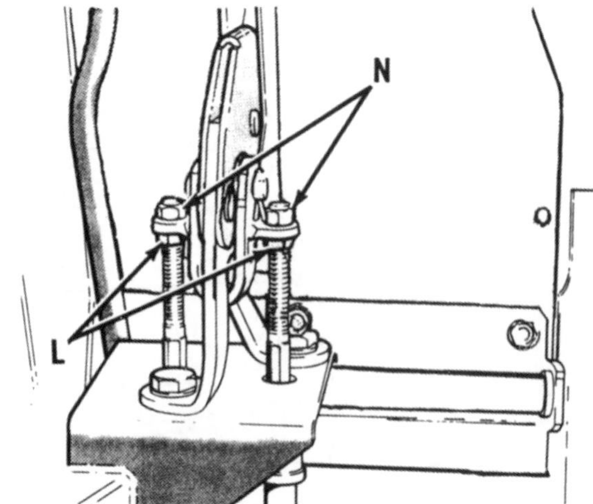

Fig. 167—View of the two hand brake cables used on some models. Refer to text.

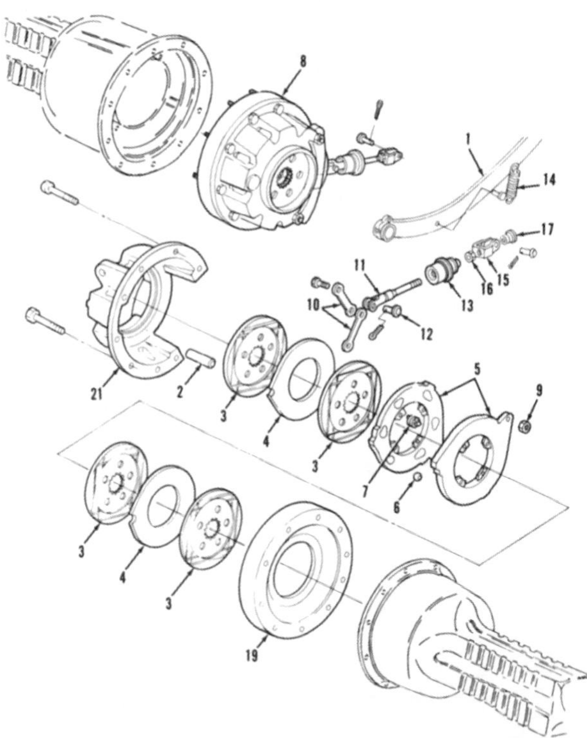

Fig. 168—Exploded view of rear wheel brake assembly used on all models. Parts (19 & 21) are also shown in Fig. 155. Components of left brake assembly (8) are similar, but all parts are not exactly alike. Some models are equipped with only one stationary disc (4) and three brake discs (3) instead of the number shown.

1. Pedal
2. Torque pin
3. Brake discs
4. Stationary discs
5. Actuating discs
6. Steel balls (6)
7. Return springs (4)
8. Left brake assembly
9. Nuts
10. Actuating links
11. Pull rod
12. Pin
13. Oil seal
14. Pedal return spring
15. Clevis
16. Locknut
17. Nut
19. Brake outer housing
21. Brake inner housing

new bearing cup is installed. Refer to paragraph 142 to adjust carrier bearings if shims are lost, if right side inner brake housing (21—Fig. 168) is renewed or if brake outer housing (19) on right side is renewed.

The brake actuating assembly can be disassembled, if necessary. Remove clevis pin (12) and detach the four return springs (7), then separate the actuating discs (5) and remove the six steel balls (6).

To reassemble, position one disc (5) on bench with inner side up and locate one steel ball (6) in each of the ramped seats. Position the other disc (5) over the first steel disc and steel balls so that lugs for the torque links (10) are about 25 mm (1 inch) apart, then install the four return springs (7). If links (10) were removed, new self-locking nuts (9) should be installed. Connect operating rod (11) to links and install pin (12) and new cotter pin.

To reinstall components in axle housing, first install outer brake housing (19) on the retaining studs. Install torque pin (2) in brake housing. Refer to Fig. 168 and install brake components. If only one stationary disc (4) is installed, the disc should be installed closest to the inner brake housing (21) and only one brake disc (3) should be located near the outer brake housing (19) instead of the two shown.

BRAKE CROSS SHAFT AND SEALS

All Models

155. REMOVE AND REINSTALL. A cross-shaft (7—Fig. 169 and Fig. 170) runs through the rear axle center housing to support brake arms of models with cab or brake pedals of models without cab. The pedal or brake arm for the left brake is attached near the right end of the cross-shaft. The shaft pivots in renewable bushings located in the center housing bores and an oil seal at the outer side of each bushing prevents oil from leaking past the shaft.

To renew the cross-shaft, bushings or seals, first drain lubricant from the rear axle center housing. Disconnect brake operating linkage from both ends of the cross-shaft, disconnect clutch linkage from left end and remove pedals or levers from right side. Remove key from keyway in right end of shaft and

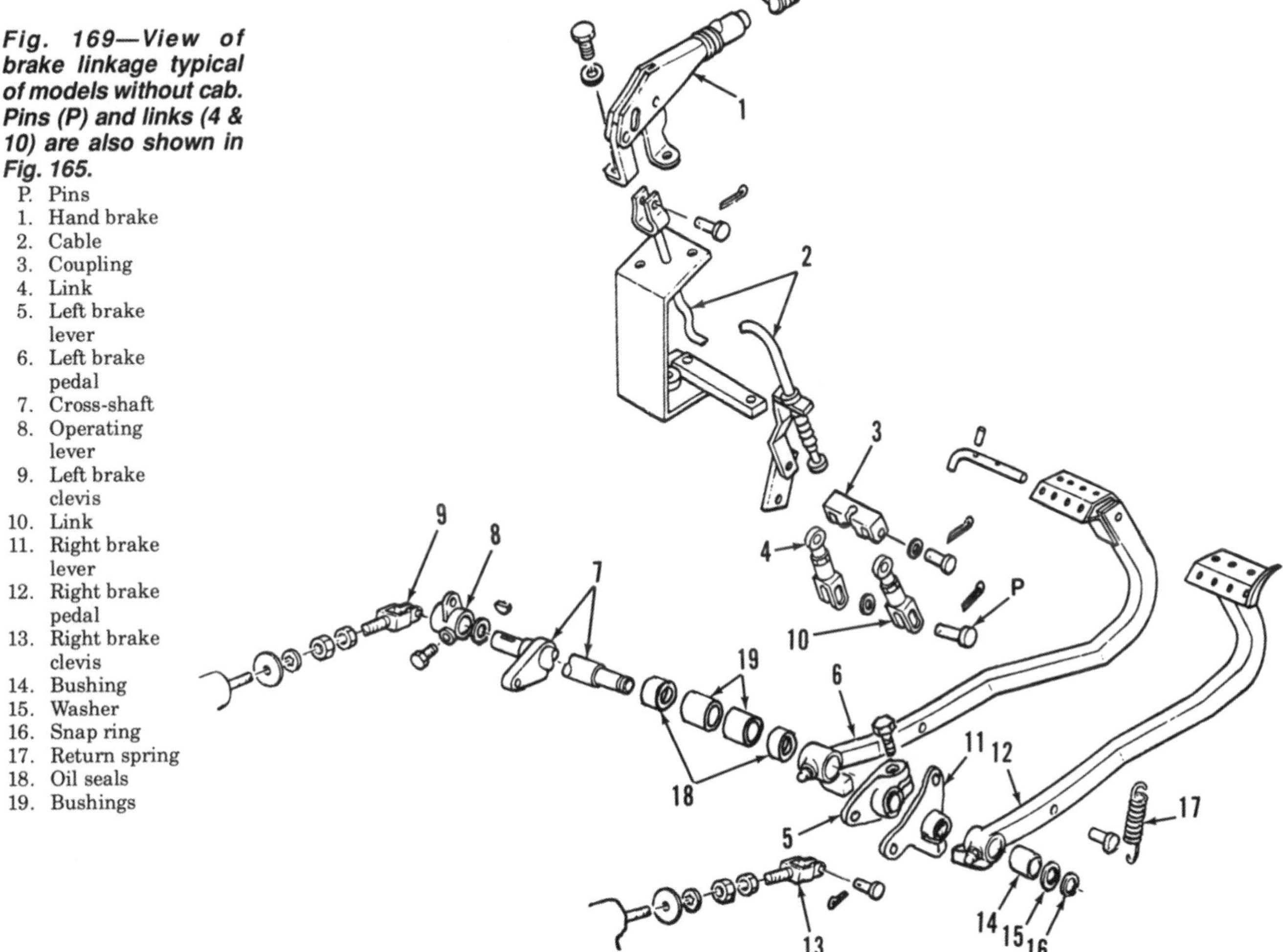

Fig. 169—View of brake linkage typical of models without cab. Pins (P) and links (4 & 10) are also shown in Fig. 165.

P. Pins
1. Hand brake
2. Cable
3. Coupling
4. Link
5. Left brake lever
6. Left brake pedal
7. Cross-shaft
8. Operating lever
9. Left brake clevis
10. Link
11. Right brake lever
12. Right brake pedal
13. Right brake clevis
14. Bushing
15. Washer
16. Snap ring
17. Return spring
18. Oil seals
19. Bushings

remove any burrs from right end of shaft. Withdraw cross shaft from left side.

Pry oil seals from each side of center housing. Inspect bushing in right brake pedal and in rear axle center housing. Bushings are presized and should not require reaming if carefully installed. Lips of seals should be toward inside. Reassembly is reverse of disassembly procedure. Be especially careful when sliding cross shaft through the right side seal. Adjust the brakes as outlined in paragraphs 152 and 153.

TRANSFER GEARBOX BRAKE

4630 and 4830 Models So Equipped

156. Refer to paragraph 16 and Fig. 22 for service to the brake assembly contained in the front drive transfer gearbox of some models.

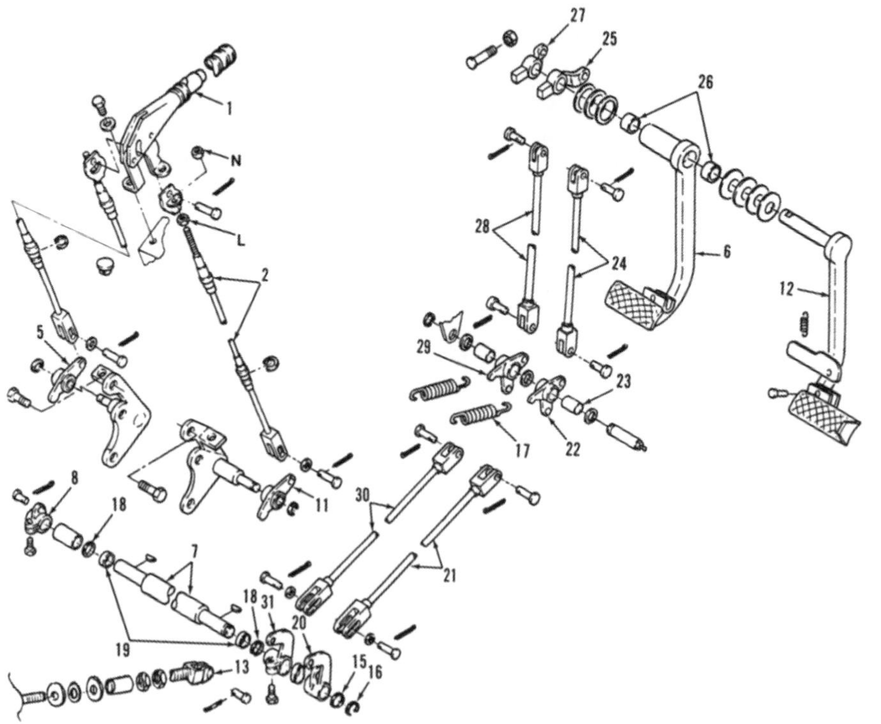

Fig. 170—Brake linkage for models with cab. Parts (L & N) are also shown in Fig. 167.

- L. Locknut
- N. Adjusting nut
- 1. Hand brake
- 2. Cable
- 5. Left brake lever
- 6. Left brake pedal
- 7. Cross shaft
- 8. Operating lever
- 11. Right brake lever
- 12. Right brake pedal
- 13. Right brake clevis
- 15. Washer
- 16. Snap ring
- 17. Return spring
- 18. Oil seals
- 19. Bushings
- 20. Right brake lever
- 21. Link rod
- 22. Bellcrank
- 23. Bushings
- 24. Link rod
- 25. Operating lever
- 26. Bushings
- 27. Operating lever
- 28. Link rod
- 29. Bellcrank
- 30. Link rod
- 31. Left brake lever

INDEPENDENT POWER TAKE-OFF

OPERATION

All Models So Equipped

157. All models are equipped with an independent-type power take-off with a hydraulically operated multiple disc clutch that can be engaged or disengaged anytime the engine is running. Standard 540 rpm pto speed is available at 1750 engine rpm for models with cab and synchronized shift (16 × 8 and 8 × 8) transmissions, and at 1800 engine rpm for other models. The hollow pto input shaft (2—Fig. 171) is splined to the engine clutch cover (1) and fitted with a drive gear which turns the driven (drop) gear (3) which is splined to the front of the pto countershaft (4). Engaging the pto clutch (6) drives the rear pto shaft (7). A gear reduction unit is located at the rear to provide proper output shaft speed. Refer to Fig. 172 for schematic of clutch engaging system.

TROUBLE SHOOTING

All Models

158. OPERATING CHECKS. Refer to the following when problems are encountered with the pto system.

PTO WILL NOT ENGAGE WITH LOAD OR STOP WITHOUT LOAD. Could be caused by:
A. Low rear axle oil level.
B. Hydraulic pump failure.
C. Incorrect operation of control lever.
D. Clutch feed tube damaged.
E. Control valve assembly damaged.
F. Pressure regulator valve damaged.

PTO WILL NOT ENGAGE WITH LOAD, BUT STOPS WITHOUT LOAD. Could be caused by:
A. Clutch housing to brake housing sealing rings leaking.
B. Control valve piston stuck or spring broken.
C. Clutch plates worn or teeth broken.
D. Clutch piston seals leaking.
E. Clutch piston or housing damaged.
F. Blocked oil passages.

PTO ENGAGES BUT WILL NOT DISENGAGE. Could be caused by:
A. Incorrect operation of control lever.
B. Control valve stuck.
C. Clutch return spring damaged.
D. Blocked oil passages.
E. Seized or warped clutch plates.

INCORRECT OPERATION OF BRAKE, SHAFT ROTATES WITHOUT LOAD AND WITH CLUTCH DISENGAGED. Could be caused by:
A. Brake lever damaged or worn.
B. Blocked oil passages.
C. Control valve assembly damaged.
D. Brake piston "O" rings leaking.
E. Brake pressure oil passages leaking.

159. CHECK PTO HYDRAULIC PRESSURE. To check pto system hydraulic pressure, first operate tractor until the hydraulic oil in rear axle housing

1. Engine clutch cover
2. Input shaft
3. Driven gear
4. Pto countershaft
5. Pto clutch input shaft
6. Pto clutch
7. Rear shaft
8. Hydraulic pump drive gear
9. Pto control valve
10. Pto brake regulating valve

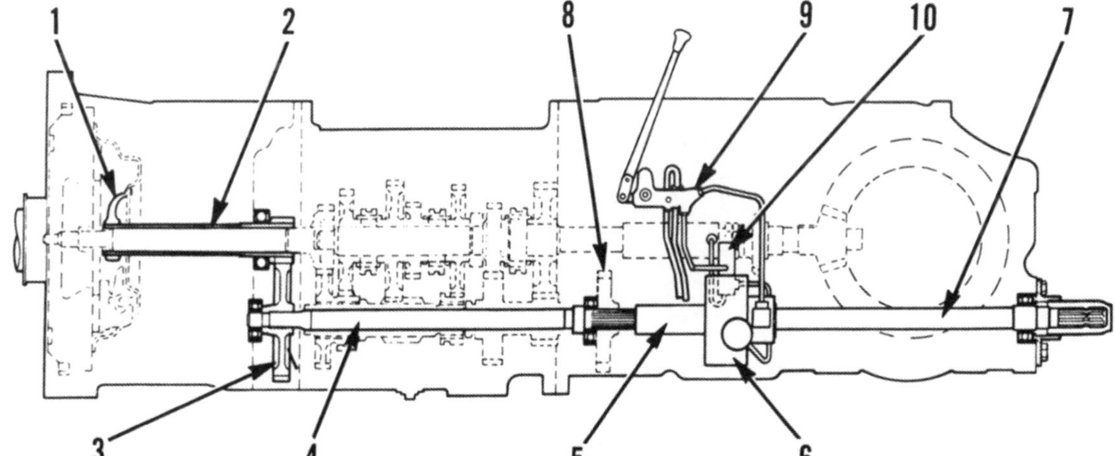

Fig. 171—Schematic of the independent pto typical of models with constant mesh transmission. Parts may be different depending upon type of transmission, but operation and power flow is shown.

reaches normal operating temperature--57-63° C (134-146° F), then stop the engine. Remove plug from rear of hydraulic pump (4—Fig. 173). Install adapter (3), tube (2) and pressure gage (1) in test opening.

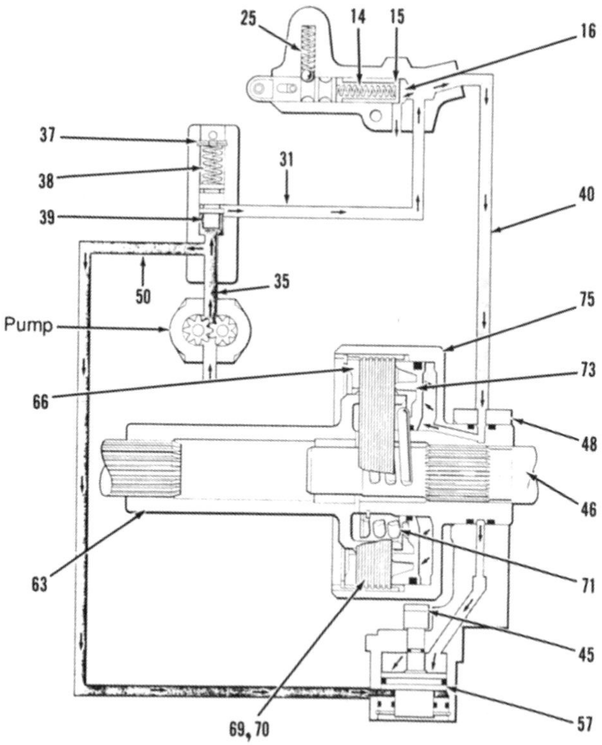

Fig. 172—Cross section drawing of pto brake and clutch showing flow of operating oil. Refer to Fig. 174, Fig. 176 and Fig. 177 for legend.

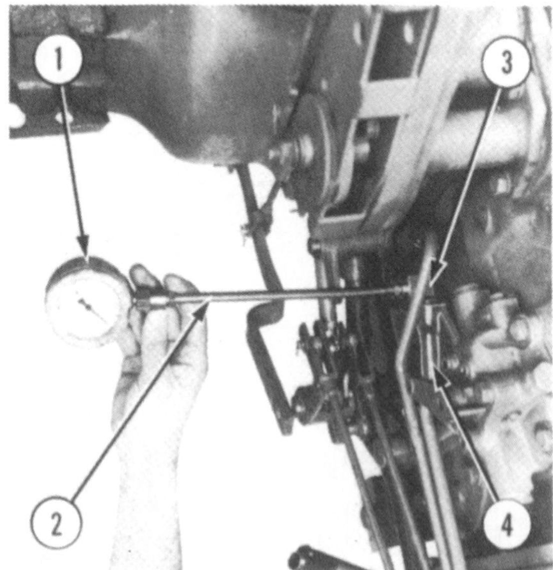

Fig. 173—Connect adapter (3), tube (2) and pressure gage (1) as shown to test port of hydraulic pump (4) to check pressure regulator valve setting.

1. Pressure gage
2. Tube (Churchill FT.4100-1 N774-2; Nuday 1552 N-774-2)
3. Adapter (Churchill FT.4100-2; Nuday 4657)
4. Hydraulic pump

To check the brake pressure regulating valve, start engine and operate at 2300 rpm. Move pto control to **disengaged** position and note gage reading. Correct pressure is 1518-1587 kPa (220-230 psi). Change pressure by varying the number of shims (37—Fig. 174) located in valve assembly. Pressure should remain within the range of 1518-1587 kPa (220-230 psi) with engine operating between 1000 and 2100 rpm.

Continue tests as follows: Check clutch engaged pressure with engine operating at 2300 rpm and clutch engaged. Pressure should be slightly higher (about 70 kPa or 10 psi) than when checking brake pressure. If pressure is incorrect, check control valve pressure as described in the following paragraph.

To check control valve pressure, remove the hydraulic lift cover as outlined in paragraph 178. Detach pressure tube (40—Fig. 176) from control valve (24) and attach an adapter (3—Fig. 175), tube (2) and pressure gage (1) as shown. Attach a hose (4) from the hydraulic lift cover feed to rear sump. Start engine and operate at 1300 rpm, engage and disengage pto clutch several times while noticing the gage pressure. If gage pressure is not within 1518-1587 kPa (220-230 psi) range, remove control valve assembly as outlined in paragraph 161 and add shims (15—Fig. 177). Shims are available in thicknesses of 0.381 mm

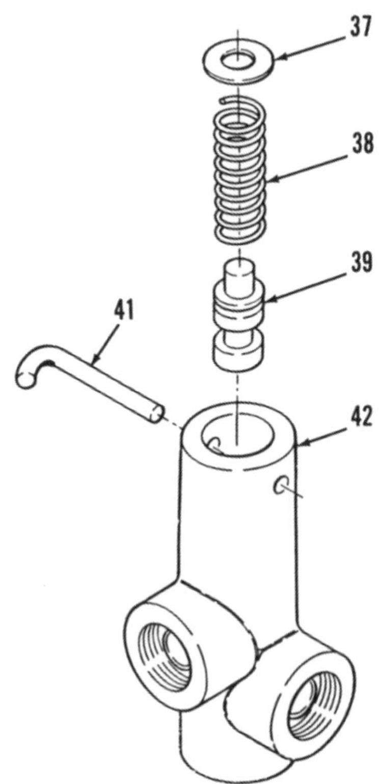

Fig. 174—View of pressure regulator valve used. Valve is also shown at (42—Fig. 176).

37. Shims
38. Spring
39. Valve
41. Cotter pin
42. Valve housing

(0.015 inch) and 0.762 mm (0.030 inch). No more than five shims should be installed.

If pressure can not be increased by adding shims, remove and overhaul hydraulic lift pump assembly as outlined in paragraph 196 or 198. If pressure is within limits, but pto clutch slips under load, remove and overhaul clutch as outlined in paragraphs 162 and 163.

PRESSURE REGULATOR VALVE

All Models

160. R&R AND OVERHAUL. Remove the hydraulic lift cover as outlined in paragraph 178. Drain lubricant from rear axle center housing until pressure regulator valve is completely exposed. Disconnect tubes from pressure regulator valve assembly, then remove the valve assembly from the rear axle center housing.

Remove pin (41—Fig. 174) and withdraw shims (37), spring (38) and valve (39) from housing (42). Clean all parts with a suitable solvent and blow dry with compressed air. Inspect housing and bore for cracks, corrosion, excessive wear or any other damage. Inspect valve for roughness, corrosion, excessive wear or any other damage. Examine spring for any type of damage. Renew components as needed.

Reassemble in reverse order of disassembly. Coat valve and housing bore with oil before reassembling. Install valve (39) in housing bore with notched land toward base of housing.

Reinstall valve assembly and tighten tube connections to 12 N•m (9 ft.-lbs.) torque. Refill rear axle center housing to specified level and check system pressures (disengaged and engaged) as outlined in paragraph 159. If pressures are not within limits, remove valve and change the number and thickness of shims (37—Fig. 174 and 15—Fig. 177) as required.

CONTROL VALVE

All Models

161. R&R AND OVERHAUL. To remove the pto clutch control valve, first remove the hydraulic lift cover as outlined in paragraph 178, then proceed as follows:

Disconnect tubes (31 and 40—Fig. 176) from control valve and lever (9) from the valve plunger. Unbolt valve, then remove from the center housing with exhaust tubes (28 and 30) attached.

Use a suitable pin punch to drive plunger retaining pin (13—Fig. 177) from valve housing, then carefully remove plunger, catching detent ball (26) and spring (25). Remove control valve spring (14), shims (15) and control valve (16) from housing. Compressed air can be used to blow valve (16) from housing if stuck. Be careful not to lose shims (15).

Clean parts in a suitable solvent, air dry and coat lightly with oil. Inspect valve, plunger and valve bore in housing for excessive wear or scoring. If housing is not suitable for reuse, install a new valve assembly. Check springs (14 and 25) for cracks or distortion.

To assemble, insert control valve (16) in housing with hollow end out, then insert shims (15) and valve spring (14) inside the valve. Position detent spring (25) and ball (26) in housing. Hold spring and detent ball compressed in housing bore with a punch while inserting plunger (12). Hold the plunger (12) in housing, compressing spring (14) and install pin (13). Attach exhaust tubes (28 and 30—Fig. 176) if removed and install in rear axle center housing. Tighten screw retaining the control valve to 45 N•m (33 ft.-lbs.) torque and tube connections to 12 N•m (9 ft.-lbs.) torque.

Connect hydraulic gage as shown in Fig. 175 and check system pressure as outlined in paragraph 159. If pressure is not within limits, remove valve, change thickness of shims (15—Fig. 177) and recheck. Reconnect pressure tube and reinstall hydraulic lift cover after pressure is correct.

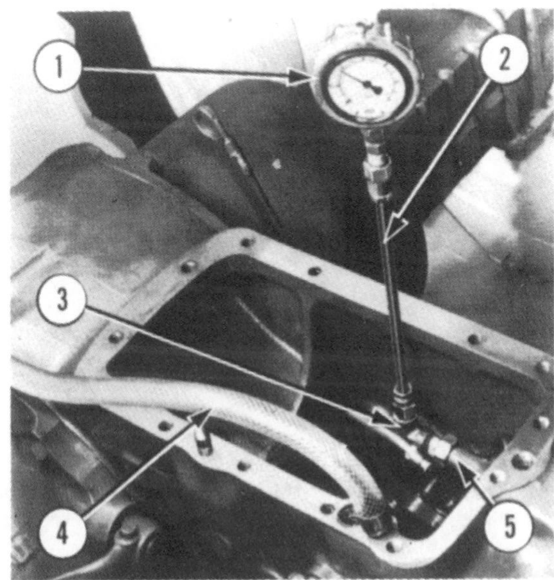

Fig. 175—View showing adapter (3), tube (2) and pressure gage (1) connected to the control valve assembly (5) for checking pressure. Hose (4) is connected from the hydraulic lift cover feed to the rear axle sump. Refer to text.

1. Pressure gage
2. Tube (Churchill FT.4100-1 N774-2; Nuday 1552 N-774-2)
3. Adapter (Churchill FT.4097 SW.17B; Nuday 1221 SW-17)
4. Hose (Churchill T.8503-4)
5. Control valve

PTO CLUTCH AND BRAKE ASSEMBLY

All Models

162. REMOVE AND REINSTALL. It is possible to remove the brake piston and install new seals without separating the tractor. Remove cover (59—Fig. 176), retainer (53) and piston (57) from tractor left side. Be sure to install "O" ring seals in proper location. The legend in Fig. 176 lists sizes.

To remove the pto clutch and brake assembly, first remove the hydraulic lift cover as outlined in paragraph 178 and hydraulic pump as outlined in paragraph 197. Remove the pto shaft as described in paragraph 165 or paragraph 166 from rear of tractor. Refer to paragraph 139 or 140 and split tractor between rear of transmission and front of rear axle center housing.

NOTE: Shims may be located on rear end of transmission pto shaft or stuck to the pto clutch (see Fig. 178) and may be lost or damaged when tractor is separated. Be careful to find these shims and not to lose or damage them.

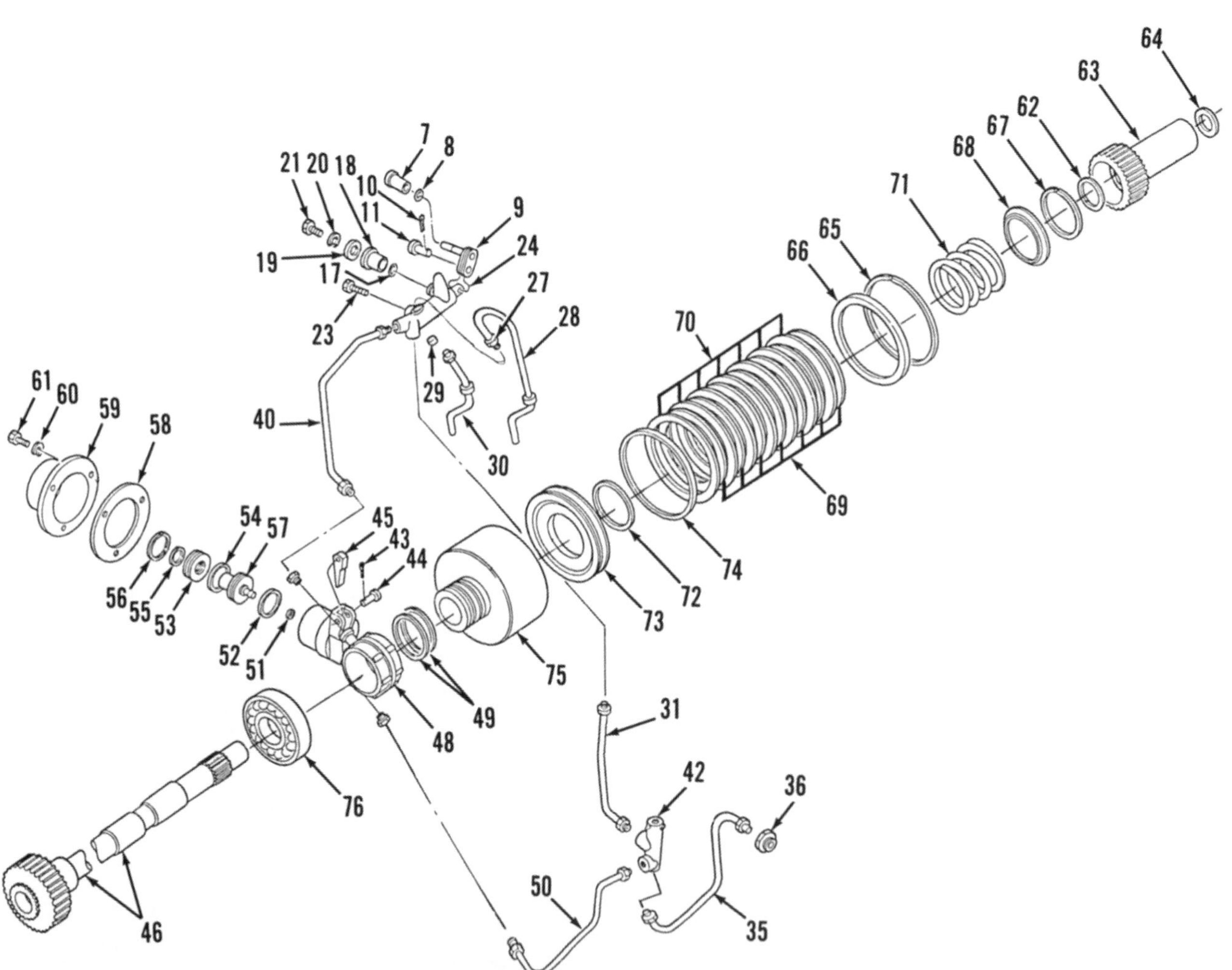

Fig. 176—Exploded view of independent pto clutch, brake, brake pressure regulator valve and control valve assemblies. Refer to Fig. 174 for exploded view of pressure regulator (42) and to Fig. 177 for control valve (24).

7. Bushing
8. "O" ring
9. Clutch valve lever
10. Cotter pin
11. Clevis pin
17. Oil seal
18. Bushing
19. Washer
20. Washer
21. Cap screw
23. Cap screw
24. Control valve
27. Tube seal
28. Tube assy.
29. Tube seal
30. Tube assy.
31. Tube assy.
35. Tube assy.
36. Connector
40. Tube assy.
42. Regulator valve housing
43. Cotter pin
44. Clevis pin
45. Brake lever
46. Pto shaft
48. Brake housing & support
49. Oil seal
50. Tube assy.
51. "O" ring (1/4 X 0.07 in.)
52. "O" ring (1.31 X 0.07 in.)
53. Retainer
54. "O" ring (0.80 X 0.07 in.)
55. "O" ring (1.44 X 0.07 in.)
56. Retaining ring
57. Brake piston
58. Gasket
59. Cover
60. Lock washer
61. Cap screw
62. Thrust washer
63. Clutch drive coupling & hub
64. Shim washer
65. Snap ring
66. Pressure plate
67. Snap ring
68. Spring retainer
69. Internally splined drive plates
70. Externally splined driven plates
71. Piston release spring
72. "O" ring
73. Piston
74. "O" ring
75. Clutch housing
76. Ball bearing

Disconnect tubes (40 and 50—Fig. 176) from housing (48), withdraw coupling and hub (63), then remove pto clutch (65 through 75), brake housing (48) and associated parts as an assembly. Refer to paragraph 163 for overhaul procedure.

To reinstall, reverse the removal procedure. Always install new "O" rings and lubricate parts liberally when assembling. Install same thickness of shims (64) as were removed unless parts (63 or 75) were renewed. Refer to paragraph 164 if selection of shims (64) must be determined. Refer to paragraph 159 and check pto hydraulic pressure before reinstalling hydraulic lift cover.

163. OVERHAUL. Slide brake housing (48—Fig. 176) from clutch housing (75) to separate assemblies. Withdraw hub (63) from clutch and remove thrust washer (62). Remove snap ring (65), then withdraw pressure plate (66) and clutch plates (69 and 70). Compress spring (71) using a suitable tool pressing against outer edge of retainer ring (68) and remove snap ring (67). Release spring slowly, then remove retainer (68) and spring (71). Compressed air can be used to blow piston (73) from housing (75).

CAUTION: Use only low-pressure air to prevent injury and damage to parts.

Remove snap ring (56) and withdraw piston (57) and guide (53). Remove cotter pin (43), clevis pin (44) and brake lever pad (45). Remove sealing rings from grooves and clean parts in a suitable solvent, air dry and carefully inspect for cracks, scoring, excessive wear or other defects, such as overheating or warped plates.

Always install new "O" ring seals and lubricate all parts liberally before assembling. Reassembly procedure is the reverse of disassembly procedure. Use a suitable tool to compress spring (71), with retainer (68) installed, while installing snap ring (67). Release compressor tool and check for seating of snap ring.

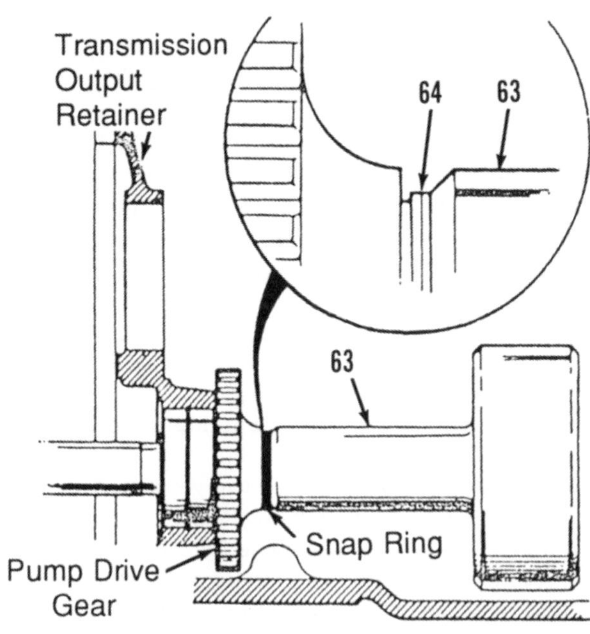

Fig. 178—Shims (64) are located between hub (63) and snap ring. Parts (63 & 64) are shown in Fig. 176. Refer to paragraph 164 to select thickness of shims for correct end play.

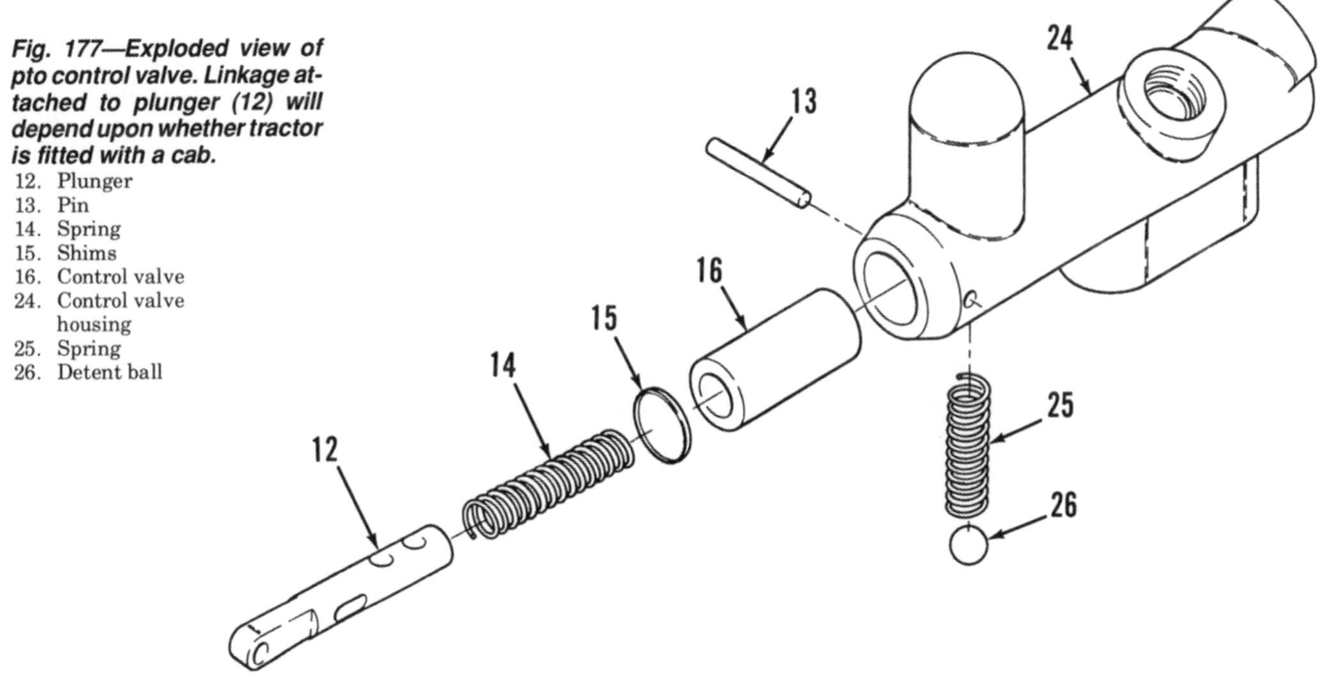

Fig. 177—Exploded view of pto control valve. Linkage attached to plunger (12) will depend upon whether tractor is fitted with a cab.

12. Plunger
13. Pin
14. Spring
15. Shims
16. Control valve
24. Control valve housing
25. Spring
26. Detent ball

Install clutch plates (69 and 70), beginning with one plate (70) with external drive lugs. Install one plate (69) with internal splines (lugs), which will engage hub (63), then alternate plates (69 and 70). Install pressure plate (66) and retaining snap ring (65). Install new sealing rings (49) in grooves of housing (75). Assemble brake lever and pad (45), pin (44) and cotter pin (43) in support (48), then slide support over housing (75) and sealing rings (49). Install new "O" rings (51, 52, 54 and 55) on piston (57) and retainer (53), assemble the retainer on piston, then insert piston and retainer into bore of support (48). Refer to paragraph 162 for reassembly of tractor.

164. ADJUST PTO CLUTCH SHAFT END PLAY. Shims (64—Fig. 176) are located between the hydraulic pump drive gear retaining snap ring and pto clutch hub (63) to limit end play within range 0-0.63 mm (0-0.025 inch). If the installation of new components would change end play or if original shims are lost or damaged, end play should be measured and new shims selected and installed. To measure end play, proceed as follows:

If pto clutch and brake assembly have not been removed, remove plate from left side of rear axle center housing and use a feeler gage to measure any gap between pto clutch housing (75) and support (48). If not already removed, remove hydraulic pump and join rear axle center housing to transmission housing using a new gasket, but without any shims (64). Work through the opening for the hydraulic pump, using a feeler gage to measure gap between pto clutch hub and snap ring as shown in Fig. 178. Add this measured gap to the gap measured previously between the rear of the clutch housing (75—Fig. 176) and support (48). Separate tractor and install shims (64) equal to total measured gap less 0-0.63 mm (0-0.025 inch).

If the tractor is separated between the transmission and rear axle center housing, reconnect rear axle center housing to transmission housing using a new gasket, but without any shims (64). Do not install hydraulic pump or hydraulic lift cover. Make sure that clutch housing (75) is fully to the rear. Work through the opening for the hydraulic pump, using a feeler gage to measure gap between pto clutch hub and snap ring as shown in Fig. 178. Add this measured gap to the gap measured previously between rear of clutch housing (75—Fig. 176) and support (48). Separate tractor and install shims (64) equal to the total measured gap less 0-0.63 mm (0-0.025 inch).

PTO OUTPUT SHAFT

Models with Constant Mesh Transmission

165. R&R AND OVERHAUL. Refer to Fig. 179 for models with constant mesh (8 × 2) transmission.

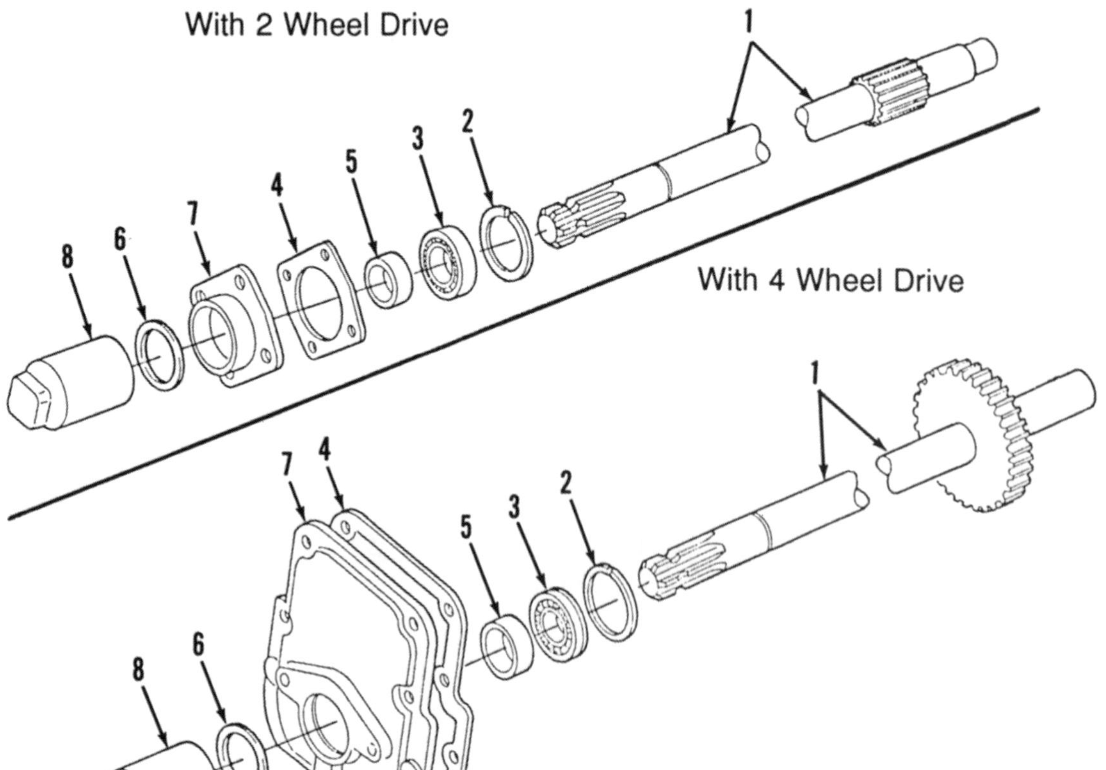

Fig. 179—Exploded view of pto rear shaft and associated parts used with constant mesh (8 X 2) transmission. Parts at top are used on 2-wheel drive tractors and parts shown in the lower drawing are for models with 4-wheel drive.

1. Pto rear shaft
2. Snap ring
3. Ball bearing
4. Gasket
5. Spacer
6. Oil seal
7. Retainer
8. Cover

Differences will be noted between parts for 2-wheel drive and models with 4-wheel drive. Remove cover (8) and any protective covers. Drain rear axle center housing. Remove screws attaching retainer (7), then withdraw shaft (1), bearing and seal.

Inspect all parts for damage and renew as necessary. Reassemble by reversing disassembly procedure. Install oil seal (6) with lip toward bearing (3). Install new gasket (4) and tighten retainer attaching screws to 68 N•m (50 ft.-lbs.) torque.

Models with Synchronized Transmission

166. R&R AND OVERHAUL. Refer to Fig. 180 for models with synchronized (8 × 8 and 16 × 8) transmission. Remove cover (8) and any protective covers. Drain rear axle center housing. Remove screws attaching cover (9) to rear axle center housing. Bearing (3) and seal (6) can be removed after unbolting and removing retainer (7). To disassemble and remove countershaft parts (13 through 19), remove cover (22) and snap ring (20).

Inspect all parts for damage and renew parts as necessary. Reassemble by reversing disassembly procedure. Install oil seal (6) with lip toward bearing (3). Install new "O" ring (4) and tighten retaining screws to 68 N•m (50 ft.-lbs.) torque.

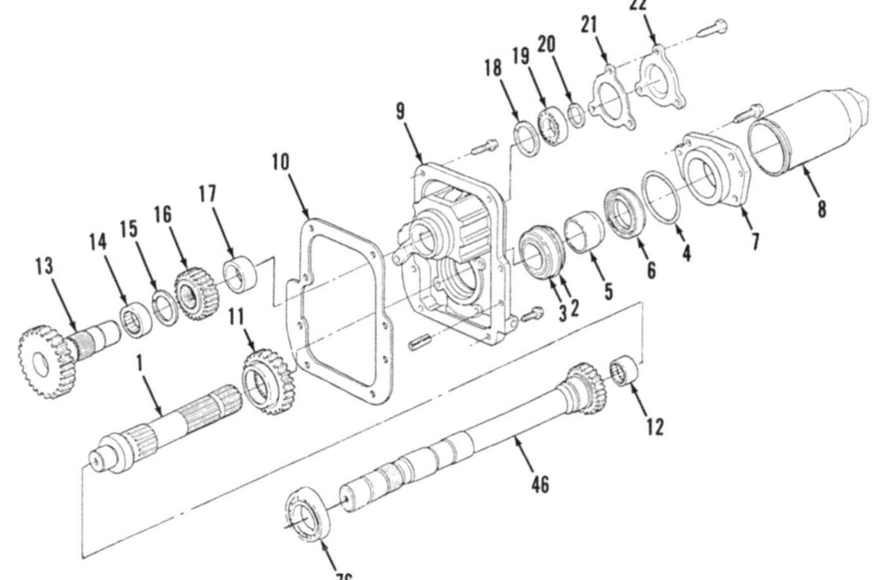

Fig. 180—Exploded view of pto rear shaft and reduction unit used on tractors with synchronized (8 X 8 & 16 X 8) transmission. Shaft (46) and bearing (76) are also shown in Fig. 176.

1. Pto rear shaft
2. Snap ring
3. Ball bearing
4. "O" ring
5. Spacer
6. Oil seal
7. Retainer
8. Cover
9. Rear cover & reduction housing
10. Gasket
11. Driven gear
12. Roller bearing
13. Pto counter-shaft
14. Roller bearing
15. Snap ring (same as 18)
16. Countershaft drive gear
17. Spacer
18. Snap ring (same as 15)
19. Ball bearing
20. Snap ring
21. Gasket
22. Cover
46. Pto drive shaft
76. Ball bearing

HYDRAULIC LIFT SYSTEM

All Models

167. The hydraulic lift system incorporates automatic draft control. Fluid for the system is common with the differential and final drive lubricant.

Tractors may be equipped with either a gear-type pump located on the left side of the engine or a different gear-type pump located inside the right side of the rear axle center housing. The pump attached to the engine is driven by a gear located on the rear of the camshaft. The pump mounted in the rear axle center housing is driven by the same shafts that provide power to the independent pto.

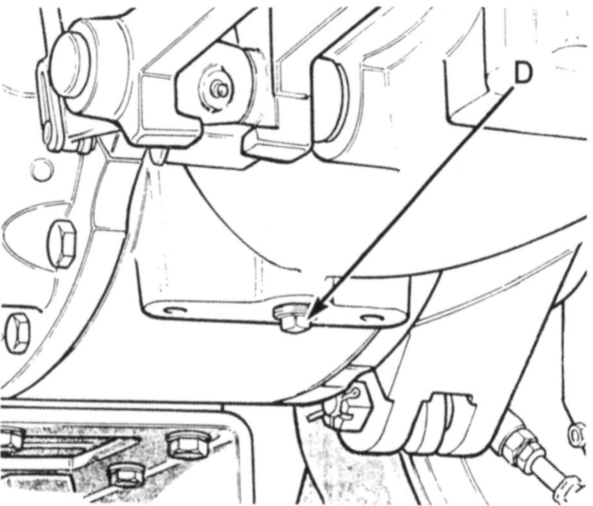

Fig. 181—View of rear axle housing drain (D) typical of models with 2-wheel drive.

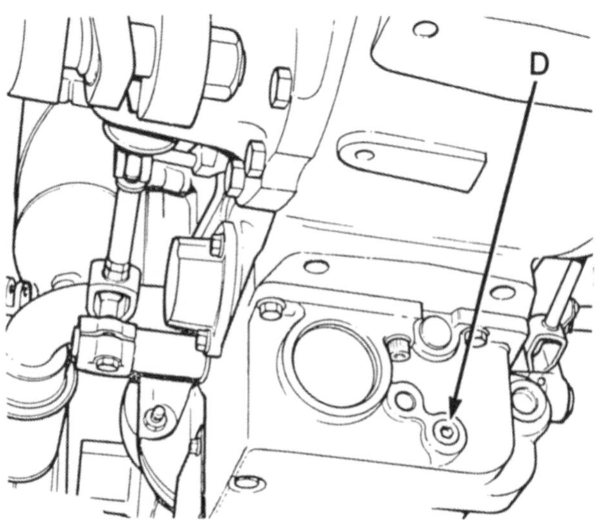

Fig. 182—View of rear axle and transfer gearbox drain (D) typical of models with 4-wheel drive.

HYDRAULIC FLUID

All Models

168. Recommended hydraulic fluid and lubricant for the rear axle and final drive is Ford M2C134-D/C or equivalent. Fluid capacity depends upon transmission type and if equipped with 4-wheel drive option. Refer to the following for rear axle, final drive and hydraulic system approximate capacity:

With Constant Mesh (8 × 2) transmission—
- With 2-Wheel Drive . . . . 45.7 L (48 qts.)
- With 4-Wheel Drive . . . . 47.1 L (49.5 qts.)

With Synchronized (8 × 8 or 16 × 8) transmission—
- With 2-Wheel Drive . . . . 32.5 L (34 qts.)
- With 4-Wheel Drive . . . . 33.9 L (35.5 qts.)

The front-wheel-drive transfer case is lubricated by oil common with the lubricant contained in the rear axle housing. On all models, lubricant should be drained from the rear axle center housing (Fig. 181 or Fig. 182) at least every 1200 hours of operation or once each year. Drain system with 3-point hitch lowered and any remote cylinders retracted. Check oil level (Fig. 183 or Fig. 184) with 3-point hitch in raised position and remote cylinders extended.

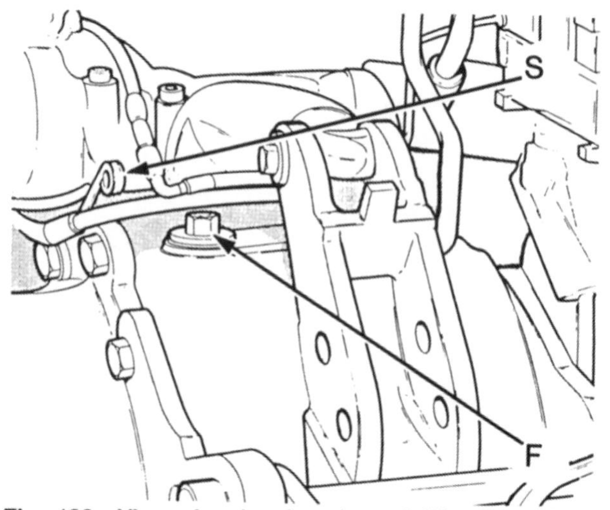

Fig. 183—View showing location of filler plug (F) and dipstick (S) typical of models without cab.

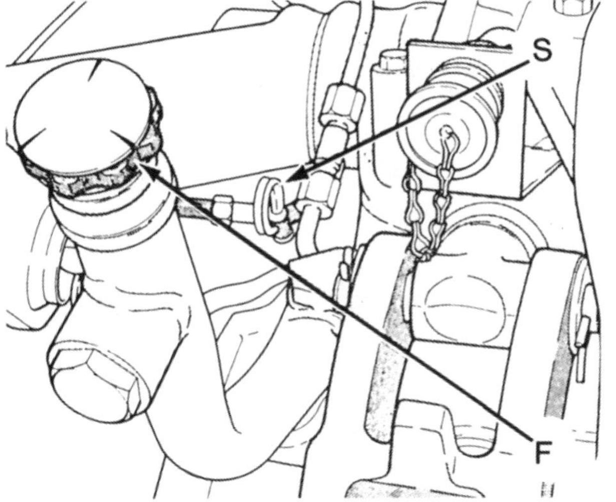

Fig. 184—View showing location of filler opening (F) and dipstick (S) typical of models with cab.

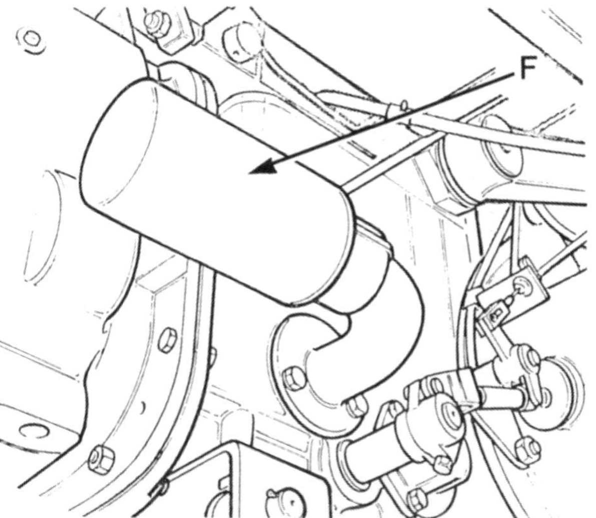

Fig. 185—View of hydraulic oil filter (F) used with hydraulic pump mounted in the rear axle center housing.

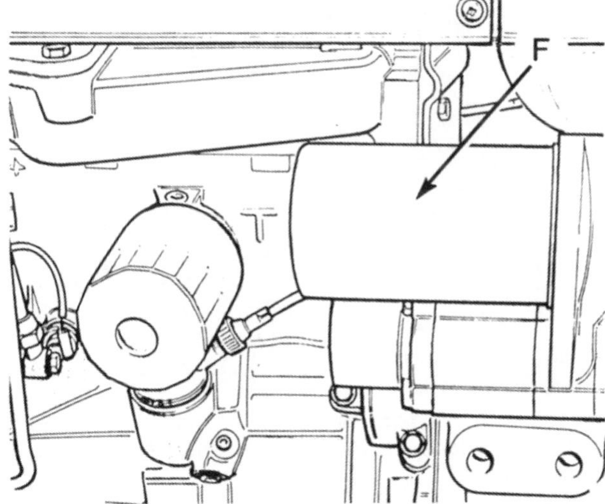

Fig. 186—View of hydraulic filter (F) used with engine mounted pump.

HYDRAULIC FILTER

All Models

169. Location of hydraulic filter (F—Fig. 185 or Fig. 186) depends upon whether pump is mounted to the rear axle center housing or the tractor engine. A new filter should be installed after every 300 hours of operation. The filter seal should be coated with clean oil and tightened $3/4$ turn past the point where filter seal just contacts housing.

TROUBLESHOOTING

All Models

170. Refer to the following when problems are encountered with the hydraulic lift system.

A. FAILURE TO LIFT UNDER ALL CONDITIONS. Could be caused by:

1. Low oil level in rear axle center housing.
2. Pressure relief valve damaged.
3. Hydraulic pump pressure low.
4. Plugged filter.
5. Pump intake leaking air.
6. Unload valve faulty.
7. Lift piston seals damaged.
8. Unload valve plug worn.
9. Lift cylinder, lift cover or pressure transfer tube leaking (cracked).

B. FAILURE TO LIFT UNDER LOAD OR SLOW LIFT. Could be caused by:

1. Hydraulic pump pressure low.
2. Damaged "O" rings between lift cylinder and lift cover or between accessory cover and lift cover.
3. Damaged "O" rings on hydraulic pump pipes.
4. Damaged lift cylinder safety valve.
5. Faulty lift piston seals.
6. Cracked or porous lift cylinder or lift cover casting.

C. EXCESSIVE CORRECTIONS (BOBBING OR HICCUPS) IN RAISED OR TRANSPORT POSITION. Could be caused by:

1. Worn or damaged check valve ball or seat.
2. Selector valve worn or damaged.
3. Lift cylinder safety valve damaged.
4. Faulty lift piston seals.
5. Control valve worn.
6. Damaged "O" rings between lift cylinder and lift cover or between lift cover and accessory cover.
7. Cracked or porous lift cylinder or lift cover casting.

PRESSURE CHECK

Models With Engine-Mounted Pump

171. Connect pressure gage and shut-off valve as shown in Fig. 187. A 0-28,000 kPa (0-4000 psi) gage should be used. After attaching test equipment, open shut-off valve fully, start engine and set engine speed at 1650 rpm. Gradually close shut-off valve while observing the gage. Hydraulic pressure should be 17,580-18,270 kPa (2550-2650 psi). **Do not close valve longer than necessary and do not allow pressure to go above 18,270 kPa (2650 psi) even if valve is not fully closed. Pump damage could occur.**

Open shut-off valve, stop engine and remove pressure testing equipment.

Models With Pump In Axle Center Housing

172. Connect pressure gage and shut-off valve as shown in Fig. 188. A 0-28,000 kPa (0-4000 psi) gage should be used. After attaching test equipment, open the shut-off valve fully, start engine and set engine speed at 1650 rpm. Gradually close the shut-off valve while observing the gage. Hydraulic pressure should be 17,580-18,270 kPa (2550-2650 psi). **Do not close valve longer than necessary and do not allow pressure to go above 18,270 kPa (2650 psi) even if valve is not fully closed. Pump damage could occur.**

Open shut-off valve, stop engine and remove pressure testing equipment.

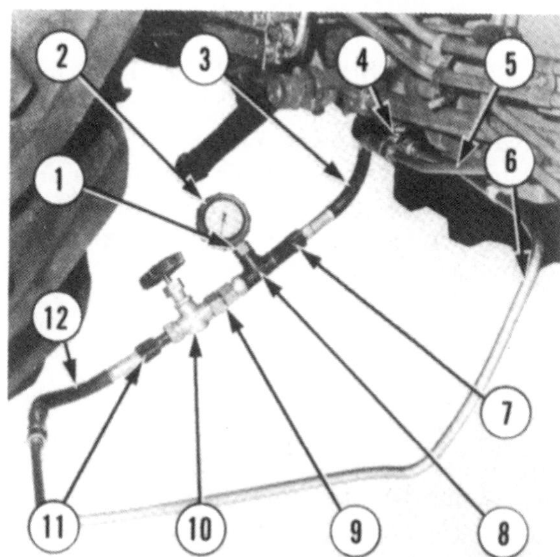

Fig. 187—View of gage installation for checking hydraulic pressure of models with engine-mounted pump.

1. Adapter
2. Pressure gage
3. Hose
4. Elbow
5. Pump pressure line
6. Tube
7. Adapter
8. "T" adapter
9. Adapter
10. Shut-off valve
11. Adapter
12. Hose

ADJUSTMENTS

All Models

173. DRAFT CONTROL MAIN SPRING. Before adjusting, unscrew yoke (1—Fig. 189), then remove housing (2) rear seat (3), draft control main spring (4) and front seat (5). Locate the rear seat (3), spring (4), original shims (5) and front seat (6) in the housing (2). Measure distance between front seat and surface of housing. The front seat should be 0.254 mm (0.010 inch) above the surface of the housing. If incorrect, add or remove shims (5) as required. Reinstall and tighten housing retaining screws to 54 N•m (40 ft.-lbs.) torque. Tighten yoke until just tight against spring, then unscrew yoke until pin hole is horizontal and reinstall pin through rocker and yoke.

If draft control is not desired, install a special sized washer ("W" in cross section) in the location shown.

Models Without Cab

174. DRAFT CONTROL SELECTOR SHAFT. The location of the draft control shaft must be adjusted to properly locate roller (R—Fig. 190) on draft lever (L). Remove lift cover as outlined in paragraph 178. Place draft control lever (C) in raised position

Fig. 188—View of gage installation for checking hydraulic pressure of models with pump mounted in the rear axle center housing.

A. Accessory valve
C. Flow control knob
G. Gage
R. Return hose
T. "T" adapter
V. Shut-off valve

and note if roller (R) is centered on draft lever (L). To adjust roller position, add or remove spacers (S). Since location of position control components is affected by spacers, be sure position control linkage operates properly after adjusting roller (R) position.

175. DRAFT AND POSITION CONTROL LINKAGE. Remove lift cover as outlined in paragraph 178. Place draft control lever (D) at bottom of quadrant as shown in Fig. 191, then place position control lever (P) 25.4 mm (1 inch) from bottom of quadrant. Lower the lift arms (L) to the fully lowered position. Remove lift cylinder baffle plate (57—Fig. 195). Loosen lock nut then turn eccentric (E—Fig. 191) so that control valve spool is 0.762 mm (0.030 inch) below face of lift cylinder. Special setting gage (Churchill FT.8600 or FT.8606; Nuday 1261 SW-508-A) may be used as shown (T) to check measurement. Retighten lock nut and install lift cylinder front plate.

Models With Cab

176. DRAFT CONTROL LINKAGE. Remove the lift cover as outlined in paragraph 178, then install quadrant assembly as shown in Fig. 192. Disconnect draft control rod (R) from lever (L) or at disc (D). Place draft control lever (L) at bottom of quadrant. Align hole in tab of disc (D) with quadrant bracket reference hole (H). Adjust rod (R) length so that ends will mate with lever (L) and disc (D) holes, then reattach rod.

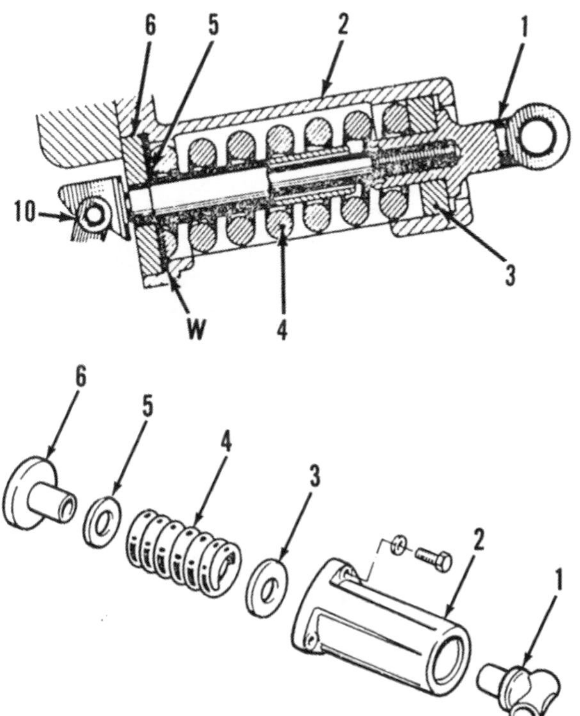

Fig. 189—A washer ("W" in cross section) can be installed between main draft control front spring seat (6) and housing (2) to prevent draft control reaction from tension on top link. Shims (5) do not need to be removed.

177. POSITION CONTROL LINKAGE. Remove lift cover as outlined in paragraph 178, then install quadrant assembly as shown in Fig. 193. Place draft control lever at bottom of quadrant then disconnect position control rod (R). Place position control lever (L) so bottom edge of lever is 21.1-22.1 mm (0.83-0.87 inch) from quadrant stop. Place lift arms in full down position. Insert an 11 mm (7/16 inch) drill bit in quadrant bracket reference hole (H). Rotate disc (D) so

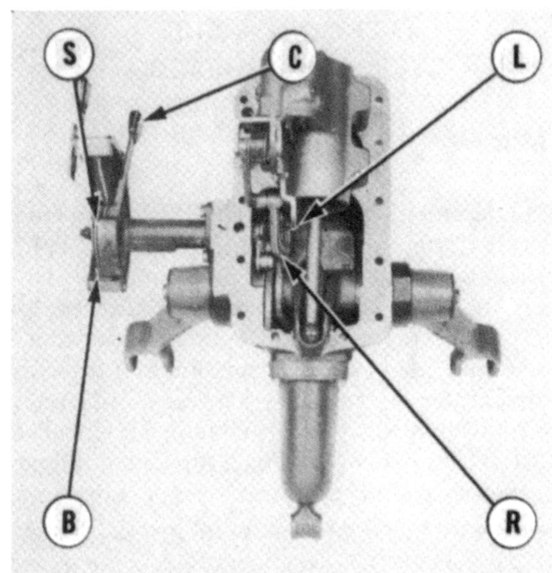

Fig. 190—Refer to text for draft control selector shaft adjustment.

B. Bolt
C. Draft control lever
L. Draft lever
R. Roller
S. Spacers

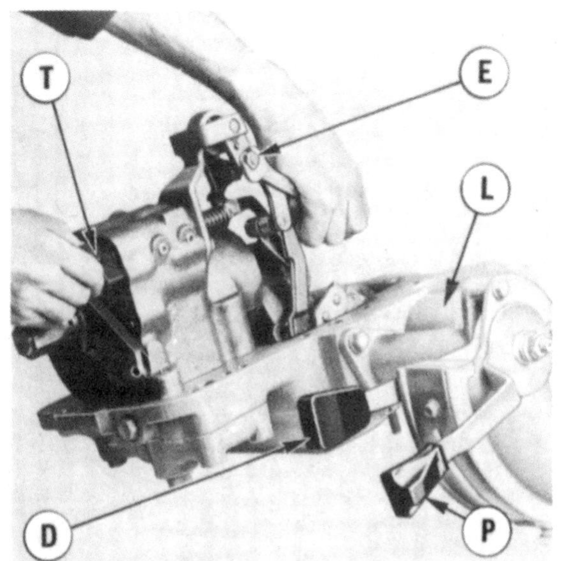

Fig. 191—Adjust draft and position control linkage as outlined in text.

D. Draft control lever
E. Eccentric
L. Lift arm
P. Position control lever
T. Gage tool

upper edge of disc tab (T) contacts drill bit. Adjust rod (R) length so ends will mate with lever (L) and disc (D) holes, then reattach rod. Remove lift cylinder baffle plate (57—Fig. 195). Loosen locknut then turn eccentric (E—Fig. 191) so control valve spool is 0.762 mm (0.030 inch) below face of lift cylinder. Special setting gage (Churchill FT.8600 or FT.8606; Nuday 1261 SW-508-A) may be used as shown (T) to check measurement. Retighten locknut and install lift cylinder front plate.

LIFT COVER AND CYLINDER ASSEMBLY

All Models

178. REMOVE AND REINSTALL. On models with cabs, refer to paragraph 211 and tilt cab for access. Remove seat from models without cab. On all models, be sure lift arms are fully lowered, then detach lift linkage from lift cylinder yoke and lift arms. Clean lift cover and surrounding area thoroughly. Disconnect auxiliary services oil line and remote control valve hoses, if so equipped. Unbolt and remove selector valve (8—Fig. 210). Unbolt lift cover and remove carefully to prevent component damage.

Reverse removal procedure to install lift cover. Place a steel rule between flow control cam adjuster and follower (see Fig. 202) so that follower cannot be forced downward when installing cover. Withdraw steel rule just prior to final positioning of lift cover. Tighten lift cover retaining screws to 106 N•m (78 ft.-lbs.) torque. Tighten accessory cover bolts to 81 N•m (60 ft.-lbs.) torque. Refer to paragraph 211 when reinstalling cab to models so equipped.

179. R&R LIFT CYLINDER. Remove lift cover and cylinder as outlined in paragraph 178 and proceed as follows. Remove snap ring (26—Fig. 194) and cylinder retaining screws. Move the lift arms to raised position and carefully separate lift cylinder from cover and linkage.

Make sure all mating surfaces are clean and free of nicks, burrs and scoring, before installing cylinder. Install "O" rings and hollow dowels between cylinder and lift cover and tighten cylinder retaining screws to 106 N•m (78 ft.-lbs.) torque.

180. OVERHAUL. Refer to the appropriate following paragraphs when servicing specific sections of the lift cylinder.

181. LIFT CYLINDER SAFETY VALVE. A safety relief valve (49—Fig. 195) is threaded into lift cylinder to protect lift cylinder from excessive hydraulic pressure due to shock loads imposed by rear mounted implements. The safety valve should open when pressure reaches 19,650-20,340 kPa (2850-2950 psi). Opening pressure of the safety valve is above normal operating pressure of the tractor hydraulic system and testing the valve requires removal and checking with hand operated pump and test equipment. If suitable test equipment is not available and condition of the valve is questioned, install a new valve. **Do not attempt to disassemble or adjust safety valve.**

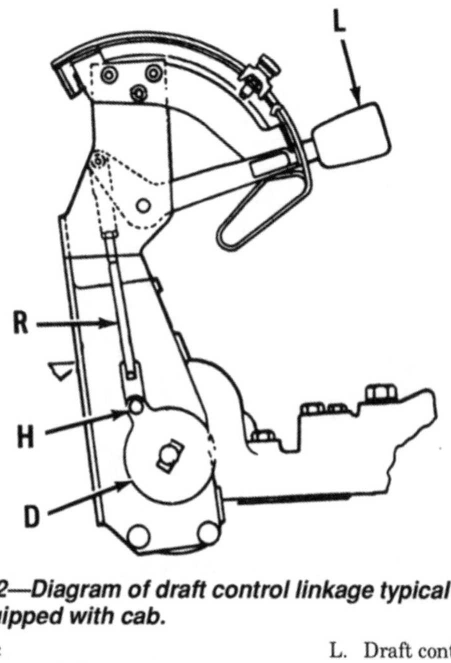

Fig. 192—Diagram of draft control linkage typical of models equipped with cab.

D. Disc
H. Reference hole
L. Draft control lever
R. Control rod

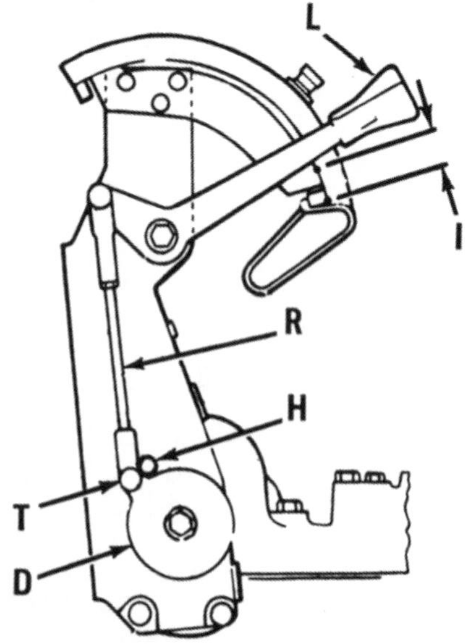

Fig. 193—Diagram of position control linkage typical of models equipped with cab.

D. Disc
H. Reference hole
I. 21.082-21.590 mm (0.83-0.85 inch)
L. Position control lever
R. Control rod
T. Disc tab

Fig. 194—Exploded view of lift control linkage. Refer also to Fig. 196.

1. Pin
2. Snap ring
3. Woodruff key
4. Flow control cam
5. Lever & roller
6. Draft control shaft
7. Cam follower roller
8. Snap ring
9. Thrust washer
10. Lift control lever shaft
11. Pin
12. Lever
13. Lift cylinder
14. Snap ring
15. Pivot pin
16. Valve actuating lever
17. Roller
18. Draft control lever & link
19. Spring
20. Stud
21. Pivot pin
22. Valve actuator
23. Lever
24. Snap ring
25. Snap ring
26. Snap ring

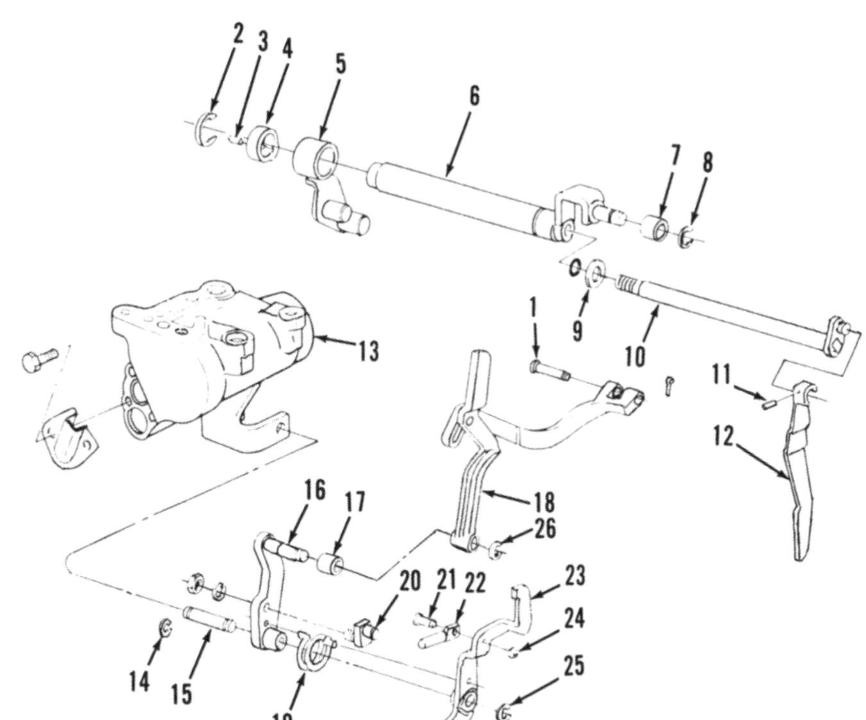

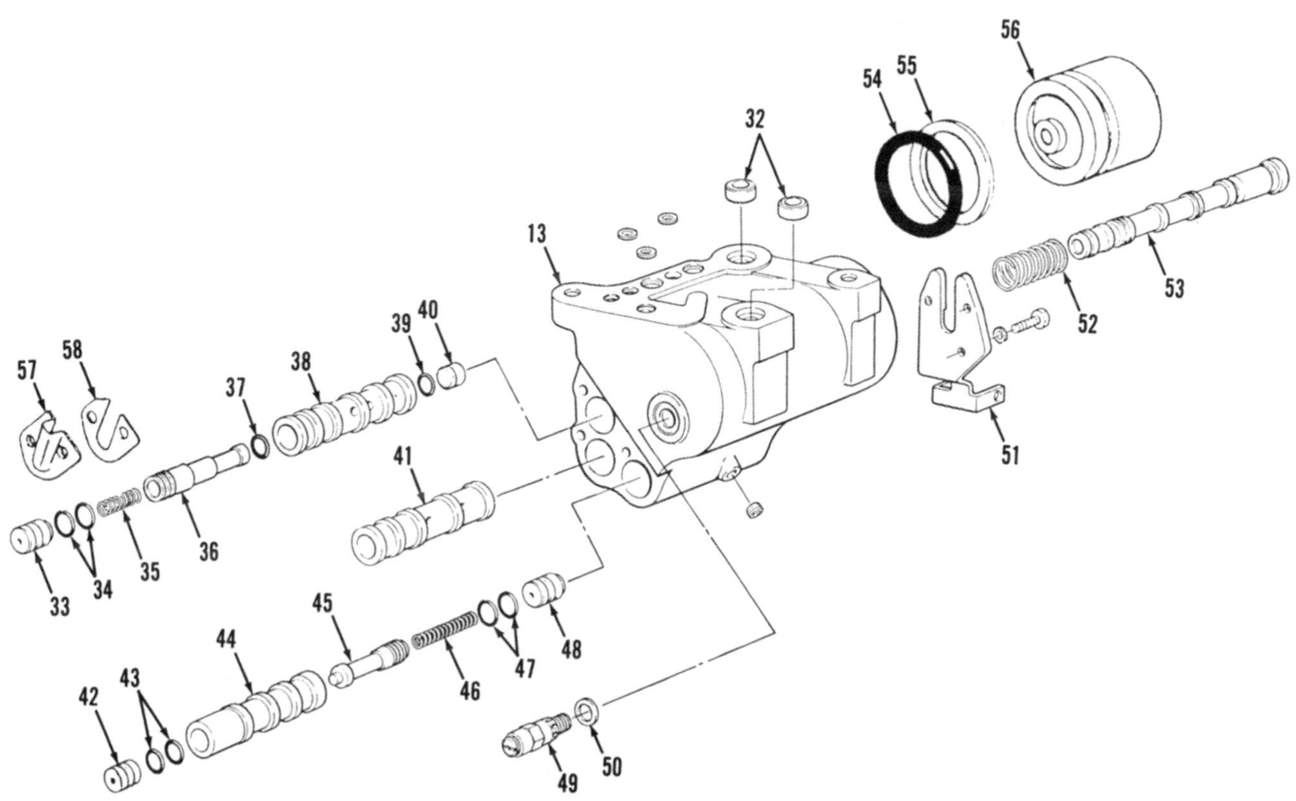

Fig. 195—Exploded view of typical hitch lift cylinder.

13. Lift cylinder
32. Hollow dowels
33. Plug
34. "O" rings
35. Spring
36. Unloading valve
37. "O" ring
38. Bushing
39. "O" ring
40. Plug
41. Bushing
42. Plug
43. "O" ring
44. Bushing
45. Feathering valve
46. Spring
47. "O" rings
48. Plug
49. Safety valve
50. Seal ring
51. Valve plate
52. Spring
53. Control valve spool
54. "O" ring
55. Back-up ring
56. Piston
57. Baffle
58. Gasket

To remove the safety valve, remove the lift cover as outlined in paragraph 178, then unscrew valve (49—Fig. 195) from the cylinder casting (13).

When reinstalling valve, use new "O" ring (50) and tighten valve to 112 N•m (83 ft.-lbs.) torque. Excessive torque may distort valve body and cause improper operation of the valve.

182. LIFT CYLINDER PISTON AND SEALS. To remove the lift cylinder piston, first remove the cylinder as outlined in paragraph 179, then remove safety valve as outlined in paragraph 181. Use a small rod inserted into valve opening and push piston from cylinder.

Inspect lift piston and cylinder bore for scoring or excessive wear and inspect cylinder casting for cracks. Renew cylinder if cracked or deeply scored. Install new piston if worn or scored.

To install leather back-up ring on piston, first soak in water for about five minutes. Install leather back-up ring on piston with rough side toward closed end of piston. Lubricate "O" ring in hydraulic fluid and install in groove at front (toward closed end of piston) of leather back-up ring. Allow leather back-up ring to shrink back to original size, lubricate piston and cylinder bore, then install piston in cylinder.

183. CONTROL VALVE AND BUSHING. To service the control valve, first remove the lift cylinder as outlined in paragraph 179. Remove screws retaining bracket (51—Fig. 195) and carefully separate control valve spool and actuating linkage from lift cylinder as shown in Fig. 196. Remove snap ring (24), pin (21) and actuator (22). Compress spring (52) and slide control valve spool (53) from bracket (51). Remove

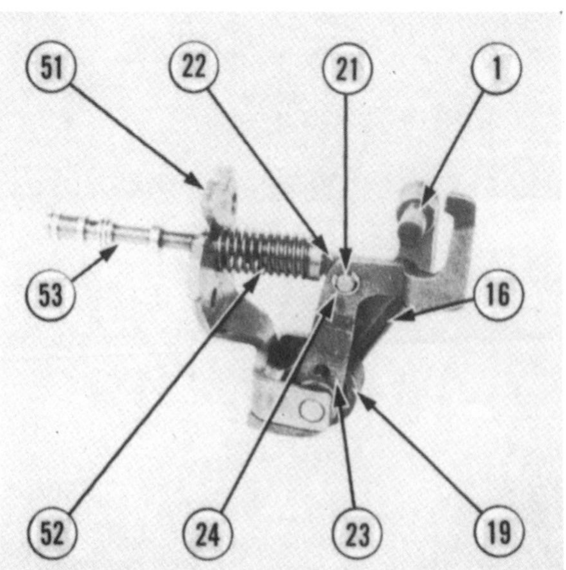

Fig. 196—View of control valve and linkage. Refer to Fig. 194 and Fig. 195 for legend.

baffle (57—Fig. 195) and gasket, then press bushing (41) out closed end of cylinder.

NOTE: Be especially careful when pressing bushing (41) from cylinder. If cylinder is cracked, scored or otherwise damaged while removing or installing bushing, cylinder must be renewed. Special removal and installing tools (Churchill T.8510 or Nuday 2191 N-508-A) are available to assist safe removal and installation.

Inspect lands on control valve spool for erosion or scoring. Renew valve spool and bushing if either is damaged in any way or if control valve leakage is indicated. Neither valve or bushing should be renewed without renewing mating part.

The control valve bushing is available in several different outside diameter ranges. Bushing is color coded to indicate size range and cylinder casting is also color coded near the bushing hole. Always renew bushing with one of same size range. Bushing outside diameter size range and color code are as follows:

CONTROL VALVE BUSHINGS

Size Range	Color Code
25.4000-25.4050 mm (1.0000-1.0002 inches)	Blue/White
25.4050-25.4101 mm (1.0002-1.0004 inches)	White
25.4101-25.4152 mm (1.0004-1.0006 inches)	Blue
25.4152-25.4203 mm (1.0006-1.0008 inches)	Yellow
25.4203-25.4254 mm (1.0008-1.0010 inches)	Green
25.4254-25.4304 mm (1.0010-1.0012 inches)	Orange
25.4304-25.4255 mm (1.0012-1.0014 inches)	Green/White

When installing new bushing, be sure that bore in cylinder and bushing are clean and free of nicks or burrs. Lubricate both bushings and bushing bores, then install bushing with oil passage holes toward closed end of cylinder.

Control valve spools are available in five different size ranges. The correct size valve can be determined by selective fit only after bushing has been pressed into cylinder casting. To check valve spool fit, lubricate valve spool and bushing, then insert spool into bushing from open end of cylinder. A slight drag should be felt on spool while moving in bushing through normal operating range of travel. If valve moves freely through bushing, select a valve spool of next larger size and check fit of the larger valve. If valve sticks or binds, select a smaller size. The correct size valve spool is the largest size that will slide freely through bushing from its own weight. The different valve ranges are color coded as follows:

CONTROL VALVE SPOOLS

Size Range	Color Code
15.0291-15.0342 mm (0.5917-0.5919 inch)	White
15.0342-15.0393 mm (0.5919-0.5921 inch)	Blue
15.0393-15.0444 mm (0.5921-0.5923 inch)	Yellow
15.0495-15.0520 mm (0.5925-0.5926 inch)	Green
15.0545-15.0571 mm (0.5927-0.5928 inch)	Orange

NOTE: Color code indicates a range only and a valve of one range may fit correctly while other valves of same color code may be too tight or too loose. Color code on bushing indicates outside diameter only, so do not attempt to match color codes of bushing and spool. It may happen that the same color code occurs on bushing and properly fitting spool, but it will be only coincidental.

Lubricate valve spool prior to installation. Refer to Fig. 197 when assembling actuating linkage. Tighten screws retaining baffle (57—Fig. 195) and screws retaining bracket (51) to 37 N•m (27 ft.-lbs.) torque. Refer to paragraph 179 for installing lift cylinder.

184. UNLOADING VALVE, BUSHING AND PLUGS. To remove the unloading valve, first remove control valve as outlined in paragraph 183. It is not necessary to remove the control valve bushing unless renewal is indicated. Proceed as follows:

Thread slide hammer adapter into unloading valve plugs (33 and 40—Fig. 195) and pull plugs from valve bores. Remove unloading valve (36) by pushing valve forward, then remove "O" ring (37). Remove unloading valve bushing (38) if renewal is necessary by pressing out toward closed end of cylinder.

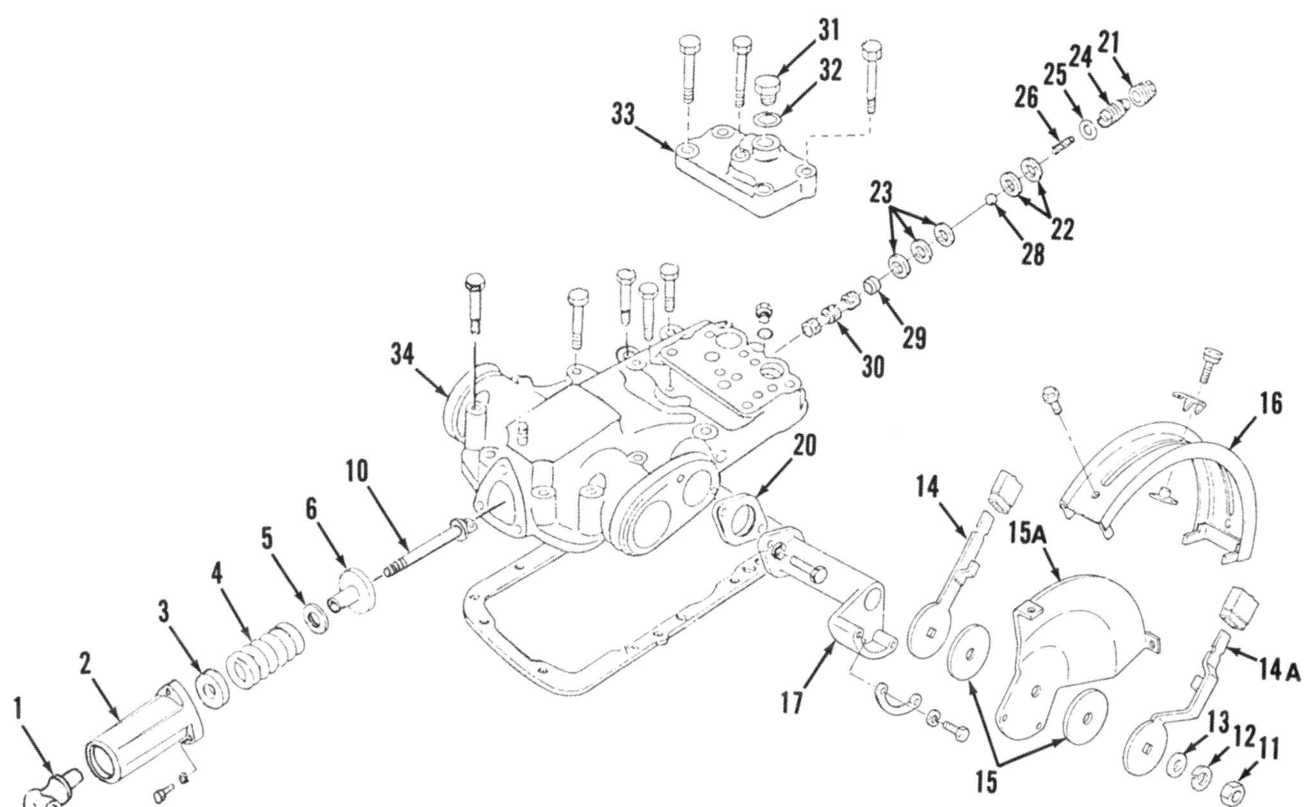

Fig. 197—Exploded view of lift cover assembly. Refer to Fig. 198 for quadrant and levers used on models equipped with cab.

1. Yoke
2. Housing
3. Rear seat
4. Spring
5. Shims
6. Front spring seat
10. Plunger
11. Self-locking nut
12. Spring
13. Washer
14. Draft control lever
14A. Position control lever
15. Friction discs
15A. Friction plate
16. Quadrant
17. Support
20. Gasket
21. Plug
22. Seal rings (0.364 in. ID)
23. Seal rings (0.489 in. ID)
24. Retainer
25. "O" ring (0.621 in. ID × 0.103 in.)
26. Spring
28. Ball (15/32 inch)
29. Seat
30. Sleeve
31. Plug
32. Seal ring
33. Accessory cover
34. Lift cover

NOTE: Be especially careful when pressing bushing (38) from cylinder. If cylinder is cracked, scored or otherwise damaged while removing or installing bushing, cylinder must be renewed. Special removal and installing tools (Churchill T.8510 or Nuday 2191 N-508-A) are available to assist safe removal and installation.

The unloading valve bushing (38) is available in eight outside diameter size ranges and new bushing should be the same size range as original. It is usually not necessary to install new unloading valve plugs unless damaged during removal procedure. Unloading valve bushing outside diameter size ranges and color codes for different size ranges are the same as listed in paragraph 183 for the control valve bushing, plus the additional larger size color-coded red/white that is 25.4355-25.4406 mm (1.0014-1.0016 inches).

When installing a new bushing, be sure lift cylinder bore is clean and free of nicks and burrs. Lubricate new bushing and bore in cylinder. Install bushing with large port holes toward open (rear) end of lift cylinder. Front of bushing should be flush with front (closed end) of lift cylinder.

Renew unloading valve (36) if scored, excessively worn or otherwise damaged. Valve is available in one size only. Lubricate valve and insert in bushing **WITHOUT** "O" ring (37). Valve should be a free sliding fit in bushing. Remove valve and install "O" ring. Lubricate valve, "O" ring and bushing bore, then reinstall valve in bushing. The "O" ring should provide a slight drag when moving valve back and forth in bushing. If valve sticks or binds, or if valve moves as freely as without "O" ring installed, check for proper sized "O" ring or other problem.

CAUTION: Do not install "O" ring of unknown quality or composition at this location. Some "O" ring material may shrink or swell when subjected to hydraulic fluid and heat, causing the unloading valve to malfunction.

Install unloading valve with cupped end toward closed end of lift cylinder. Install plugs with threaded hole out so outer face of plug is flush with machined surface of lift cylinder.

185. FEATHERING VALVE, BUSHING AND PLUGS. First remove control valve as outlined in paragraph 183. It is not necessary to remove the control valve bushing unless renewal is indicated. Proceed as follows:

Thread slide hammer adapter into feathering valve plugs (42 and 48—Fig. 195) and pull plugs from bushing. Feathering valve (45) can be removed after plugs. Bushing (44) can be removed from rear (open) end of lift cylinder if renewal is required.

NOTE: Be especially careful when pressing bushing (44) from cylinder. If cylinder is cracked, scored or otherwise damaged while removing or installing bushing, cylinder must be renewed. Special removal and installing tools (Churchill T.8510 or Nuday 2191 N-508-A) are available to assist safe removal and installation.

The feathering valve bushing (44) is available in seven different outside diameter size ranges and new bushing should be same size range as original. Feathering valve bushing outside diameter size ranges and color codes for different size ranges are the same as listed in paragraph 183 for the control valve bushing.

When installing a new bushing, be sure that lift cylinder bore is clean and free of nicks and burrs. Lubricate new bushing and bore in cylinder. Install bushing with larger diameter end toward open (rear) end of lift cylinder. Front of bushing should be flush with front (closed end) of lift cylinder.

Renew feathering valve (45) if scored, excessively worn or otherwise damaged. Valve is available in one size only. Lubricate valve and bushing bore, then insert valve in bushing with cupped end toward rear (open) end of lift cylinder. Valve should be a free sliding fit in bushing. Install plugs (42 and 48) with threaded hole out and outer face of plugs flush with machined surface of lift cylinder.

LIFT CONTROL LINKAGE AND SHAFT

All Models

186. HYDRAULIC SYSTEM CHECK VALVE. The check valve for the hydraulic lift system is located in front edge of lift cover.

Remove plug (21—Fig. 197). Use needlenose pliers to remove retainer (24), spring (26) and steel check valve ball (28). Withdraw nylon check valve seat (29) using a hooked wire.

NOTE: Sleeve (30) is pressed into lift cover and cannot be withdrawn for service.

Inspect check valve seat and valve ball for worn spots, nicks, burrs and any other damage. Renew any defective parts. Install new check valve spring if rusted, cracked or worn. Always renew sealing "O" ring (25) and seal rings (22).

Reassemble in reverse order of disassembly. Install and tighten retainer plug (21) to 68 N•m (50 ft.-lbs.) torque.

187. R&R INTERNAL LINKAGE AND SHAFT. Remove lift cover as outlined in paragraph 178 and lift cylinder as outlined in paragraph 179. Remove and disassemble the lift control linkage and lift shaft as follows:

Unscrew yoke (1—Fig. 197) from rear end of draft control plunger (10). Unbolt draft control main spring housing (2) from rear end of lift cover and slide the housing, seat (3), spring (4), shims (5) and spring seat (6) from draft control plunger.

NOTE: Special washer (W—Fig. 189) is available and may be installed in location shown if top link tension is not desired.

Place lift arms in lowered position and extract plunger (10—Fig. 197) along with draft control lever and link (18—Fig. 194) through rear opening in lift cover. Unbolt and remove lift arms from lift shaft and disconnect piston rod from lift ram arm. Note thrust washer (61—Fig. 199) on right side of lift arm. If not previously removed, refer to Fig. 197 or Fig. 198 and disassemble support, quadrant and lever components. With control lever shaft assembly removed as shown in Fig. 200, refer to Fig. 194 for disassembly and reassembly of control lever shaft assembly.

Need and procedure for further disassembly will be evident on inspection of removed parts. To reassemble and reinstall, reverse removal and disassembly procedure. Tighten control lever shaft support retaining cap screws to 56 N.m (41 ft.-lbs.) torque. Be sure that control lever is a free sliding fit over shaft. Install flat washer, spring washer and retaining nut, then tighten retaining nut until a force of 22-26 kg. (10-12 lbs.) is required to move the control lever. Tighten cap screws retaining lift arm to lift shaft evenly so all end play is removed, but without causing arms to bind. Lift arms should fall of their own weight. Bend tangs

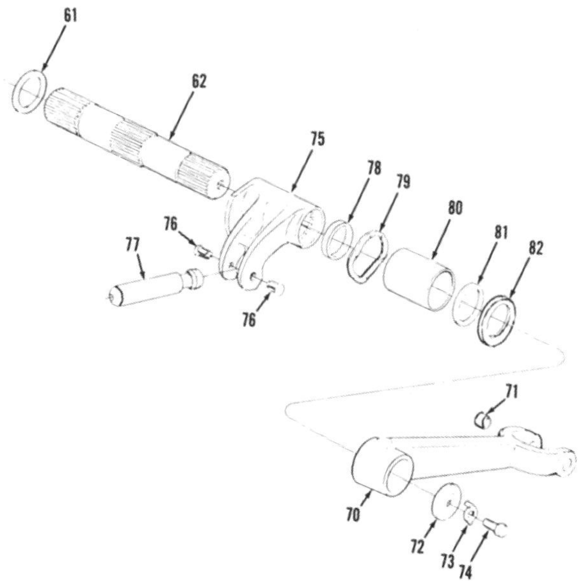

Fig. 199—Exploded view of hydraulic lift shaft assembly. Shaft (62), ram arm (75) and lift arms (70) have master splines that permit installation in one position only.

61. Spacer
62. Lift shaft
70. Lift arm
71. Bushing
72. Retainer
73. Lock
74. Bolt
75. Ram arm
76. Pins
77. Piston rod
78. Sleeve
79. Wave washer
80. Bushing
81. Dust seal
82. Washer

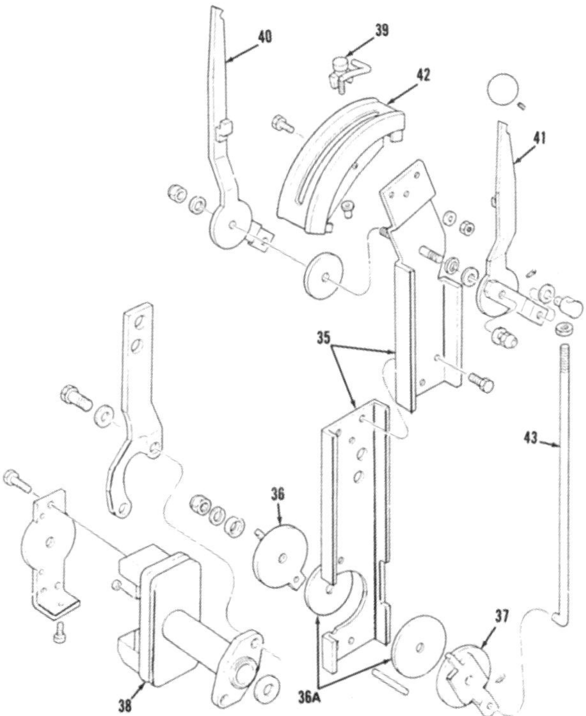

Fig. 198—Exploded view of quadrant, levers and linkage used on models equipped with a cab.

35. Bracket
36. Inner disc
36A. Friction discs
37. Outer disc
38. Support
39. Lever stop
40. Position control lever
41. Draft control lever
42. Quadrant
43. Control rod

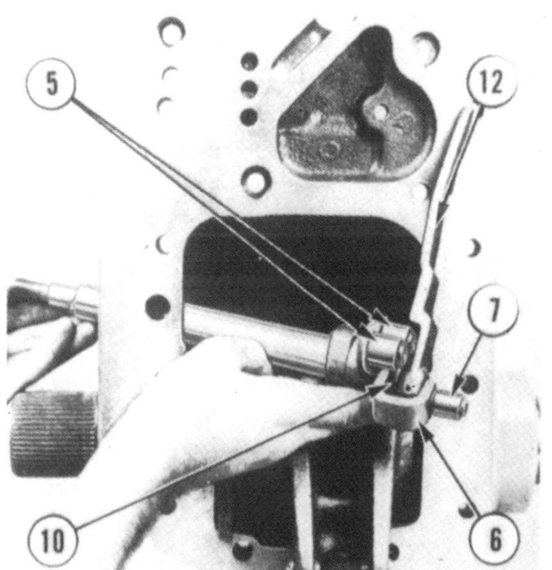

Fig. 200—View showing hydraulic lift control linkage removal. Refer to Fig. 194 for legend.

of retainers (73—Fig. 199) against cap screws when lift shaft end play is properly adjusted.

Refer to the appropriate paragraphs for adjustment of draft control main spring, draft control selector shaft, draft and position control linkage and flow control valve linkage.

FLOW CONTROL VALVE AND OPERATING LINKAGE

Models So Equipped

188. OPERATING PRINCIPLES. Turning flow control knob (30—Fig. 201) or lever (47) adjusts position of restrictor (9) in pressure line from hydraulic pump. Position of flow control valve spool (20) is controlled by full pump pressure against inner end of spool and by reduced pressure via drilled ports from passage at upper side of restrictor aided by pressure from spring (19). According to position of restrictor, control valve spool moves in bore to dump some oil back into sump via tube (43). When the 3-point hitch control lever is moved to top of quadrant (raising position), a cam on control lever shaft forces plunger (6, 7 and 8) downward and moves restrictor to full-flow position through connecting linkage (3 and 10).

189. ADJUST FLOW CONTROL VALVE LINKAGE. The flow control restrictor is returned to full flow position by moving lift control lever to top of quadrant (raising position). To properly adjust linkage between cam on control lever shaft and restriction, proceed as follows:

Remove hydraulic lift cover and cylinder assembly as outlined in paragraph 178. Move lift control lever to bottom of quadrant and, using a depth gage, measure distance (A—Fig. 202) between lower machined face of lift cover and nearest point of cam on lift control shaft.

Turn the flow control restrictor knob or lever fully clockwise and insert flow control override adjuster

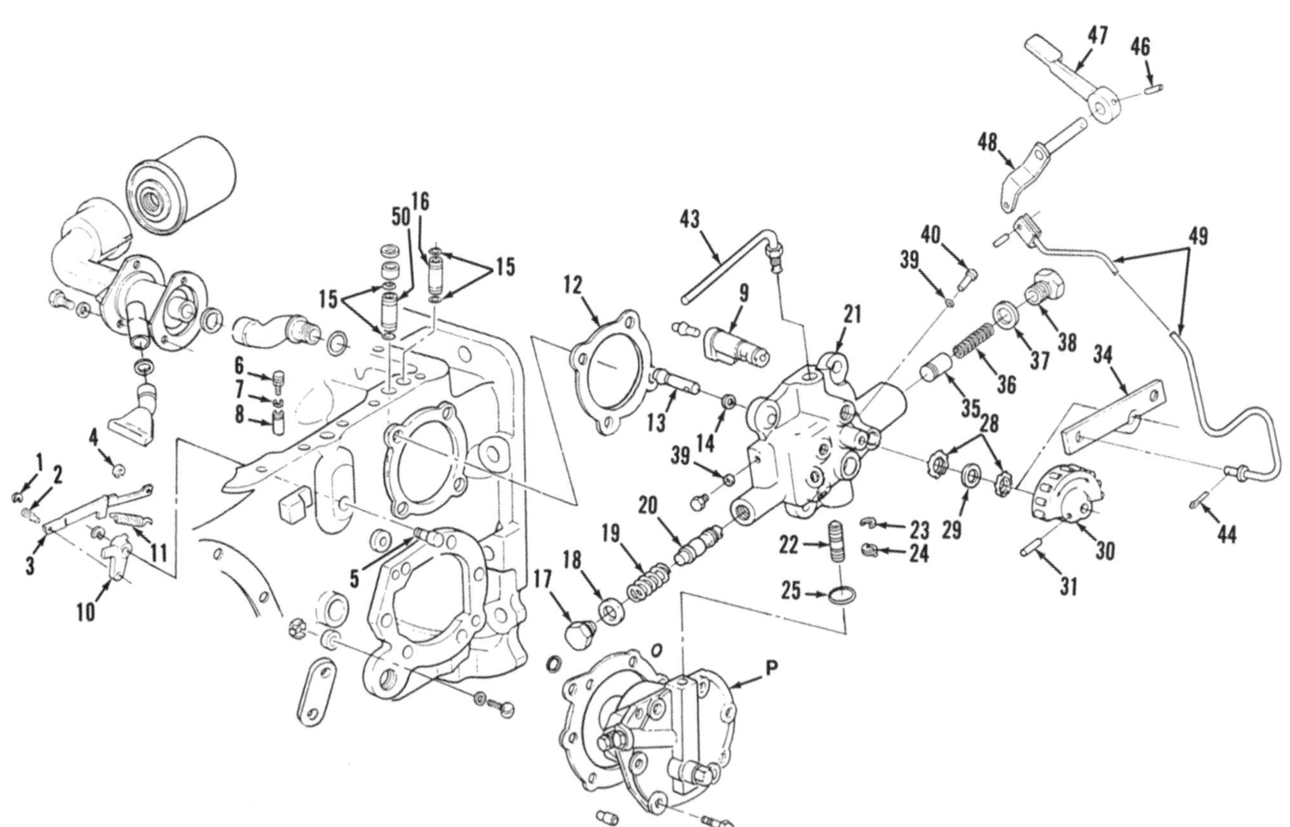

Fig. 201—Exploded view of flow control valve and linkage. Pump mounted in rear axle center housing is shown at (P).

1. Retaining ring
2. Pin
3. Connector link
4. Retaining ring
5. Pivot pin
6. Cam follower
7. Shims
8. Adjuster
9. Flow control restrictor valve
10. Lever
11. Spring
12. Gasket
13. Shaft
14. "O" ring (0.344 × 0.070 in.)
15. "O" rings (0.500 × 0.070 in.)
16. Pressure tube
17. Plug
18. Seal ring
19. Spring
20. Flow control valve
21. Valve plate
22. Pressure tube
23. Snap ring
24. Seal washers (2)
25. "O" rings (same as 15)
28. Spring washers (2)
29. Flat friction washers (2)
30. Flow control knob
31. Roll pin
34. Lever (w/cab)
35. Oil cooler valve
36. Spring
37. Washer
38. Plug
39. Seals (2)
40. Retaining pin
43. Tube
44. Pin (w/cab)
46. Roll pin (w/cab)
47. Lever (w/cab)
48. Arm & shaft (w/cab)
49. Control rod (w/cab)
50. Return tube

into bore of rear axle center housing, then measure protrusion of adjuster above machined surface of rear axle center housing. The adjuster should protrude 0.254 mm (0.010 inch) more than measurement from face of lift cover to cam on control lever shaft dimension (A—Fig. 202) plus 0.254 mm (0.010 inch).

If necessary, unscrew cam follower (6—Fig. 201) from top of adjuster (8) and add or remove shims (7) as required to obtain correct protrusion of adjuster. Shims are available in thicknesses of 0.254 mm (0.010 inch) only.

To adjust linkage on models with a cab, detach control rod (49—Fig. 203) and turn flow control valve lever (34) and hand lever (47) to 3 o'clock position. Adjust control rod (49) length to connect with levers.

190. REMOVE AND REINSTALL. To remove flow control valve, first drain hydraulic system (rear axle center housing) and remove the hydraulic lift cover and cylinder assembly as outlined in paragraph 178. Detach oil cooler line from flow control valve housing and loosen the four screws retaining the flow control valve housing. Extract the return feed pipe (50—Fig. 204), disconnect spring (11) and withdraw flow control cam follower assembly (6). Remove snap ring (4) and detach link from restrictor pin. Disconnect oil exhaust elbow (41) from flow control valve housing and extract upper tube (16). On models with pump (P—Fig. 201) mounted in the rear axle center housing, the lower pressure pipe (22) is external. Remove snap ring (23) and push or drive pipe upward into valve until clear of hydraulic pump flange. On models with only an engine-driven pump, lower pressure pipe (26—Fig. 204) is internal as shown. Remove snap ring and push or drive lower pressure pipe (26) downward until pipe is clear of flow control valve. On all models, remove retaining screws, then remove the flow control valve.

To reinstall flow control valve, reverse removal procedure. Tighten valve retaining screws to 56 N•m (41 ft.-lbs.) torque. Adjust flow control valve linkage as outlined in paragraph 189. Refer to paragraph 178 to reinstall hydraulic lift cover. Refill axle center housing with lubricant (see paragraph 168).

191. OVERHAUL. Refer to Fig. 201, then unscrew plug (17) and remove spring (19) and valve spool (20). Inspect spring for wear, cracks, distortion and free

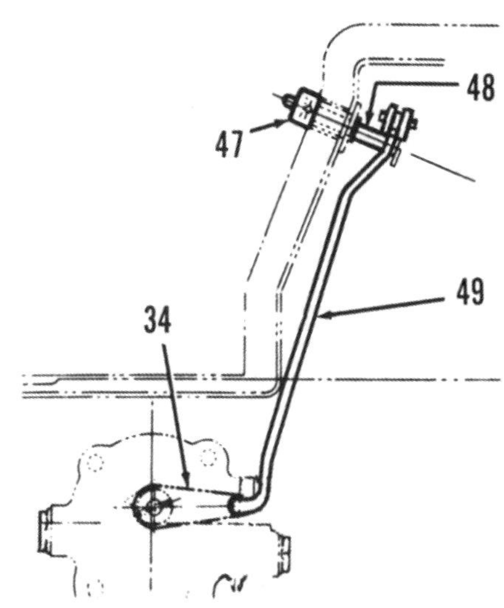

Fig. 203—Drawing of flow control valve linkage typical of models with cab. Refer to Fig. 201 for legend.

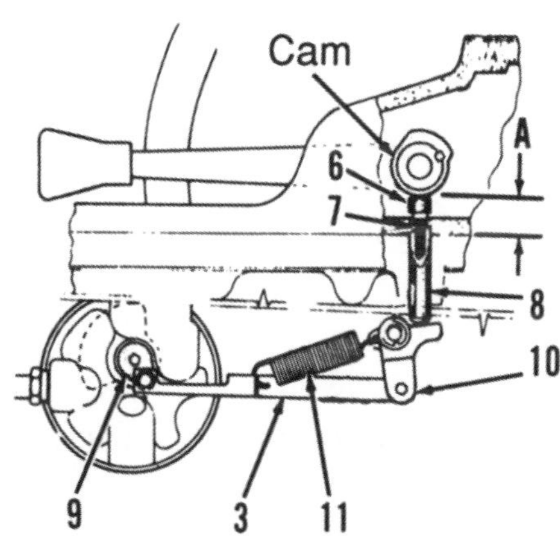

Fig. 202—Drawing of flow control override linkage and actuating cam. Dimension "A" is from machined surface of lift cover to cam on lift control lever shaft.

3. Connector link
6. Cam follower
7. Shims
8. Adjuster
9. Flow control restrictor valve
10. Lever
11. Spring

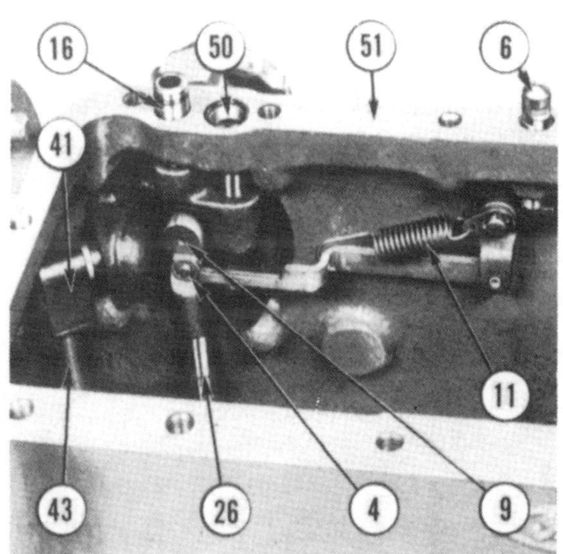

Fig. 204—View of flow control assembly for models with engine driven hydraulic pump. Other models are similar. Refer to Fig. 201 for legend except for lower pressure pipe (26), exhaust elbow (41) and center housing (51).

length. Install new spring if length is less than a new spring or if spring is damaged in any way.

Flow control valve spool (20) should be free sliding fit in bore of housing. Renew spool if worn or scored. A new valve spool with the same color coding as old valve is usually installed, but the next size can be installed if it slides freely in bore. Valve should not bind or drag. Valves are color coded as follows:

Valve Outside Diameter	Color Code
16.9418-16.9468 mm (0.6670-0.6672 inch)	Red
16.9468-16.9519 mm (0.6672-0.6674 inch)	Yellow
16.9519-16.9570 mm (0.6674-0.6676 inch)	Blue
16.9570-16.9621 mm (0.6676-0.6678 inch)	Green
16.9621-16.9672 mm (0.6678-0.6680 inch)	White

If flow control valve bore is scored or worn excessively, complete valve and housing must be renewed.

Remove restrictor retaining pin (40) and pull restrictor (9) from bore. Renew restrictor if scored or worn; pin is available separately. Renew flow control valve and housing assembly if restrictor bore is worn excessively or scored. Inspect tip of retainer pin (40) and renew if worn or damaged. Unscrew plug (38), then remove oil cooler valve spool (35) and spring (36).

Refer to exploded view (Fig. 201), cross section (Fig. 202) and assembled view (Fig. 204) when disassembling and reassembling operating linkage. Renew bent or otherwise damaged components.

PRIORITY VALVE

Models So Equipped

192. OPERATING PRINCIPLES. The priority (unload) valve is mounted on the hydraulic lift cover and is installed when remote control valves and dual pump are installed. This valve regulates oil flow in remote control valve circuit and combines auxiliary pump flow with main pump flow to maintain constant pressure. A pressure relief valve is installed and an opening of 18,615-18,960 kPa (2700-2750 psi) protects auxiliary pump and remote valve circuit.

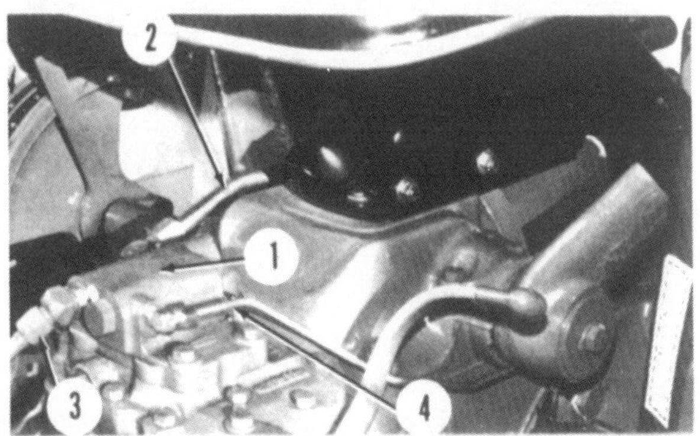

Fig. 205—View showing location of priority (unload) valve typical of all models. Exploded view of priority valve is shown in Fig. 206.

193. REMOVE AND REINSTALL. Disconnect out line (2—Fig. 205), auxiliary pump supply line (3) and pressure sensing pilot line (4). Plug all openings to prevent entrance of contaminants. Remove mounting screws and separate priority valve from lift cover.

Notice location of oil passage "O" rings in bottom of valve assembly. Remove and discard "O" rings.

Reinstall valve assembly in reverse of removal. Position "O" rings in counterbores in bottom of valve housing using light grease. Tighten retaining screws to 81 N•m (60 ft.-lbs.) torque.

194. OVERHAUL. Refer to paragraph 193 and remove valve assembly. Remove plug (16—Fig. 206) and withdraw spring (14), guide (13) and ball (12). Remove plug (9) and withdraw spring (7) and valve assembly (2 through 6).

Clean all parts in a suitable cleaning solvent and blow dry with clean compressed air. Place compo-

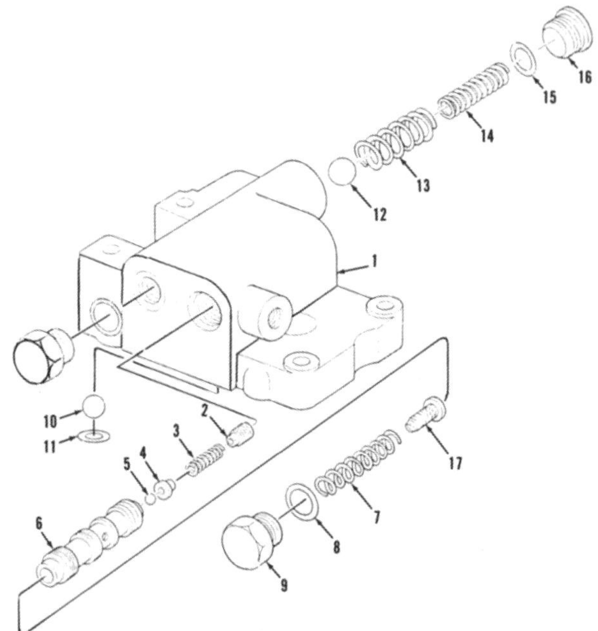

Fig. 206—Exploded view of priority (unload) valve assembly.

1. Housing
2. Plug
3. Spring
4. Seat
5. Ball
6. Valve
7. Spring
8. "O" ring
9. Plug
10. Ball
11. Ring
12. Ball
13. Guide
14. Spring
15. "O" ring
16. Plug
17. Filter

nents on clean, lint-free cloth. Inspect valve (6) for excessive wear, scoring, burrs and for freedom of movement in bore. Inspect ball (12) for corrosion and excessive wear. Inspect springs for cracks, weakness and other damage. Examine valve housing for cracks or restricted passages and bores for scoring and excessive wear. Renew parts as needed.

Reassemble in reverse order of disassembly. Renew filter (17) and all "O" rings. Lubricate valve components with hydraulic oil during reassembly. Tighten plugs securely.

HYDRAULIC PUMP

Engine-Mounted Pump

195. REMOVE AND REINSTALL. A gear-type hydraulic pump is mounted on the rear left side of the engine and is driven by a gear attached to rear of engine camshaft.

To remove pump, first thoroughly clean area around pump and hydraulic lines. Disconnect pump outlet tube (30—Fig. 207), unbolt pump housing (8) from engine, then lift pump from engine, pulling inlet tube (31) from inlet bore. Cover all openings to prevent entrance of dirt.

Reverse removal procedure to install pump. Install new "O" rings on pump lines and new gasket between housing and engine. Tighten retaining screws to 41 N•m (30 ft.-lbs.) torque.

196. OVERHAUL. Mark the drive housing (8—Fig. 207), body (16) and cover (18) before disassembling so they can be reassembled in original positions. Unscrew filter (19), then remove retainer (1) and plug (2). Straighten tabs of lock washer (5), then remove nut (4) and lock washer. Remove the four through-bolts and use a soft mallet to separate housing, body and cover. Tap end of shaft to release drive gear (6) from taper of pump drive shaft. Note position of bearings and gears before removing from body. Re-

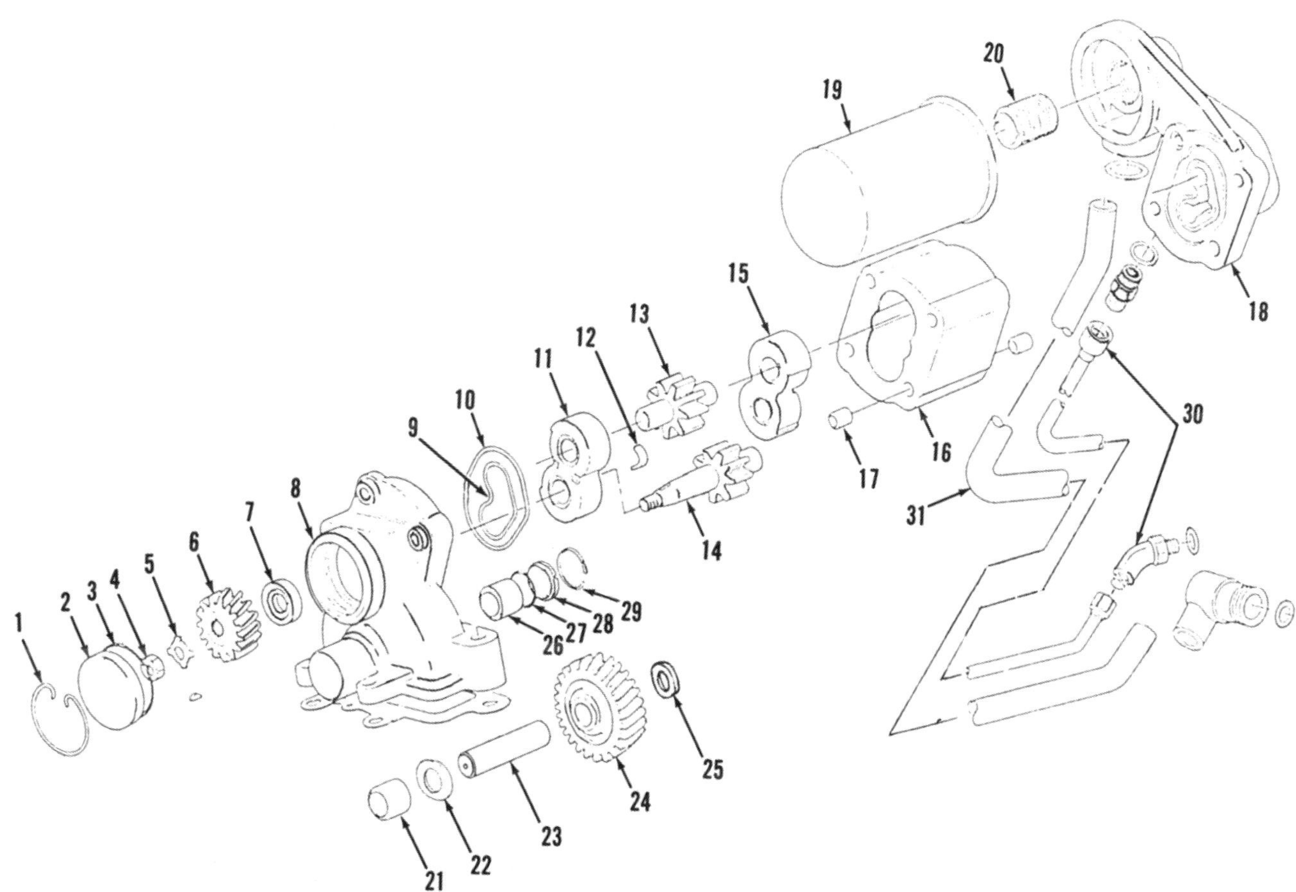

Fig. 207—Exploded view of engine-mounted hydraulic pump.

1. Clip
2. Plug
3. "O" ring
4. Nut
5. Tab washer
6. Gear
7. Seal
8. Housing
9. Seal ring
10. Seal ring
11. Bearings
12. Stuffer strip
13. Driven gear
14. Drive gear
15. Bearings
16. Body
17. Dowels
18. Cover
19. Filter
20. Adapter
21. Bushing
22. Washer
23. Idler shaft
24. Idler gear
25. Washer
26. Bushing
27. "O" ring
28. Plug
29. Snap ring
30. Pressure tube
31. Inlet tube

move snap ring (29), plug (28) and "O" ring (27). If plug is not easily removed, compressed air can be directed behind plug through bleed hole in pump mounting surface. Remove idler shaft (23), idler gear (24) and washers (22 and 25).

Clean all parts and inspect carefully. Install new pump body if scored or if gear track is worn more than 0.101 mm (0.004 inch) deep into inlet side of body. Maximum runout across gear face to tooth edge (measured at 90 degrees to gear centerline) should not exceed 0.025 mm (0.001 inch) and bearing journals on each side of gear should be within 0.012 mm (0.0005 inch) of the other side. There must be no more than 0.005 mm (0.0002 inch) difference in widths of drive gear (14) and driven gear (13). Remove light score marks from gear bearing faces with "0" grade emery paper on flat lapping plate. "0" grade emery paper also can be used to remove light scoring from gear shaft journals. Bearings (11 and 15) and gears (13 and 14) are available only as matched sets.

Clean parts thoroughly, air dry, then lubricate with clean hydraulic fluid before assembling. Reassemble by reversing disassembly procedure. Install bearings (11 and 15) so notches in bearing faces are toward gears and relieved area on outside of bearing is toward outlet side of body. Pack high-temperature grease between lips of shaft seal (7). Tighten through-bolts to 50 N•m (37 ft.-lbs.) torque.

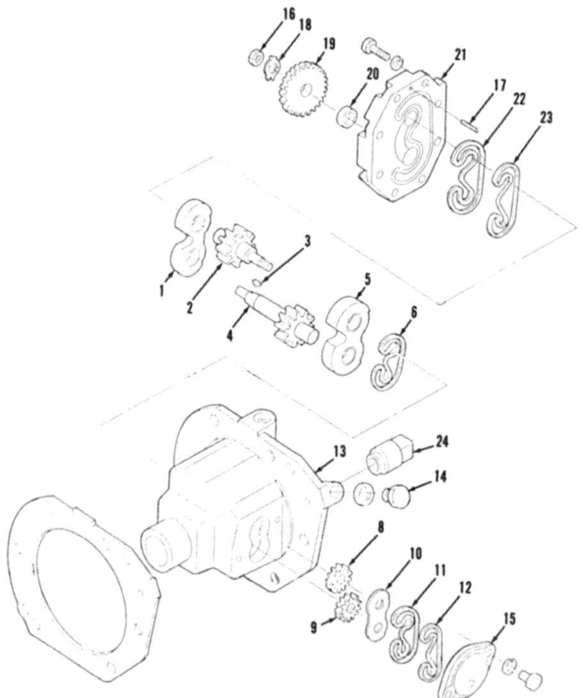

Fig. 208—Exploded view of hydraulic pump mounted in the rear axle center housing.

1. Bearing
2. Driven gear
3. Key
4. Drive gear & shaft
5. Bearing
6. Seal
8. Drive gear
9. Driven gear
10. Pressure plate
11. Outer seal
12. Inner seal
13. Pump body
14. Plug
15. Rear cover
16. Nut
17. Roll pin (1/8 × 0.437 in.)
18. Lock washer
19. Gear
20. Seal
21. Front cover
22. Outer seal ring
23. Inner seal ring
24. Pressure relief valve

Pump Mounted In Axle Center Housing

197. REMOVE AND REINSTALL. To remove the hydraulic pump from right side of rear axle center housing, drain lubricant from center housing and remove lift cover as outlined in paragraph 178. Disconnect brake return springs and remove right platform (step) from models without cab. On models with cab, disconnect brake return rods. On all models, remove brake levers that interfere with removal of pump. Pry snap ring (23—Fig. 201) from groove in pipe (22) between hydraulic pump (P) and flow control valve housing (21) and drive pipe (22) upward until clear of pump flange. Remove retaining screws, detach internal oil lines and remove pump.

Reverse removal procedure when installing pump. Tighten mounting screws to 68 N•m (50 ft.-lbs.) torque. Refer to paragraph 178 when installing lift cover and to paragraph 168 for filling rear axle center housing with correct lubricant.

198. OVERHAUL. Bend lock washer (18—Fig. 208) away from nut (16), then remove nut. Remove gear (19) and key (3) from pump drive shaft. Unbolt front and rear covers (15 and 21), then remove components. Note original location of bearings (1 and 5) so they can be reinstalled in same position.

Clean and carefully inspect all parts. Install new pump body (13) if scored or if gear track is more than 0.101 mm (0.004 inch) deep on inlet side. Maximum runout across gear face to tooth edge (measured at 90 degrees to gear centerline) should not exceed 0.025 mm (0.001 inch) and diameter of bearing journals on each side of gear should be within 0.012 mm (0.0005 inch) of the other side. There must be no more than 0.005 mm (0.0002 inch) difference in widths of drive gear (4 or 8) and driven gear (2 or 9). Remove light score marks from gear bearing faces with "0" grade emery paper on flat lapping plate. "0" grade emery paper also can be used to remove light scoring from gear shaft journals. If pump gears (2, 4, 8 and 9) are damaged, complete pump must be renewed.

Clean parts, air dry, then lubricate with clean hydraulic fluid before assembling. Reassemble by reversing disassembly procedure. Install bearings (1 and 5) so notches in bearing faces are toward gears and relieved area on outside of bearing is toward outlet side of body. Install pressure plate (10) with notches in outer edge toward outlet side of pump. Pack high-temperature grease between lips of shaft seal (20).

REMOTE CONTROL VALVE

AUXILIARY SELECTOR CONTROL VALVE

All Models So Equipped

199. OPERATION. The auxiliary selector control (A.S.C.) valve provides operation of single-acting remote hydraulic cylinders, either in conjunction with or independent from the 3-point lift cylinder, from the tractor system control valve. Valve is operated by knob (2—Fig. 209 or 1—Fig. 210) or lever (20—Figs. 209 and 210).

When spool (3 or 14—Fig. 210) is fully in, control valve directs pressure to the 3-point hitch only. When valve is pulled fully out, pressure is directed to selector valve port (10) only. When the valve is located at mid-position, pressure is directed to both 3-point lift cylinder and remote cylinder port of selector valve.

NOTE: The lift arms of 3-point hitch must be slightly lowered before pulling selector valve spool out to operate remote cylinder independently of 3-point hitch.

200. R&R AND OVERHAUL. To remove the selector valve, first place selector lever on lift cover in draft control position, move control lever to bottom of quadrant and stand on lift arms to force all oil from 3-point hitch lift cylinder. Remove valve from top of lift cover and "O" rings from counterbores in bottom of valve body.

To disassemble valve on models without a cab, proceed as follows: Remove plug (5—Fig. 210), spring

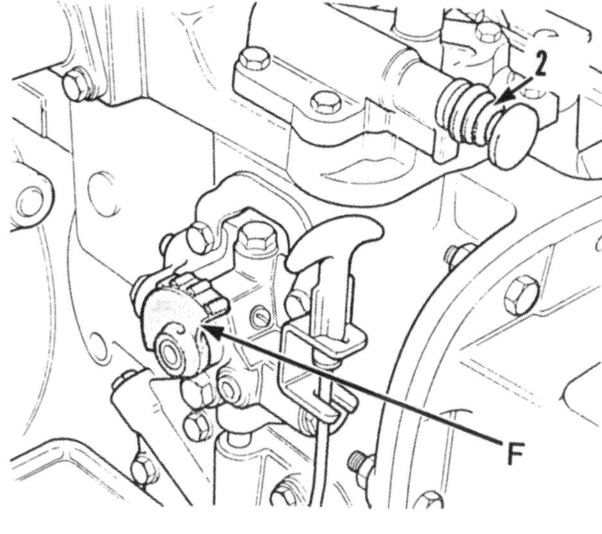

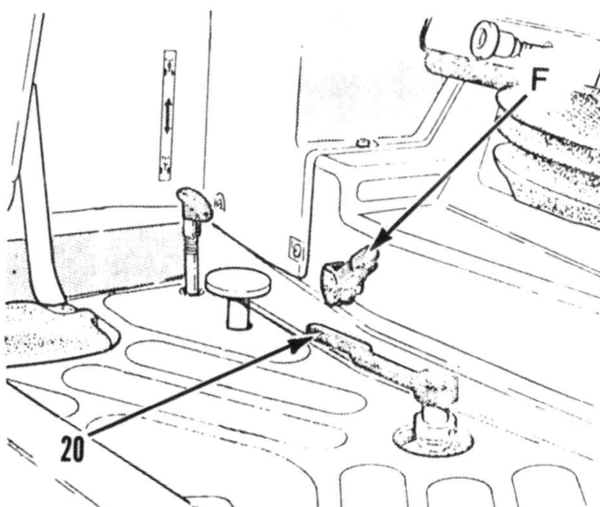

Fig. 209—View of installed Auxiliary Service Control (A.S.C.) selector valve installed on models without cab, at top and with cab as shown in bottom illustration.

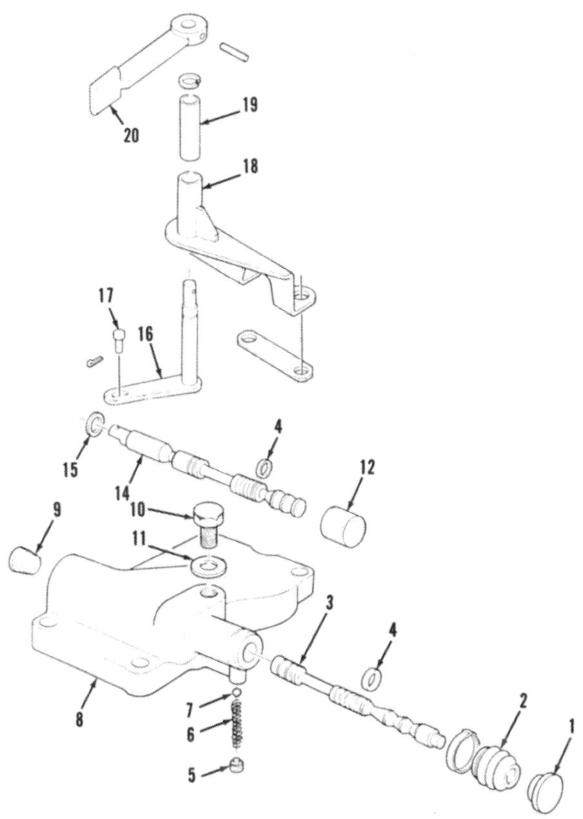

Fig. 210—The A.S.C. selector valve used on models with cab and without cab. Spool, linkage and related parts (12-20) are used in place of parts (1, 2, 3 & 9), which are used only on models without cab.

1. Knob
2. Boot
3. Valve spool
4. "O" ring (7/16 × 0.070 inch)
5. Plug
6. Spring
7. Detent ball
8. Valve body
9. Plug
10. Port sealing plug
11. Seal washer
12. Boot
14. Spool
15. Seal (0.564 × 0.139 inch)
16. Lever & shaft
17. Clevis pin
18. Bracket
19. Bushing
20. Lever

(6) and ball (7). Loosen or remove clamp, then pull boot (2) from boss of valve body and withdraw spool (3). Remove "O" ring (4) from groove in valve spool, unscrew knob (1) and remove boot. Remove plug (9) for cleaning.

To disassemble valve on models with a cab, proceed as follows: Remove cotter pin and withdraw clevis pin (17), then separate operating controls (16 through 20) from valve and spool assembly. Remove plug (5), spring (6), ball (7) and boot (12), then withdraw valve spool (14). Remove "O" ring (4) from groove in valve spool and remove seal (15) from valve body.

On all models, inspect spool and bore in body for damage. Complete valve assembly must be renewed if either is worn or otherwise unserviceable.

To assemble, install new "O" ring (4) on valve spool, lubricate valve spool liberally and insert into bore of valve body (8). Remainder of assembly is reverse of disassembly procedure.

Clean mounting surfaces and install new "O" rings in body counterbores. Tighten retaining screws to 56 N•m (41 ft.-lbs.) torque. Plug (10) and seal (11) block port to control valve for remote hydraulic cylinder.

SINGLE- AND DOUBLE-SPOOL REMOTE CONTROL VALVES

All Models So Equipped

201. OPERATION. Single- or double-spool valves may be used and may or may not be equipped with detents. Valve body and spool are a select fit and are not available separately. Spools for double spool valves must be identified when disassembling so that they can be returned to original bore.

Quick-drop valve (38 through 41—Fig. 212) allows rapid retraction of single-acting cylinders. Quick-drop valve may be installed in place of float valve on single spool control valves. Turn quick-drop valve out for use with single-acting cylinders or in for double-acting cylinders.

Disassembly of all valves will be the same except double spool valve will have approximately twice as many parts and the non-detent valves do not include detent assemblies.

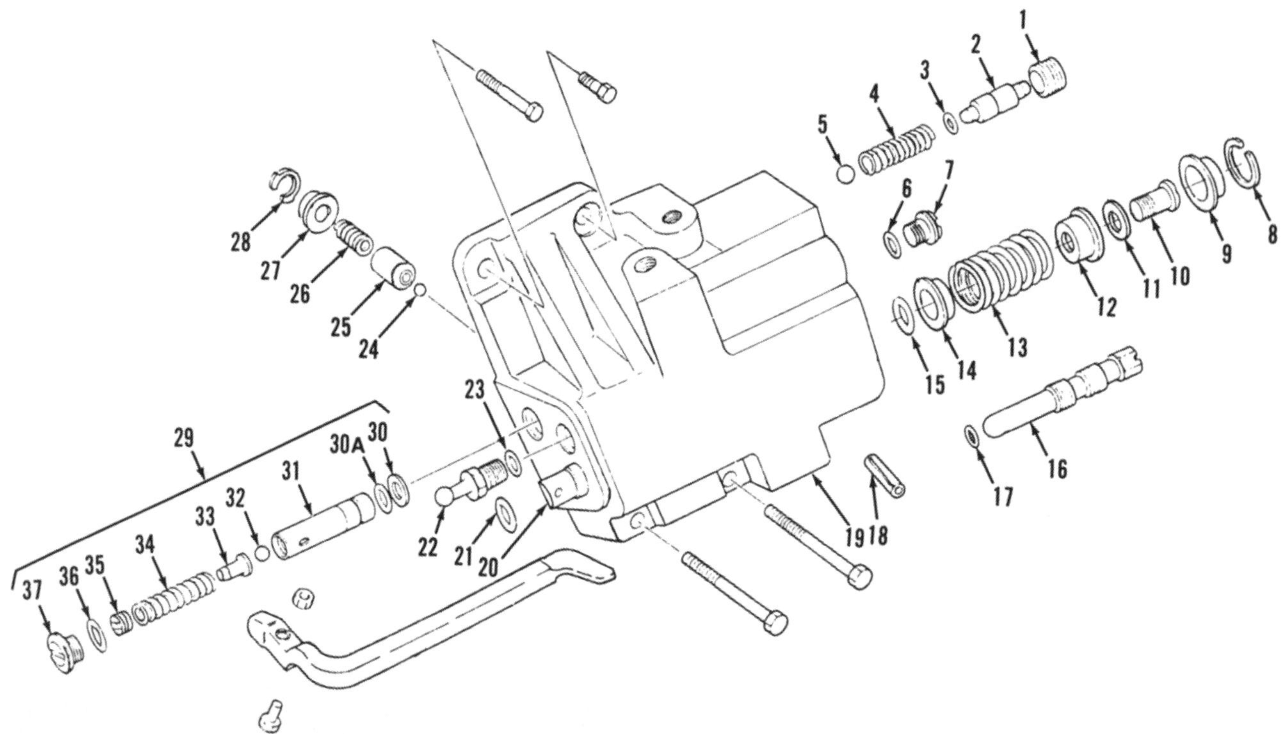

Fig. 211—Exploded view of single spool remote control valve. Valve shown has detent (items 24 through 28). Non-detent valves, except for detent assembly, remain the same. Detent valve can be made to function as non-detent valve by removing only detent valve spring (26).

1. Plug (5/8 - 18 × 5/8 inch)
2. Check valve plug
3. Seal
4. Spring
5. Check valve ball (3/8 inch)
6. Seal
7. Plug
8. Retaining ring
9. End cap
10. Screw (1/4 - 20 × 1/2 inch)
11. Washer (1/4 inch)
12. Spring retainer
13. Centering spring
14. Spring retainer
15. "O" ring (0.625 × 0.103 inch)
16. Float valve stem
17. Seal
18. Spring pin (1/8 × 5/8 inch)
19. Valve body
20. Valve spool
21. "O" ring (0.625 × 0.103 inch)
22. Handle pivot
23. Seal
24. Detent ball
25. Poppet
26. Spring
27. Spring retainer
28. Snap ring
29. Detent regulating valve assy.
30. Seal (0.468 × 0.078 inch)
30A. Back-up ring
31. Valve body
32. Steel ball (1/8 inch)
33. Spring retainer
34. Spring
35. Adjusting plug
36. "O" ring
37. Plug

202. ADJUST DETENT REGULATING VALVE. The valve (29—Fig. 211 or Fig. 212) should be adjusted to release spool detent poppets at a pressure slightly higher than that of normal cylinder operating pressure. When so adjusted, control valve spool will be returned to neutral position without excessive pressure buildup when remote cylinder reaches end of stroke.

To check opening pressure, connect a suitable pressure gage and loading valve to lift coupling and route testing hose into rear axle center housing filler opening. Open loading valve and run engine at about 1700 rpm. Warm hydraulic oil by operating system. Move remote control valve to raise position, then slowly close loading valve while watching pressure gage. Note pressure reading when valve moves remote control lever to neutral position. Pressure should be 13,790-14,670 kPa (2000-2100 psi). If pressure is incorrect, remove valve plug (37), then turn plug (35) with the proper size Allen wrench. Turn plug IN to increase release pressure or OUT to lower release pressure. Reinstall plug with new "O" ring (36).

203. REMOVE AND REINSTALL VALVE. Remove access cover over valves, then remove clevis pins that attach control rods to the remote control valve spools of models with cab. On all models, disconnect hoses from valve, remove valve attaching screws, then carefully remove valve assembly.

Reinstall by reversing removal procedure. Be sure to clean all mating surfaces and tighten retaining screws to 56 N•m (41 ft.-lbs.) torque.

204. OVERHAUL. Remove valve as outlined in paragraph 203 and disengage or remove lever arm from spool if still attached. Remove snap rings (28—Fig. 211 or Fig. 212), spring retainer (27), detent spring (26), poppet (25) and detent ball (24).

NOTE: On double spool detent valves, keep spools identified with their bores as each spool and bore is a matched fit.

Remove retaining ring (8) and end cap (9). Push on handle end of spool to remove spool and centering spring assembly from valve body (19). Remove screw

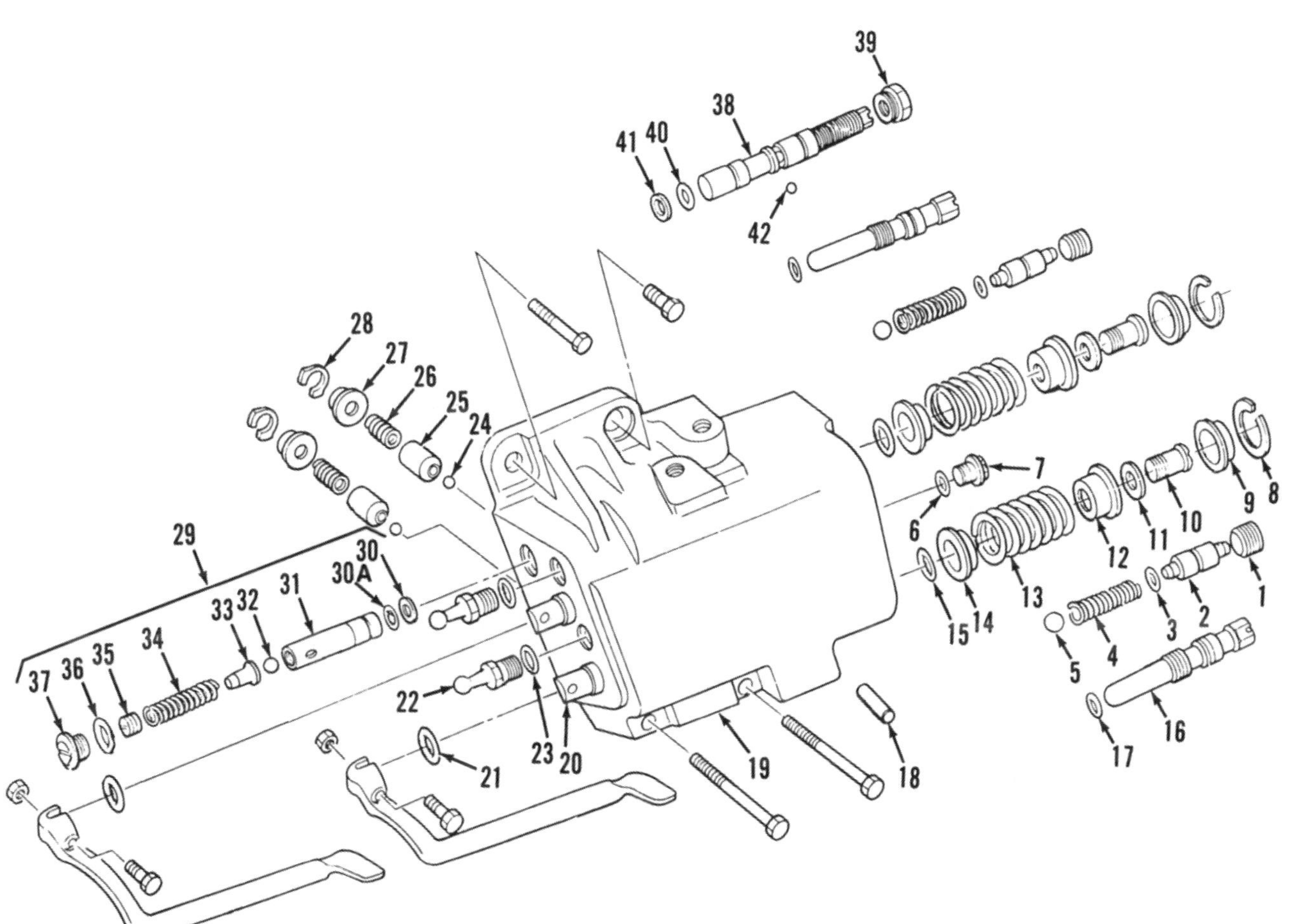

Fig. 212—Exploded view of double spool remote control valve. Except for detent assembly, (items 24 through 28), non-detent valves remain the same. The inboard spool, without detents, is always double acting and cannot be converted to single acting operation. Refer to Fig. 211 for legend, except the following: (38) Quick-drop valve, (39) Nut, (40) Seal (0.426 inch), (41) Seal (0.364 inch) and (42) Ball (5/16 inch).

(10), washer (11), spring retainer (12), centering spring (13) and spring retainer (14) from spool, then remove "O" rings (15 and 21). Remove plug (37) and "O" ring (36), then pull regulating valve (29) from body. Remove adjusting plug (35), regulating valve spring (34), spring retainer (33) and ball (32) from regulating valve body (31).

Remove socket head plug (1), then remove check valve plug (2) with seal (3), spring (4) and check valve ball (5) from valve body. To remove float valve (16), drive pin (18) out, then unscrew float valve. Unscrew quick-drop valve retaining nut (39) and withdraw quick-drop valve (38). **Nut (39) and matching threads on valve (38) are left-hand and require turning in clockwise direction to remove.**

Thoroughly clean all parts and inspect carefully. If spool or spool bore are excessively worn or otherwise damaged, complete valve must be renewed. Free length of new centering spring is 54.787 mm (2.157 inches). Renew centering springs if substantially different from new free length, if distorted or if any fracture is evident. Inspect check valve spring (4) and regulator spring (34) for distortion or fractures, then renew as necessary.

Reassemble in reverse of disassembly procedure, using new "O" rings and seals. Lubricate all parts liberally while assembling. Refer to paragraph 203 when reinstalling and to paragraph 202 for adjusting release detent, if so equipped.

CLOSED-CENTER REMOTE CONTROL VALVES

All Models So Equipped

205. Optionally available closed-center remote control valves may be either single spool or double spool and may be installed in combinations of one, two, three or four spools. Double spool valves contain twice as many components as a single spool valve and a shuttle check valve (47 through 50—Fig. 213) is installed in the internal pilot passage connecting the two spools. A flow control restrictor (10 through 25) is used on each remote control valve spool and may be adjusted to a minimum of 11.4 liters/min. (3 gpm) flow and a maximum of 57 liters/min. (15 gpm) flow. When flow control restrictor is set to less than maximum output, excess fluid from pump is returned to the sump by the combining valve located in the priority valve assembly.

Closed-center valves may be used to operate single or double-acting cylinders. When used to operate single-acting cylinders, a single hose is connected to lift port of quick-release coupling. The cylinder is extended by moving control lever to float position.

206. ADJUST LINKAGE. Control valve levers should be centered and aligned (if more than one is installed) when all valves are in neutral position. Operating linkage is adjustable at clevis attached to valve spool. Loosen locknut and turn clevis until lever

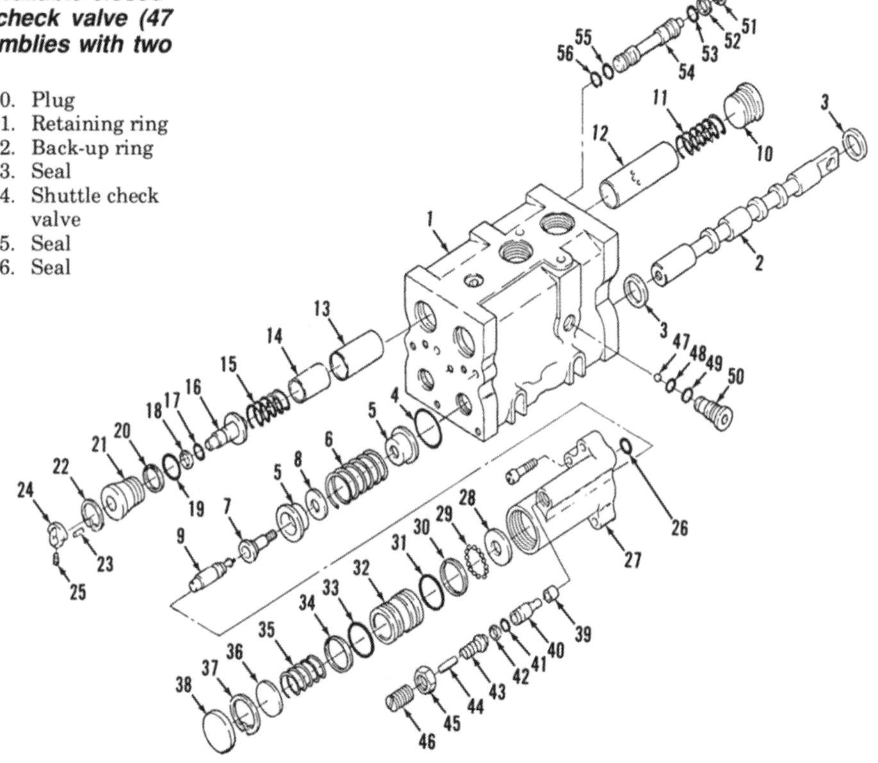

Fig. 213—Exploded view of optionally available closed-center remote control valves. Shuttle check valve (47 through 50) is used only on valve assemblies with two spools as described in text.

1. Valve body
2. Valve spool
3. Seals
4. Seal
5. Spring retainers
6. Spool centering spring
7. Pivot detent coupling
8. Detent spacer
9. Detent shaft
10. Plug
11. Spring
12. Flow control spool
13. Flow control restrictor
14. Load check valve
15. Spring
16. Shaft
17. Seal
18. Back-up ring
19. Seal
20. Back-up ring
21. Shaft retainer
22. Retaining ring
23. Spring pin
24. Knob
25. Set screw
26. Seal
27. Detent housing
28. Washer
29. Detent balls
30. Back-up ring
31. Seal
32. Detent spool
33. Seal
34. Back-up ring
35. Spring
36. Retainer
37. Retaining ring
38. Expansion plug (1.18 in.)
39. Seat
40. Poppet
41. Seal
42. Back-up ring
43. Spring
44. Dowel pin
45. Locknut
46. Detent adjusting screw
47. Ball
48. Seal
49. Seal
50. Plug
51. Retaining ring
52. Back-up ring
53. Seal
54. Shuttle check valve
55. Seal
56. Seal

is correctly positioned, then tighten locknut. Slot in clevis should be vertical when adjustment is complete and locknut is tightened.

207. ADJUST DETENT REGULATING VALVE. Detent regulating valve (46—Fig. 213) should be adjusted to release control valve spool at a pressure slightly higher than normal cylinder operating pressure. When correctly adjusted, control valve spool will return to neutral without excessive pressure buildup when remote cylinder reaches end of stroke.

To adjust, loosen locknut (45) and turn adjusting screw (46) as required. Turn screw IN to increase pressure required to return control valve to neutral. Turn screw OUT to lower release pressure.

208. REMOVE AND REINSTALL. Clean area around remote control valve assembly and disconnect linkage. Identify and tag hoses, then detach hydraulic lines from control valves. Cover openings to prevent entrance of dirt. Unbolt and remove remote control valve(s) and quick-release coupling(s) as an assembly. Separate quick-release coupling(s) from control valve(s) after removing.

Reverse removal procedure to reinstall. Tighten retaining screws to 56 N.m (41 ft.-lbs.) torque. Refer to paragraph 206 for adjusting linkage.

209. OVERHAUL. Remove screws retaining detent housing (27—Fig. 213) to valve body (1), then withdraw detent housing. Extract valve spool (2) and centering spring (6) as an assembly from detent housing end of valve body.

NOTE: Slide valve spool and centering spring assembly from body slowly, without applying undue pressure, to reduce chances of damage.

Remove plug (10) from end of flow control spool bore and withdraw spring (11) and spool (12). Remove set screw (25) and knob (24). Remove snap ring (22), then withdraw retainer (21), shaft (16), spring (15), load check valve (14) and restrictor (13). Remove spiral retaining ring (51), then use a suitable tool to pull shuttle check valve (54) from body. On double spool valves, a third shuttle check valve seat (50) and ball (47) is located in side of valve body.

Separate components as needed, clean all parts thoroughly and inspect all parts carefully before reassembling. Valve spools and body are not available separately and complete assembly must be renewed if damaged. Inspect springs for signs of fracture or distortion and renew if necessary.

Reassembly is the reverse of disassembly. Use new "O" rings and seals. Lubricate all parts liberally while assembling. Refer to paragraph 208 when reinstalling and to paragraph 206 for adjusting linkage.

QUICK-RELEASE COUPLING

All Models So Equipped

210. OVERHAUL. Unscrew sleeve (8—Fig. 214) from housing (1). Remove roll pin (5) or clip and washers from handle shaft, then remove handle. Withdraw coupling from housing, then refer to Fig. 214 and Fig. 215 to complete disassembly.

Clean components thoroughly and inspect for excessive wear or other damage. Reassemble in reverse of disassembly procedure, using new "O" rings and seals. Lubricate all parts liberally while assembling.

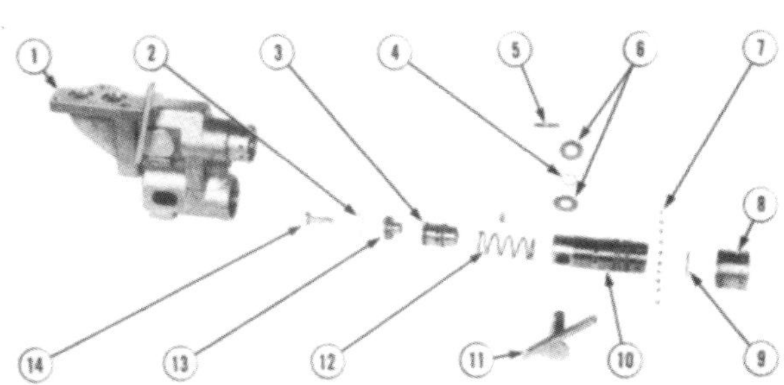

Fig. 214—View of typical quick-release coupling assembly for closed-center remote control valves.

1. Housing
2. Clip
3. Check valve assy.
4. Spring washer
5. Roll pin or clip
6. Washers
7. Balls
8. Sleeve
9. Clip
10. Body
11. Lever assy.
12. Spring
13. Guide
14. Plunger

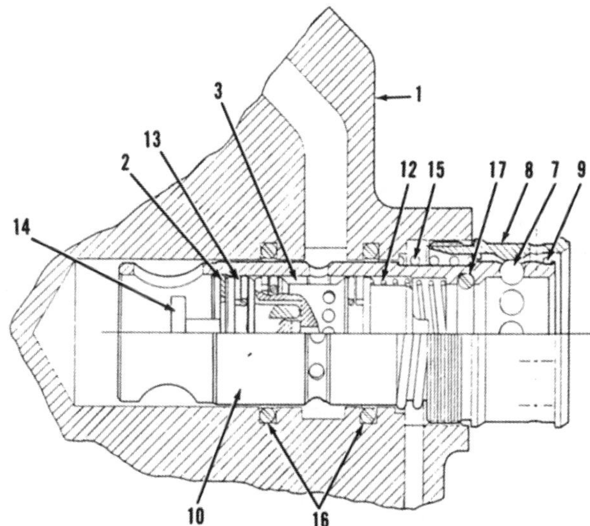

Fig. 215—Cross section of quick-release coupling. Refer to Fig. 214 for legend except the following:

15. Spring
16. Seal & back-up ring
17. Seal ring

CAB

Refer to the following paragraphs to remove or tilt the tractor cab forward. The following procedures apply only to tractors equipped with a Ford cab.

All Models So Equipped

211. TILT CAB FORWARD. It should be noted that extensive damage to the cab is extremely likely at all stages of this operation. Do not raise the cab more than necessary; about 77 mm (3 inches) is generally considered the limit because the windshield may break. Not supporting the raised cab evenly can also result in stress that may bend or break parts that are not easily repaired. Do not attempt to tilt the cab to the rear.

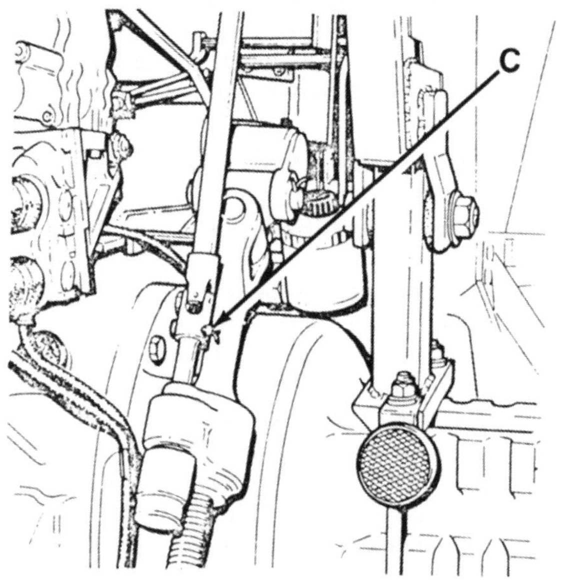

Fig. 216—The lift linkage controls in the cab are connected to the external linkage by clevis pin (C).

Disconnect the ground strap between the cab and the hydraulic lift cover and disconnect the in-cab hitch leveling linkage (C—Fig. 216). Disconnect controls (Fig. 217) for the remote valve, hydraulic lift control (6), auxiliary control valve levers and system control lever (5). Disconnect spring (4) from the lift control cross shaft and remove the cross shaft support bracket. Disconnect pto lever from upper end of actuating rod. Disconnect differential lock linkage (2) and remove spacer (8).

Remove retaining screws and plates from the cab floor, to reveal the cab front mounting stud access covers, then pry the access covers from the floor. Loosen the front cab retaining stud nuts until the nuts are flush with the stud ends. Remove the four bolts that attach the cab to the rear axle housings. Remove the cab roof retaining bolts from the sides near the rear and attach suitable lifting equipment (L—Fig. 218). Cab can be tilted by raising rear of cab from bottom using a wooden beam positioned under cab. Blocks (made of 2x4x3 inch boards) can be positioned between rear axle housings and cab rear mounting supports as shown in Fig. 219 to hold the cab up. **Do not attempt to raise rear of cab more than 90 mm (3½ inches).**

When reassembling, tighten the front mounting stud nuts to 340 N•m (250 ft.-lbs.) torque, rear cab to mounting bracket bolts to 380 N•m (280 ft.-lbs.) torque and rear mounting bracket to rear axle hous-

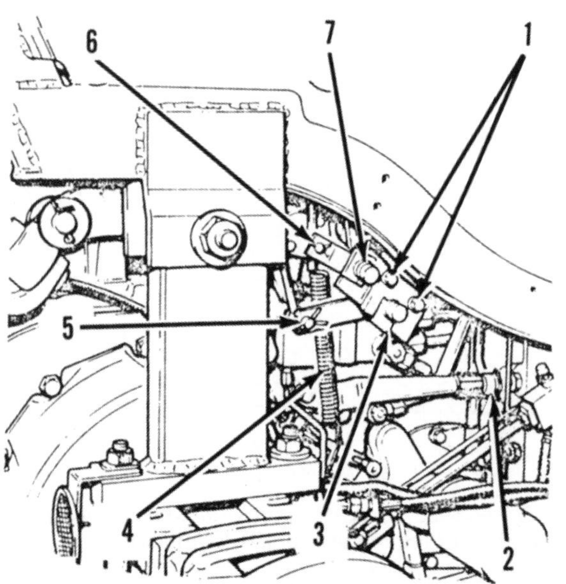

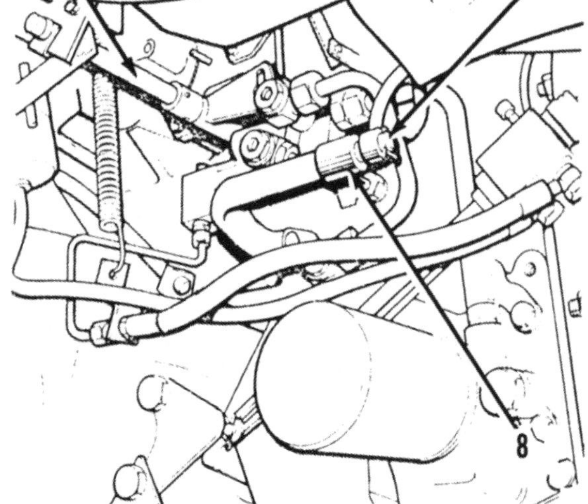

Fig. 217—View of various linkages that must be disconnected before raising or removing tractor cab.

1. Bracket retaining bolts
2. Differential lock linkage
3. System selector cross-shaft
4. Spring
5. System selector linkage
6. Lift control linkage
7. Lift control cross-shaft
8. Spacer

ing bolts to 270 N•m (200 ft.-lbs.) torque. If rear axle housing has assist ram, cab rear mounting bracket bolts should be tightened to 340 N•m (250 ft.-lbs.) torque. Access panel bolts and gearshift gaiter retaining bolts should be tightened to 9 N•m (7 ft.-lbs.) torque. The cab ground bolt should be tightened to 48 N•m (35 ft.-lbs.) and the roof retaining screws should be tightened to 23 N•m (18 ft.-lbs.) torque.

212. REMOVE AND REINSTALL. It should be noted that extensive damage to the cab is extremely likely at all stages of this operation. Do not attempt to lift the cab unless proper lifting equipment is available. Improper/uneven lifting or improper/uneven supporting of the removed cab can result in stress that may bend or break parts that are not easily repaired. The cab is especially vulnerable when it is removed because support is not in the stable locations selected by the manufacturer.

If cab is equipped with air conditioning, disconnect refrigerant lines between the engine and the cab. Self sealing couplings are usually located in lines to facilitate removal of the cab. Make sure that air conditioner drain hose is free to be raised with the cab. If cab is equipped with heater, drain cooling system and disconnect heater hoses (1 and 5—Fig. 224).

On all models, remove the access panel over the transmission and remove the floor mat. Unbolt and remove the gearshift console and gaiter. Unbolt and remove the brake access panel (1—Fig. 220) and pry the cab front mounting stud access covers (2) from the floor. Remove the front cab retaining stud nuts located under access covers (2). Remove cotter pins (2—Fig. 221) and pins (1 and 3) to disconnect the brake operating rods. Loosen locknuts (1 and 2—Fig. 222) and remove hand brake adjusting nuts (3). Push hand brake operating cables out to the underside of the cab. Disconnect the engine stop cable (2—Fig. 223) and speed control rod/cable (3) from the fuel injection pump. Detach the proofmeter drive cable. Detach cable housings from brackets and pull cables back so they hang freely from tractor cab. Disconnect wiring harnesses (W—Fig. 224). Disconnect the clutch operating linkage and mark the four hydrostatic steering hoses on the left side to facilitate identification when reconnecting. Separate the four lines at hose fittings and cover all openings to prevent the entrance of dirt. Disconnect the ground strap

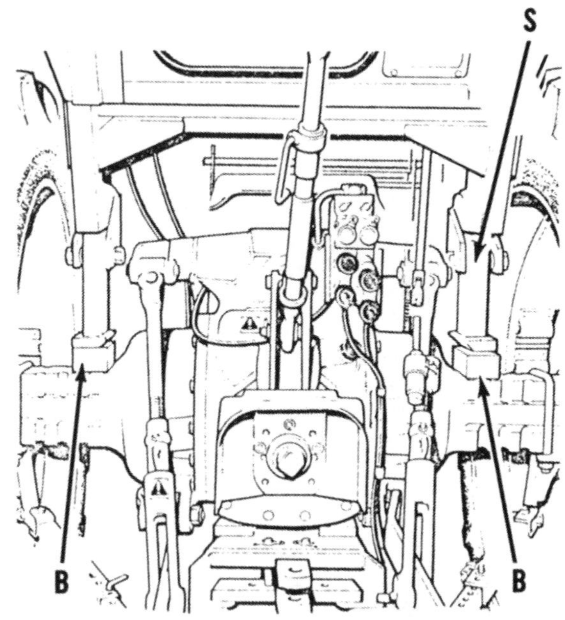

Fig. 219—Wooden blocks (B) can be positioned as shown, between cab rear supports (S) and rear axle housings.

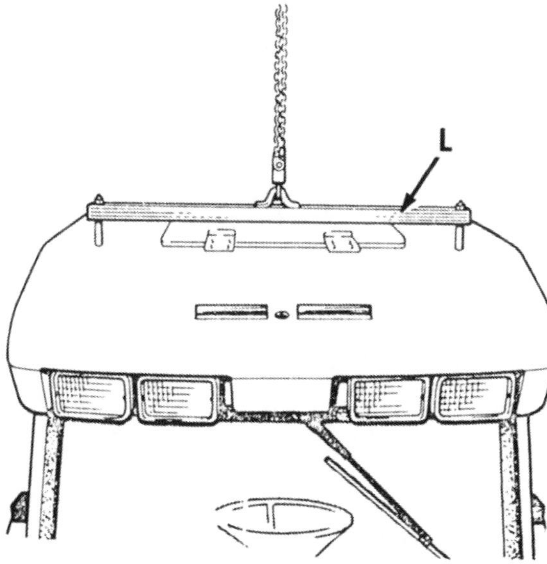

Fig. 218—Lifting equipment can be attached at locations where roof is attached.

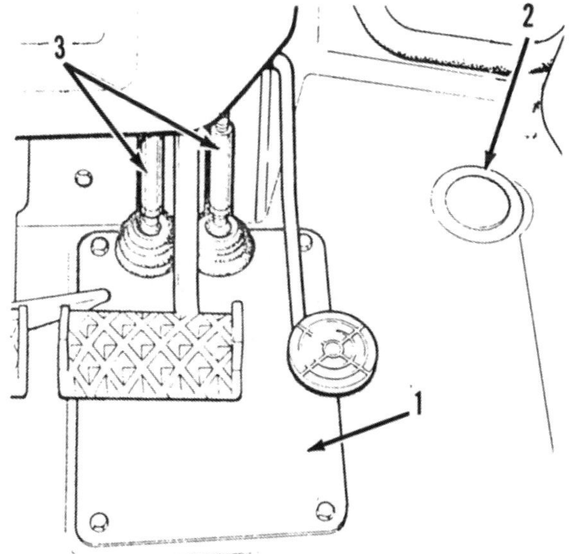

Fig. 220—View showing location of brake access panel (1), one of the two front cab mount stud covers (2) and the brake pedal linkage (3).

between the cab and the hydraulic lift cover and disconnect the in-cab hitch leveling linkage (C—Fig. 216). Disconnect controls (Fig. 217) for the remote valve, hydraulic lift control (6), auxiliary control valve levers and system control lever (5). Disconnect spring (4) from lift control cross-shaft and remove cross-shaft support bracket. Disconnect pto lever from upper end of actuating rod. Disconnect differential lock linkage (2) and remove spacer (8). Remove the four bolts attaching the cab to rear axle housings.

Remove cab roof retaining bolts from the sides near midpoint of roof and attach suitable lifting equipment (L—Fig. 218). Raise cab slowly, checking to be sure that cables, lines, linkage and wires are nct catching. Cab can remain suspended if properly secured to prevent damage or injury. If cab is lowered, make sure to provide adequate support to prevent damage.

When reinstalling, tighten the front mounting stud nuts to 340 N•m (250 ft.-lbs.) torque, rear cab to mounting bracket bolts to 380 N•m (280 ft.-lbs.) torque and rear mounting bracket to rear axle housing bolts to 270 N•m (200 ft.-lbs.) torque. If rear axle housing has assist ram, cab rear mounting bracket bolts should be tightened to 340 N•m (250 ft.-lbs.)

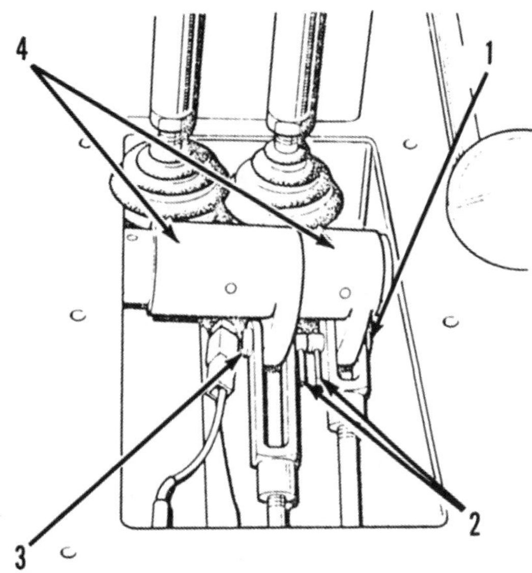

Fig. 221—Brake pedal linkage is located under panel shown in Fig. 220.

1. Right hand clevis
2. Cotter pins
3. Clevis pin for left brake linkage
4. Brake pedal pivots

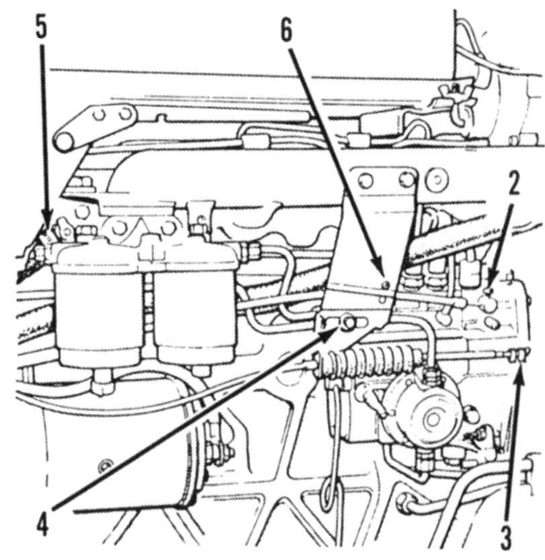

Fig. 223—View of engine right side showing some of the items to be disconnected before the cab can be removed.

2. Engine stop cable
3. Engine speed control cable
4. Throttle bracket retaining screw
5. Rear heater hose
6. Engine stop cable retaining clip

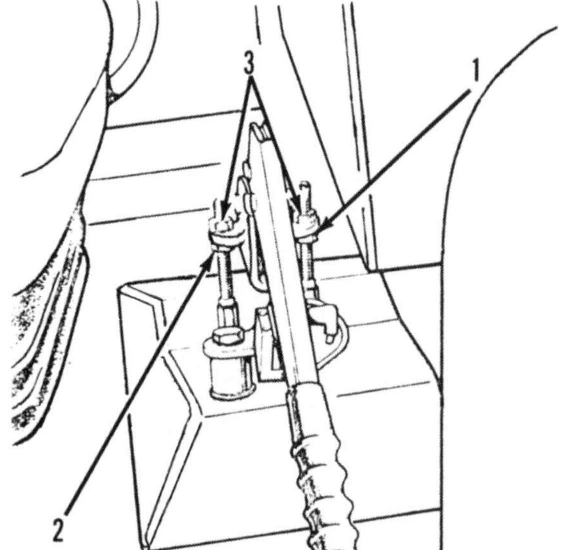

Fig. 222—View of hand brake cables. Adjustment nuts are shown at (3) and locknuts are shown at (1 & 2).

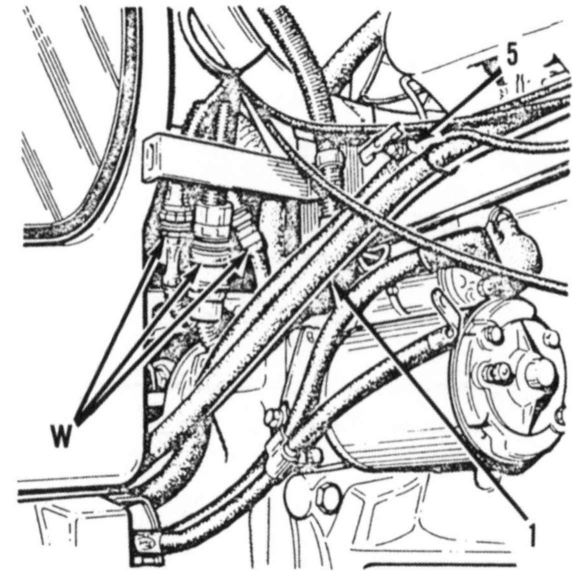

Fig. 224—View of right side showing location of wiring harness connectors (W) and heater hoses (1 & 5).

torque. Access panel bolts and gearshift gaiter retaining bolts should be tightened to 9 N•m (7 ft.-lbs.) torque. Tighten the gearshift console retaining screws to 3.5 N•m (30 in.-lbs.) torque. The cab ground bolt should be tightened to 48 N•m (35 ft.-lbs.) and the roof retaining screws should be tightened to 23 N•m (18 ft.-lbs.) torque.

ELECTRICAL COMPONENT LOCATIONS

213. Refer to Fig. 225 or Fig. 226 for tractor electrical components.

Fig. 225–Electrical components typical of models without a cab.

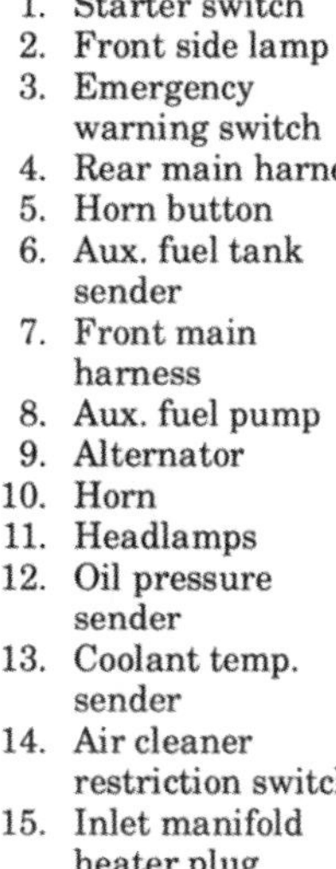

1. Starter switch
2. Front side lamp
3. Emergency warning switch
4. Rear main harness
5. Horn button
6. Aux. fuel tank sender
7. Front main harness
8. Aux. fuel pump
9. Alternator
10. Horn
11. Headlamps
12. Oil pressure sender
13. Coolant temp. sender
14. Air cleaner restriction switch
15. Inlet manifold heater plug
16. Starting motor/solenoid
17. Battery temp. sensor
18. Battery
19. Rear fender harness switch
20. Safety start switch
21. Flasher unit
22. Stoplight switch
23. Fuse box
24. Front fender harness switch
25. Light switch
26. Right fender harness
27. Implement lamp
28. Main fuel tank sender
29. Instrument cluster
30. Flasher switch
31. Trailer socket
32. License plate lamp
33. Rear, stop and flasher lamp
34. Left fender harness

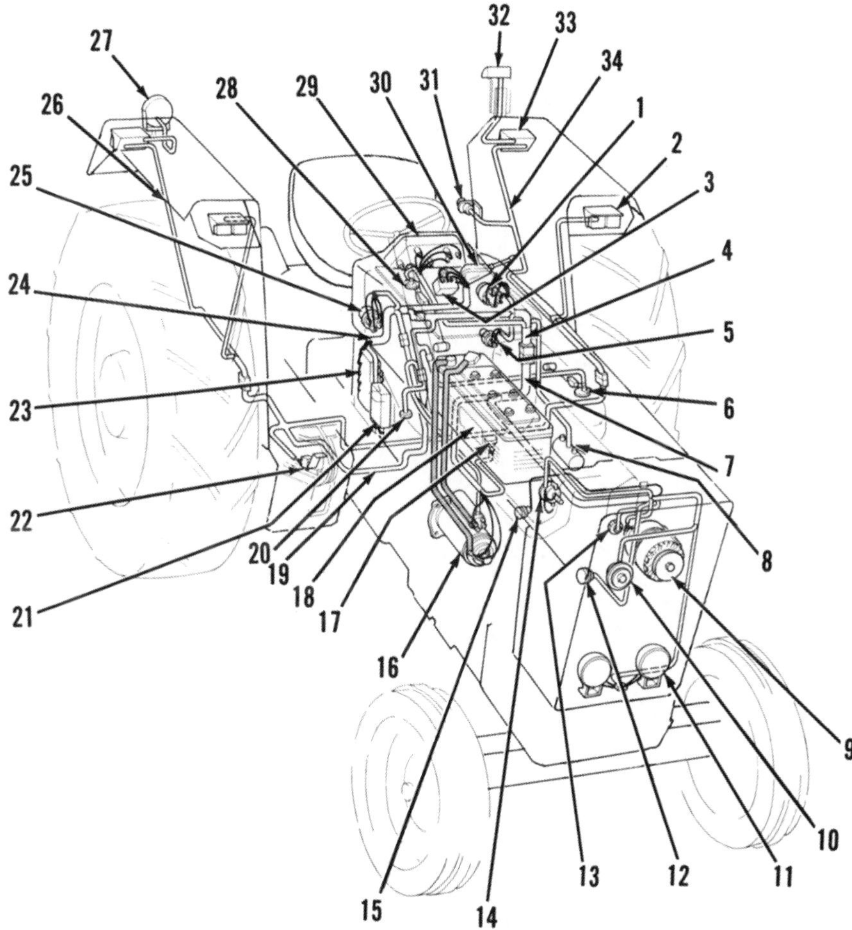

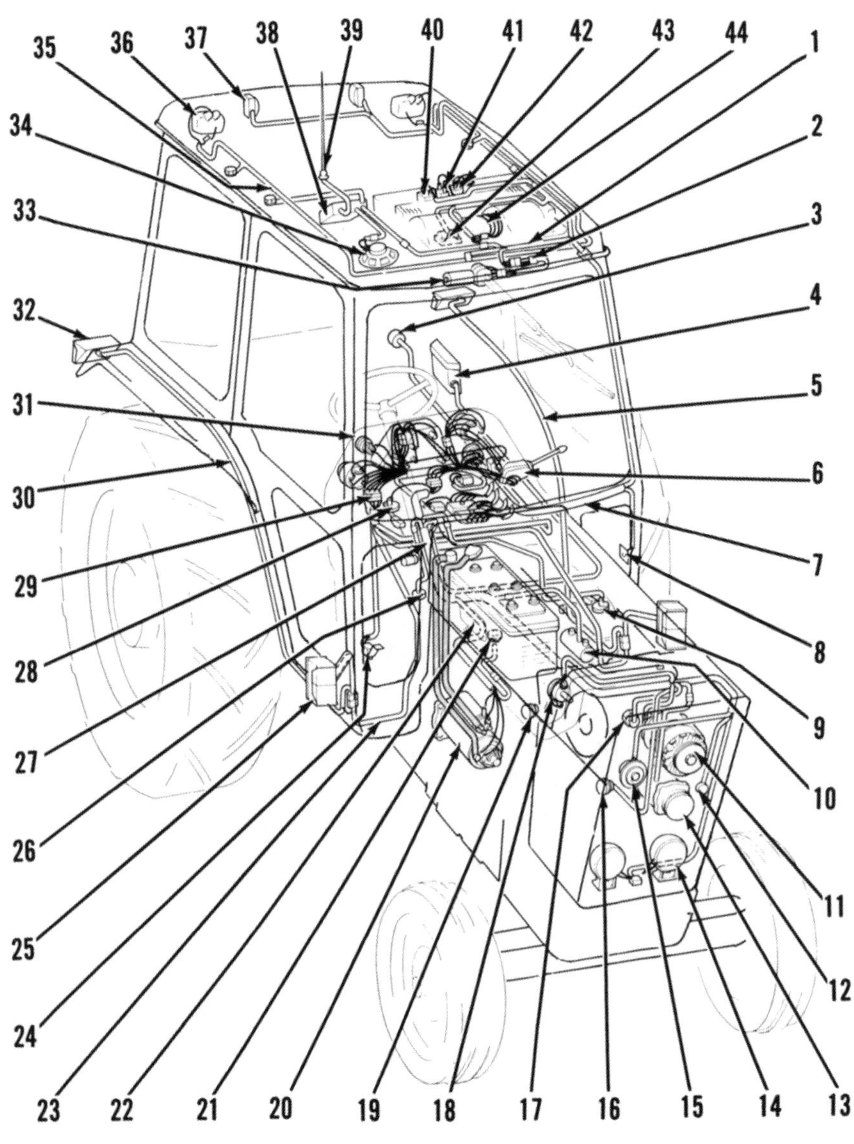

Fig. 225–Electrical components typical of models equipped with a cab.

1. Heater control harness
2. Radio suppression filter
3. Trailer socket
4. Windshield washer motor
5. Left fender harness
6. Flasher switch
7. Rear main harness
8. Interior lamp switch
9. Aux. fuel tank sender
10. Aux. fuel pump
11. Alternator
12. Thermostat
13. Air conditioning compressor
14. Headlamps
15. Horn
16. Oil pressure sender
17. Coolant temp. sender
18. Air cleaner restriction switch
19. Inlet manifold heater plug
20. Starting motor/solenoid
21. Battery temp. sensor
22. Battery
23. Fender rear harness
24. Stoplight switch
25. Front side lamp
26. Safety start switch
27. Front main harness
28. Main fuel tank sender
29. Fender rear harness
30. Right fender harness
31. Instrument console
32. Rear, stop and flasher lamp
33. Windshield wiper motor
34. Loudspeaker
35. Rear lamp harness
36. Rear lamp
37. License plate lamp
38. Radio
39. Antenna
40. Heater blower switch
41. Air conditioner evaporator switch
42. Windshield wiper/washer switch
43. Interior lamp
44. Heater blower motor

NOTES

NOTES

NOTES

NOTES